HEALTH
AND
SOCIETIES

HEALTH
AND
SOCIETIES

Changing Perspectives

SARAH CURTIS

Queen Mary and Westfield College, University of London

and

ANN TAKET

South Bank University

A member of the Hodder Headline Group
LONDON • NEW YORK • SYDNEY • AUCKLAND

© 1996 Sarah Curtis and Ann Taket

First published in Great Britain 1996 by
Arnold, a division of Hodder Headline PLC,
338 Euston Road, London NW1 3BH

Co-published in the USA by Halsted Press,
an imprint of John Wiley & Sons, Inc.,
605 Third Avenue,
New York, NY 10158-0012

British Library Cataloguing in Publication Data
A catalogue record for this book is available from the British Library

Library of Congress Cataloging-in-Publication Data
Curtis, Sarah,
    Health & societies: changing perspectives / Sarah Curtis and
Ann Taket.
        p. cm.
    Includes bibliographical references and index.
    ISBN 0-470-23576-4.—ISBN 0-470-23577-2
    1. Medical policy—Cross-cultural studies.   2. Medical
care—Cross-cultural studies.   3. Public health—Social aspects.
4. Social medicine—Cross-cultural studies.   5. World health.
I. Taket. A. R. (Ann R.)   II. Title.
RA394.C88   1995
362.1—dc20
                                                    95-35298
                                                       CIP

ISBN 0 340 62512 0 HB
ISBN 0 470 23576 4 HB (in the USA only)
ISBN 0 340 62513 9 PB
ISBN 0 470 23577 2 PB (in the USA only)
1 2 3 4 5   95 96 97 98 99

Typeset in 10/12 pt Sabon by
Phoenix Photosetting, Lordswood, Chatham, Kent
Printed and bound by
J.W Arrowsmith, Bristol

Contents

Acknowledgements

We always enjoy reading the acknowledgement sections in books and tracing out the interconnections they add to the text itself. We are conscious that it is impossible to name all the individuals and groups who have shaped the work and the writing of this text – even though they may never have read a word of what is written here (and indeed some of them might wish to distance themselves from its content!). We would like to thank the undergraduate students who have influenced our thinking through their work on courses we have taught at Queen Mary and Westfield College, University of London, and in various medical schools. We have been fortunate to supervise a number of postgraduate students and we would like to thank them for the many interesting discussions we have shared and for the comments and suggestions which some of them have given on drafts of this book (especially Liz Allen, Yolande Coombes, Gavin Daker-White, Mark Exworthy, Sophie Hyndman, Ian Jones, Chizoma Onouha, Vicky Roberts, and David Woodhead). We would also like to acknowledge the inspiration and stimulation provided by various other colleagues, including the staff in the Geography Department at Queen Mary and Westfield College, and many others in health professions, voluntary organizations, and academic institutions who have provided us with ideas. Special thanks go to members and workers of the Tower Hamlets Health Strategy Group, who have provided the inspiration and the location for some of the research described here. Finally, but not least, many thanks to Brian Blundell and Leroy White for moral and intellectual support, to Susan Richards in the inter-library loan section at QMW library for tirelessly pursuing numerous references, and to Laura McKelvie at Arnold for her help over the production of this book.

The authors and publishers would like to thank the following for permission to use copyright material in this book:

British Medical Association for the figure 'Trends in mortality from amenable and non-amenable causes' from Boys, R., Foster, D., and Jozan,

P., 'Mortality from causes amenable and non-amenable to medical care: the experience of Eastern Europe', *British Medical Journal* 303 (1991), p. 881; Business Education Publishers Ltd for figure 8.7 from Ottewill and Wall, *The growth and development of the community health services* (1990), p. 402, Cambridge University Press for Omran, A. 'The epidemiological transition: a theory of the epidemiology of population change', *Milbank Memorial Fund Quarterly* 64 (1971), pp. 355–91; table 1 from Puentes-Markides, C., 'Women and access to health care', reprinted from *Social Science and Medicine* 35 (1992), p. 622 and extract plus figure 8 from Curto de Casas, S., 'Geographical inequalities in mortality in Latin America', reprinted from *Social Science and Medicine* 36(10) (1993), pp. 1349–1355, with kind permission from Elsevier Science Ltd, The Boulevard, Langford Lane, Kidlington OX5 1GB, UK; Health Affairs for the extract from Kundig, D. and Yan, G. (1993) *Physician supply in rural areas with large minority populations*, p. 170, Exhibit 1; HMSO, London for the extract from Rodine, J.M., Blanchet, M. and Dowd, J. 'Health Expectancy. First Workshop of the International Healthy Life Expectancy Network (REVES)'. OPCS *Studies on Medical and Population Subjects, no. 54*; Medical Association of South Africa for the extract from Ransome, O. *et al.* Facilities for children in state and provincial hospitals, *South African Medical Journal* 69, p. 612, Table 1. Oxford University Press for figure 2 from Brannstrom, I., Persson, L., and Wall, S., 'Towards a framework for outcome assessment of health intervention: conceptual and methodological considerations', *European Journal of Public Health* 4 (1994), pp. 125–30; Professor G. Pyle for figure 7 from 'Regional inequalities in infant mortality within North Carolina, USA', *Espace Populations Societes* 3 (1990), p. 444.

Preface

Changing perspectives on health and societies

The study of health, health services and their relation to society represents a rapidly proliferating research field, of topical relevance to many groups of people: social studies undergraduates and postgraduates, health service professionals in training and continuing education, and many interested people in the voluntary sector. As research on health and health services expands, it is undergoing constant change and development. This change in academic perspectives is associated with developments in the issues being researched and in the way that they are viewed in wider society.

Change in health, services, and policy

Throughout the world, perspectives on health, health services, and health policy are shifting and evolving. Many factors are contributing to change in population health and the way that health is perceived and understood. We are becoming more conscious that views of what constitutes good health or poor health are diverse. The risk factors and diseases which threaten good health in populations are also subject to change. Health service delivery is going through a period of rapid reorganization in many countries, especially in terms of the roles of the private and state sectors of health service production and consumption. Health service policy also shows changes, as the health service professions, under a variety of influences as diverse as the World Health Organization (WHO), right-wing governments, and feminist and other liberation movements, are in the process of renegotiating and reconstructing their approaches to health and illness, increasing the emphasis on health promotion and the prevention of illness rather than on the care and/or cure of those deemed 'ill', and redefining the role of individuals, the state, and private and voluntary agencies in action to improve health. These processes of reorientation necessitate a much stronger focus on the relationships between health and social, economic, and environmental factors, and on building intersectoral links between different agencies to achieve the new goals of health promotion and illness prevention. In many cases there is tension between the stated *aims* of the parties involved and the *effects* of their programmes or actions. The rhetoric and realities do not always match. In order to understand why this is so, we need to consider the social and political processes which drive the processes of change and which set the 'hidden agendas' for health and health services.

An interdisciplinary perspective

This book considers health, health services, and health policy to be socially constructed, and it examines the study of changes in health and societies mainly from the position of disciplines which make aspects of society their object of study (we do not, for example, include much discussion of perspectives from the natural sciences, and our view of biomedicine is mainly taken from an external viewpoint). However, within the broad purview of the social sciences, we aim to take an interdisciplinary perspective, since part of the change in academic views of health and society is the realization that the issues cannot be fully understood from a single disciplinary framework. The discussion here draws on and integrates examples of the work of medical geographers, sociologists, anthropologists, epidemiologists, and other social scientists. We aim to show how ways of looking at health and health services are constantly evolving. As well as a growing tendency to work across disciplinary boundaries, we find a wider diversity of approaches within each discipline. One aim of this book is to use examples of issues relating to health which give rise to debate in societies and which in turn stimulate changing perspectives within different disciplines. Some of the discussion also shows how changing academic perspectives may influence debate in wider society.

We wish to be very clear that we do not aim to give a full account of the theoretical positions or the methodological approach of any one discipline, and we have not attempted to adapt the discussion of the issues we consider in order to make coverage of each of the different disciplines completely balanced. Indeed, we are conscious that the illustrations we have chosen draw more heavily on some disciplines than others. This book should not therefore be taken as intending to cover the whole content or the state of the art in any one of these disciplines in particular.

Themes covered in the book

We are also forced to be selective in the substantive issues which we have chosen to discuss, and there are certainly many topics relating to health, health services, and health policy which are not considered here. The themes and issues covered in the book are ones which we find illuminating examples of perspectives on health, health services, and health policy. Issues relating to health are covered particularly in Chapters 2, 3, and 4. The discussion in Chapters 5, 6, and 7 relates especially to changes in health services. Health policy issues are particularly the focus of Chapters 8 and 9. Two specific themes which run through all of the discussion in this book are the exploration of *diversity* and the importance of *context*. These are taken up in a number of different ways in the chapters that follow.

Chapter 1 examines the evolution and diversity of approaches to health and society in social studies, using the development of strands of research in medical geography as an illustration of these processes of change. These different strands show varying emphasis on alternative theories of health in society and on different types of methodology. We aim to show how these different approaches within one discipline complement each other, and also how they reflect the influence of ideas from other disciplines.

Chapter 2 explores the diversity that exists in concepts of health, disease, and illness. We use examples to illustrate how diverse interpretations of ideas of 'health', 'ill health', 'disease', 'sickness', and 'illness' interact with views on the causation of health and ill health. Such studies are an important component in understanding the basis for medical practice, and in understanding the wider relationships between the health service systems and the social context. The chapter explores both 'lay' and 'professional' views. Examples of professional views are taken from the biomedical and the socio-ecological models of health. We also consider some early studies of lay views of health, arguing that these were not particularly successful in capturing the complex nature of lay responses to health and ill health. We finish by examining more recent studies of lay views which emphasize the importance of context in understanding individuals' attempts to make sense of themselves and their experiences of health and ill health.

Chapter 2 also demonstrates how the various different discourses around health are closely linked with the structure of the society concerned, i.e. that they are all socially constructed, and that, furthermore, these discourses are subject to constant contestation in different arenas, leading to a continual redefinition of their meanings and usage. This is further illustrated in Chapter 3, which contains three case studies of contested definitions of health and disease, involving different lay and professional views, and their implications. These deal with: women and mental illness; the construction and deconstruction of homosexuality as a mental illness; and finally the proliferation of discourses on HIV and AIDS. All the examples have in common the use of a medical diagnosis of ill health, combined with the use of moral or cultural censure and sanctions, in ways that reinforce existing power relations within the societies concerned. At the same time, through these examples, the active struggles of the disempowered can also be identified.

Chapter 4 moves on to consider the evidence for differences in health between populations in different places and social groups (using measures largely derived from professional perspectives of health and ill health). These can be distinguished at all geographical scales, from global differences between groups of countries to local differences between small areas. This chapter concentrates particularly on what can be learned about health inequalities from social and geographical variations in health of populations.

In Chapter 5 we shift from discussion of perspectives on *health* to a more focused consideration of *health services* and, in particular, of the differences in health service systems which are evident at the national scale. This chapter also takes the specific examples of the health care systems of the USA, Britain, and Russia to illustrate the changing role of the state in the provision of services. For each of these countries we review the historical differences in the organization of medical services, the problems leading to change, and the recent moves for reform of national health services. We consider how far these developments provide evidence of international convergence in the ways that health services are organized.

Keeping the focus on health services, Chapter 6 examines issues of equity, efficiency, and effectiveness. It begins by exploring the problems of defining and measuring need for services at local level, illustrated with examples from resource allocation in the British NHS. The chapter then considers the measurement of equity in resource provision and use, including examination of service accessibility, before moving on to examine health service effectiveness. We examine issues related to the choice between qualitative and quantitative approaches to evaluating effectiveness, arguing that both are necessary. We also review the use of institutionalized processes such as standard setting, performance review, health service indicators, clinical audit, and consumer satisfaction surveys.

Chapter 7 begins to shift the perspective towards strategies aiming to improve population health. We discuss the public health component of health and health services, examining various aspects of the 'new' public health movement and some examples of ways in which it has been implemented locally. We show how these implementations are linked to perspectives such as disease ecology, ideas about public participation in public health alliances, the prioritization of primary health care services, and to strategies for action such as 'Healthy Cities' initiatives.

We then move on in Chapter 8 to discuss the formulation and implementation of health policy. The chapter introduces a range of different approaches used to theorize the policy process, illustrated with a discussion of some relevant studies. We distinguish between different components of health policy: policy goals; the underlying framework of values/principles; strategies and mechanisms for policy implementation. We consider policy development and implementation at international, national/federal, regional, and local levels. The chapter concludes by discussing two linked themes which are globally important in health policy: decentralization, and lay participation in health policy. These two policy themes are particularly interesting to us for a number of reasons. First, they represent two of the policy thrusts evident in many of the most recent health service reforms (although both are by no means new). Second, they illustrate many of the difficulties arising from the earliest efforts to theorize the policy process, and the necessity for more sophisticated frames for analysis. They also exemplify the importance of attention to the level at which analysis is carried out.

Chapter 9 discusses the growing influence of international organizations concerned with health in the field of health policy and health service system restructuring. The chapter begins by offering a brief historical perspective on the origins of international organizations with specific roles in relation to health, before moving on to focus on the structure, organization, and functions of the WHO. We examine some of the more recent orientations in WHO's work, in particular the 'Health for All' strategy and related programmes of policy development. Another recent entry into international discourses on health and health policy is provided by the World Bank's (1993) *World Development Report*, which takes as a central focus the notion of 'investing in health'. The implications of this report are considered and contrasted with the 'Health for All' strategy. The final section argues that important challenges remain for the future in terms of the role of international organizations in health and health policy.

In Chapter 10 we draw on what has been presented in earlier chapters to highlight some of the outstanding issues that remain to be explored through research on health, health services, and health policy. We identify globalization and interdependence, the structure of health services systems, professional and methodological pluralism, and the changing understanding of the nature of health and the role of health services as examples of topics which are likely to be important as foci for changing perspectives and research on health and societies in the future.

Glossary

Key terms and concepts

There are a number of key terms and concepts which we would like to introduce. The literature is replete with the use of these terms in slightly different ways. In part this is inevitable, as it results from the continually contested nature of key concepts like health and ill health (which are the explicit subject of Chapters 2 and 3), contested concepts in health policy (like decentralization and participation, discussed in Chapter 8), or from the evolution of concepts in health policy, as discussed in Chapters 8 and 9. However, it can be confusing and unhelpful to the reader if the different usages are due to a simple difference in the definitional scope of a term. For example, consider the term 'health services'. There are a number of different usages in the literature: it may refer to professionally provided services only; it may include self-help, self-care, and informal care; it may include the voluntary sector; it may include biomedical professionals only. In this section we shall discuss a number of key terms and indicate how we intend to use them throughout this book.

Health policy

'Health policy' is used to encompass any policy which includes strategies and actions undertaken with the aim of maintaining or improving health and providing for the care, treatment, or cure of ill health. Within the socio-ecological paradigm of health discussed in Chapter 2, this encompasses actions taken to address environmental, heredity, lifestyle, or health service system factors – all of which affect health. Health policy should be understood to refer to actions taken in *any* sector of society; in particular it is not limited to actions by health service professionals, and is not restricted to consideration of the health service system. We shall reserve the term 'health services policy' for when we wish to talk *solely* about the health service system. Our use of the term 'health policy' in this broad sense is similar to the term 'healthy public policy' originated by Milio (1987), and used more recently in the work of the WHO; this is discussed further in Chapter 9.

The health sector

We use the term 'health sector' to denote the sector within a society whose main concern is with health care. The sector comprises all the different kinds of activity undertaken with the purpose of protecting or promoting health, and providing care, treatment, cure, or support to those in ill health. Obviously, other sectors (like transport, education, housing, agriculture, and industry) also have effects on health (see the discussion of health policy above), but these are not the main concern or focus of their activities.

Within the health sector we can further distinguish between the formal and informal sectors. The formal sector is what we will refer to as the 'health services', provided through the organization of a 'health service system'; this may include any or all of the following components: public services provided by, or on behalf of, the state; privately provided services; organized voluntary services. What the formal sector has in common is a formal organizational structure and staff who can be described as 'health professionals' (to be understood to cover a wide range of different types of training and qualification). The informal sector, on the other hand, provides care which is organized informally. It includes two elements: self-care and lay care, the latter to be understood as that provided by family or friends on an informal basis. Table 0.1 summarizes the usage of these terms. In practice, of course, the boundaries between the different terms may not be sharply delineated and the terms should not be considered as always mutually exclusive; for example, the boundaries between private non-profit-making organizations, voluntary organizations, and charitable organizations may in certain settings be non-existent. In addition, there are inevitable

Table 0.1 The health sector

Health sector (health care)	
Health services (formal sector)	Informal care (informal sector)
private statutory voluntary	self-care lay care

problems with setting the boundaries to terms such as 'health professional' – at what point does experience (in caring for a relative or friend) replace, or even overtake, the possession of a formal qualification?

The final point to be made in discussing the health sector is the use of the word 'care', in particular in phrases such as 'health care' or 'primary health care'. In discussing the activities of the health sector, we have emphasized that these activities are certainly not restricted solely to the care (in a narrow sense) of those in ill health: they include activities that are promotive or protective of health, as well as curative and treatment activities. In phrases such as 'health care' and 'primary health care', therefore, 'care' should be understood in a broader sense to encompass promotion, protection, prevention, treatment, etc.

Medical pluralism: systems of medical thought and knowledge

Through history a wide variety of systems of medical thought and knowledge can be distinguished. All such systems are cultural systems, having their origins in a particular set of cultural beliefs, including those about the processes by which health and ill heath are identified, labelled, and explained (this will be a particular topic of discussion in Chapter 2). Currently, health services are still dominated by one particular medical system: 'biomedicine', which will also be referred to at times by the synonyms 'Western medicine' and 'allopathic medicine'. This book cannot discuss in detail all the different medical systems, and in places it will be convenient to use as a general term, 'non-Western medicine' (or the synonym 'non-allopathic medicine') to refer to the multiplicity of other such systems. This should *not* be taken to imply that such systems form a homogeneous group. Similarly, the use of the terms 'complementary medicine' and 'alternative medicine' should not be taken to imply that a homogeneous group of systems is being referred to.

Different parts of the world

We have deliberately chosen not to talk about different countries of the world using classifications based on notions of development (e.g. to distinguish developed from developing or underdeveloped countries). Such classifications seem unsatisfactory to us, owing to the (usually unacknowledged) assumptions they make about the nature of the development process, and about what constitutes desirable development. For different, though related, reasons we find a division into 'first', 'second', and 'third' world equally problematic. Instead, we focus at some points on different regions of the world (for example, Europe or Eastern Europe) and, at other points, we contrast aspects of the health situation in groups of countries which we will refer to as 'low-income', 'middle-income', and 'high-income' countries (in order to distinguish between relatively disadvantaged and relatively advantaged groups of countries). The exceptions to this are in Chapters 5 and 9, where we review the work of the World Bank, and where we use (and explain) the definitions it uses.

1

Changing perspectives on health and societies: the example of medical geography

Different disciplines offer varying perspectives on health and society. Views from within sociology, anthropology, politics, philosophy, economics, and geography, for example, will place varying emphasis on, and offer differing interpretations of, topics such as: the ways in which health and health services are linked to social relations; the social construction of health and cultural differences in the meaning of health for societies; efficiency, effectiveness, and equity in health and resources for health services; social justice and ethics as they relate to health; and the importance of place, environment, and context for health. At the same time, there is cross-fertilization of ideas between different disciplines, and each has an influence on the other. Furthermore, the perspectives presented within particular disciplines vary between academic communities in different societies and also over time. This is because the nature of health and societies is variable and changing, but also because of the ways in which academic thinking evolves. We aim in this book to explore selected illustrations of this diversity and change in perspectives on health and societies, using examples which particularly strike us from our own experience in research activity and in teaching a variety of different audiences, including health service professionals and social science students in various disciplines. Although this book adopts an interdisciplinary perspective, we begin the discussion by exploring developments within one discipline, in order to illustrate how an academic perspective can evolve over time.

We have chosen in this chapter to focus on changing perspectives in medical geography. We have several reasons for making this selection. We are both involved in research and teaching in medical geography, and thus we have a particular interest in exploring a variety of geographical themes. In the later chapters, strongly geographical issues emerge: the spatial patterning of health, ill health, and service provision; the interaction between different spatial levels of policy formulation and service organization; the study of the role of particular environmental factors in disease causation;

the comparative study of different societies and/or different communities or social groups in terms of their experience of health and health services; and the importance of notions of place and community in understanding health issues. The examples discussed later in the book, therefore, show how ideas from medical geography have contributed to our understanding of health and society and how they relate to ideas originating in other disciplines. While there is an extensive literature on health and societies written from the perspective of other disciplines, such as sociology and economics, relatively few publications specifically address these geographical aspects. In this chapter, therefore, we provide an overview of the issues covered in the book through the perspective of changes in medical geography. This demonstrates how, within a particular discipline, the perspectives adopted have shifted in response to changing emphases on health in societies, and exemplifies how academic perspectives are themselves influenced by change in the societies of which they are a part.

It is important to state at the outset that this chapter is *not* intended to provide a history of medical geography; that would be far too large a task for a single chapter. Those who wish to pursue the history of medical geography, especially the early origins of the discipline, might peruse Riley (1987) or the forthcoming book by Barrett. Nor does this chapter attempt to provide a comprehensive review of medical geography in different countries/continents. Those interested in bodies of work carried out this century are referred to McGlashan and Blunden's (1983) edited collection, which contains reviews of work in tropical Africa, India, the United States, Canada, the United Kingdom, Germany, France, and the Benelux countries, covering varying periods up to the early 1980s. Likewise, this book does not aim to develop a comprehensive account of theory and methodology from within any one discipline; for a full discussion of these in the context of medical geography, the reader is directed to Joseph and Phillips (1984), Eyles and Woods (1983), Jones and Moon (1987), Learmonth (1988), or Meade *et al.* (1988). However, this chapter does highlight key developments in the subject which have been covered in more detail in those texts; it also provides examples of the more recent developments, bringing the account of changing geographical perspectives more up to date.

The concerns of medical geography

The *Dictionary of human geography* provides us with a deceptively simple definition of medical geography: 'the application of geographical perspectives and methods to the study of health, disease and health care' (Johnston *et al.* 1994: 374). This alerts us immediately to two components or domains for study. The first of these, the study of health and disease, encompasses the analysis of spatial variations in human health or, more often, lack of health, i.e. death/mortality and disease/morbidity and the search for environmental and social conditions which may be causally related to health or ill health. The second component, then, is the study of health care, to be understood as comprising all formal or informal activities or services which are concerned with the care, cure, or treatment of ill health, with disease prevention, and with health promotion. This component includes studies of formal (statutory, public, or private) health service organization, provision, and use, as well as studies of the voluntary and informal sectors. Within this domain we find a strong concern with the analysis of equity and accessibility. To these two domains might be added a third domain which deals with the interaction of health and health services within society; this third strand, which touches on some of the most recent concerns of medical geography, is not well encompassed in the definition.

In many of the writings on the subject it has been customary to distinguish between two major groups of approach: the traditional and the contemporary (e.g. Phillips 1981); however, such a simple distinction fails adequately to reflect the most recent developments within medical geography. Indeed, *any* particular categorization of types of medical geographical study (as in any discipline) is inevitably imperfect for some reason or other. Thus the organization of material adopted in this chapter is a convenient device for exposition rather than a work of precise classification. We review the history and current practice of medical geography under five different headings, representing five strands that can be found woven together in different combinations within any particular study in medical geography.

The first two of the five strands are discussed together here as 'traditional medical geography', dealing first with the spatial patterning and diffusion of disease and death and second with the spatial patterning of health service provision and use. These first two strands are epistemologically distinct from the others: in their adherence to positivism, relying on strong assumptions about the existence of absolute standards of truth/correctness against which knowledge can be measured, they adopt a 'hard scientific' approach, and by and large concentrate on the application of quantitative methods.

The other three strands, all labelled 'contemporary medical geography', developed later, and adopt rather different epistemological stances, arising out of a variety of critiques of positivism, and making more use of qualitative methods. The third strand, the humanistic turn, concerns itself with studies examining, in one way or another, human awareness, agency, and creativity, for example in researching beliefs about health or health-related behaviour and service utilization; there are links here with the work of the Chicago School of Sociologists, and the work is carried out from a variety of essentially phenomenological perspectives. Many of the studies in this strand take as their focus a concern with the socio-cultural construction of health and illness.

The fourth strand is labelled 'structural/materialist/critical' turn, the alternatives illustrating that for those seeking more detailed categorization, a further subdivision might be attractive. Here we find studies informed by a variety of different perspectives on the organization of social life, sharing perhaps a common concern with highlighting the effects of broad social forces, although differing in the particular type of social theory they might draw on. So that some work, defined here as a structural substrand, might be informed by Parsonian structural functionalism or by Giddens's structuration theory, while a materialist substrand includes studies drawing specifically on marxist approaches to social dynamics, focusing on the primacy of the material basis of social life, and a critical substrand includes studies informed by the critical theory of the Frankfurt School or of Habermas.

Finally, we can identify the fifth strand as a cultural one, which draws on the new cultural geography; here there are links with poststructuralist critiques of/in social theory. In this chapter we illustrate the work within this strand by discussing studies concerned with the notion of therapeutic landscapes, and with concepts of space and place and how these influence and affect health, health policy, and health services provision.

The brief review of medical geography contained within the chapter demonstrates how the emphases of the research carried out by medical geographers has changed, reflecting – after some time lag – the concerns of mainstream geography and most particularly theoretical developments in the social theory used within geography. Similar shifts are also evident in the application of other social science disciplines to the study of health. The examples discussed in this chapter also demonstrate a shift in emphasis from 'traditional' *medical geography* towards perspectives better described as the *geography of health and health services*. In the sections that follow, each of these different strands in medical geography is examined and its concerns and limitations discussed. We do not subscribe to the Kuhnian notion of paradigm shift; all of the strands discussed here are still present in current geographical research, and they are often complementary rather than competing. Thus the chapter should not be read as a simple sequential journey through history, although the strands found earlier in the chapter do, by and large, have a longer history than those towards the end.

Traditional medical geography: the roots of a discipline

The first of the two strands considered under this heading can be linked to environmentalism in mainstream geography, while the second draws on behaviourism. Historically, such studies stand at the roots of the identification of medical geography as a distinct field of knowledge and inquiry, and are still continuing today.

Strand 1: spatial patterning of disease and death

Whoever wishes to pursue the science of medicine . . . must first investigate the seasons of the year and what occurs in them. Observe and reflect upon winds, drinking waters, site, elevation, soil, climate, astrological features, and diet, the better to understand diseases and treatments. (Hippocrates, *Airs, waters and places*)

The recognition of the relevance of geographical factors and of the nature of particular places in the study of disease is of long standing, although a separate discipline of 'medical geography' did not emerge until much later. According to Riley (1987: 5), it was during the eighteenth century that the first studies setting systematic aggregation of disease data against quantification of climatic and habitational factors emerged; in part this was made possible by the earlier developments of suitable instruments for the measurement of climate and weather (p. 7). Many of the early studies were carried out by physicians, and sought to characterize the medical geography of a particular city or locale rather than to undertake comparative analysis; the focus was heavily on geographical and topographical features, rather than meteorological. Into this category, for example, comes the work of the physician Lind with his book *Essay on the incidence of disease in hot climates*, published in 1768 (Barrett 1991).

Riley and Barrett both identify Finke, a German, as a particularly significant figure for his production of *An attempt at a general medical-practical geography* (Barrett 1993: 701). This was published over the period 1792–5 in three volumes, and covered the entire world; it also included the formulation of general methodological principles, chief of which was the plea for a move to more heavily quantitative measurements (Riley 1987: 40). Although this work was one of the first, if not the first, to use the specific term 'medical geography', Finke was not a geographer by training, but a medical doctor (Barrett 1993), and his work is particularly interesting for its wide conception of the factors influencing health: 'my book . . . must however, list everything responsible for the healthy or unhealthy nature of each and every country: climate, form of government, religious customs, morals, habits, diet, education and the like are the main requirements for the discovery of the origin of disease' (Finke (1792), quoted by Barrett 1993: 703–4); this, however, was not really lived up to by the analysis he was able to offer (based entirely on other written sources), nor in the subsequent work of medical geographers until considerably later. The year 1792 also saw the publication of a medical geography of North America in William Currie's *An historical account of the climates and diseases of the United States* (Riley 1987).

The early emphasis of the work of medical geographers in the nineteenth and early twentieth centuries was medical cartography, the mapping of occurrences of disease and/or death. These studies are best characterized as descriptive, although in many cases they could be used to suggest associations with possible causative factors. There was an almost total dependence on availability of data at appropriate level, and studies were limited by the quality of data available, since most of these works relied on data collected by other people for other purposes, rather than undertaking any special data collection exercises. There were often difficulties of interpretation in the findings, particularly with regard to causation, and in many cases the results were dependent on peculiarities of data definition. In addition, the early studies were often unsophisticated or inappropriate in their choice of the level of spatial disaggregation and in their use of cartographical techniques. More recent examples of this strand are Howe's exhaustive studies of mortality in the UK (Howe 1963, 1972), the atlases of mortality by Gardner *et al.* (1983, 1984), sections within Cliff and Haggett (1988), and the *Atlas of cancer mortality in the People's Republic of China* (Li 1989).

The most basic medical cartographical studies confined themselves to providing a description of the location of events, in terms of either deaths or diseases, across space and in some cases through time as well. However, another part of this strand of traditional medical geography comprises disease ecology studies (also referred to as 'ecological medical geography' by some and 'studies of geographical pathology' by others), which take as their specific purpose the study of geographical patterns of disease in order to suggest possible environmental links with factors causing the particular disease. Cliff and Haggett (1988: x) identify the 1830s as the start of the use of maps for the testing of causal hypotheses. The example they give is that of Malgaigne's maps of hernia amongst military recruits in France (published 1839): combination of these maps with data on consumption of olive oil and regions where wine-drinking was exceeded by cider-drinking enabled the elimination of both these factors as causes of high hernia rates.

A classic example of the disease ecology approach is John Snow's study of cholera in the 1850s (Snow 1855). Snow was a general practitioner working in the district of Soho in London; within ten days in late August and early September 1854, over 500 people died of cholera in the district of Soho. Snow plotted the distribution of cholera deaths according to where people lived, and showed that the vast majority lived in an area clustered around a particular water pump, on Broad Street. Snow went on to show that most of the deaths had occurred in people who drank water from this pump, whereas those living in the same neighbourhood but using different pumps escaped; further examination revealed that the water at the Broad Street pump had become infected by sewage from a leaking cesspool or drain. Snow had the handle of the pump removed, and thereafter the number of new cases declined. Sources differ as to whether this was a largely symbolic gesture, as the epidemic was already limiting itself, or whether it represented a decisive step in halting new cases. However, it remains an excellent demonstration of the association between cholera and contaminated water, and also illustrates how the mapping of mortality or morbidity can help point to environmental factors that might underlie the particular patterns observed. It further demonstrates the potential of such studies for instigating suitable preventive public health action.

In this case Snow already had his hypothesis about the causation of cholera, and so in some sense was using the map as a device to seek confirmation. He also went on to find out, wherever he could, the source of the water drunk by those who had died of cholera, as well as those who had not, to build up the evidence for his supposition, resulting in a careful and painstaking examination of individual cases:

> Dr Fraser also first called my attention to the following circumstances, which are perhaps the most conclusive of all in proving the connexion between the Broad Street pump and the outbreak of cholera. In the 'Weekly Return of Births and Deaths' of September 9th, the following death is recorded as occurring in the Hampstead district: 'At West End, on 2nd September, the widow of a percussion-cap maker, aged 59 years, diarrhoea two hours, cholera epidemica sixteen hours.'
>
> I was informed by this lady's son that she had not been in the neighbourhood of Broad Street for many months. A cart went from Broad Street to West End every day, and it was the custom to take out a large bottle of the water from the pump at Broad Street, as she preferred it. The water was taken on Thursday, 31st August, and she drank of it in the evening, and also on Friday. She was seized with the cholera on the evening of the latter day, and died on Saturday, as the above quotation from the register shows. A niece, who was on a visit to this lady, also drank of the water; she returned to her residence, in a high and healthy part of Islington, was attacked with cholera, and died also. There was no cholera at the time, either at West End or in the neighbourhood where the niece died. (Snow 1855: 44–5)

Throughout the work of Snow, as in other early medical geographical works, we find notions of the healthiness of particular places, exemplified above by his comments on the 'high and healthy part of Islington' in which the niece lived, reinforcing the intimate connection seen between characteristics of places and health or disease.

Snow's studies used individual as well as area-level data. Others are concerned only with data measured at area level; these are often referred to as '*ecological* associative studies', where the concern is to estimate relationships between aggregate-level data on incidence or prevalence and variables representing aspects of the physical and social environments. These studies use statistical techniques of varying sophistication. Giggs's study (1973) of schizophrenia in Nottingham in the 1960s and 1970s provides an interesting example. Giggs analysed the addresses of the 464 persons diagnosed as schizophrenic and admitted to hospital for the first time from within Nottingham over the period 1963–9. He initially analysed the attack rates for people living at different distances from the centre of the city. This showed a very interesting spatial pattern: the standardized attack ratio (standardized for the different numbers of populations at risk in the different census enumeration districts) was much higher for people living nearest to the city centre. Similar patterns were found for different population subgroups: males, females, married, single, etc. Giggs went on to demonstrate that these same areas with a high standardized attack ratio had distinct social and environmental characteristics: high levels of unemployment, low social cohesion, and poor housing; i.e. they were areas with unfavourable social and environmental characteristics and high levels of social deprivation. One major problem with the study is in the interpretation of the results found. One possible interpretation is that 'pathogenic' areas within cities have been identified – the so-called 'breeder' hypothesis – where increased incidence of illness is caused by particular factors in the adverse conditions; this was the conclusion advanced by Giggs. An alternative explanation is provided by the 'drifter' hypothesis, which argues that high rates are due to inward migration by sufferers from conditions such as schizophrenia (into areas of cheap housing, etc.); this is what Gudgin (1978) argues is a more persuasive explanation for the spatial patterns found by Giggs. Here the argument is that schizophrenia sufferers move into areas of cheap housing, etc. after they have developed the disease, perhaps because they have difficulty staying in paid employment, and thereby inflate local hospitalization rates in these areas, as well as possibly affecting indices of social deprivation. The data in the study by itself does not allow for conclusive adjudication between these hypotheses. A later study (Dauncey *et al.* 1993), carried out at the individual level, which examined the lifelong residential mobility of a cohort of schizophrenia sufferers in Nottingham, was not subject to the same limitations. The authors conclude that the patterns of mobility found are best explained by a combination of the breeder and drifter hypotheses.

Ecological associative studies use aggregate measures of health and risk factors for the populations of different geographical areas, rather than linking ill health to environmental risk factors at the level of the individual. These studies are therefore subject to the dangers of the so-called 'ecological fallacy', which involves making erroneous assumptions about the health of, and the causes of illness in, individuals based on information for aggregated populations (this is discussed further in Chapter 6). Causal relationships cannot be proved in such studies, only disproved. However, such ecological associative studies do remain useful for generating hypotheses about possible causes of ill health and health difference, and they may help to show how the combined attributes of individuals in particular geographical settings can have implications for health (as discussed in Chapter 5).

The period after the Second World War until the early 1960s saw the majority of medical geographical studies concerned with spatial patterns of specific diseases and of mortality, and their relationship to particular environmental factors. For example, cancer, coronary artery disease, cerebrovascular disease, and bronchitis received much attention, and studies of the effects of specific environmental agents included work on cigarette smoke, background radiation, and water supplies (see Jones and Moon 1987; Learmonth 1988).

In some instances the examination of spatial differences identifies a cause for concern with implications for health services and health policy. In the UK, for example, the Black inquiry (Townsend and Davidson 1982) into inequalities in health set up in 1977 was prompted by identification of differences in infant mortality between countries, and by concern that the British population was not improving on some indicators as fast as a number of other industrialized countries. A whole range of investigations in the UK into links between nuclear installations and cancer (COMARE 1986, 1988, 1989) was initiated by the identification of clusters of deaths from cancer in the neighbourhood of nuclear installations; and surveillance of disease clusters is recognized as a powerful tool for early response to health problems. Schneider *et al.* (1993), using US data from the New Jersey Cancer Registry for childhood and young adult cancers, illustrate how examination of incidence data at multiple geographical scales can answer most questions about whether clusters are significant or not without the need for costly epidemiological studies, and can help in appropriate spatial and demographic targeting of epidemiological work where necessary. Li (1989) discusses how the *Atlas of cancer mortality in the People's Republic of China* has been used as a basis for the appropriate organization of anticancer campaigns, for the investigation of cancer etiology, and for the evaluation of the quality and impact of cancer prevention and control.

Closely related to disease ecology studies, which take a 'cross-sectional' view of the spatial pattern of disease at some point in time, are disease diffusion studies, which examine the spread of disease over space through time. These are often used to examine the spread of infectious diseases. Haggett (1994) suggests that diffusion studies can be usefully subdivided into descriptive, predictive, and interdictive studies. Descriptive studies of the progress of past epidemics can be used to derive quantitative predictive models, which can be used to anticipate the future course of epidemics. Such models can also be adapted to develop interdictive scenarios, testing out the likely impact of different forms of intervention to control epidemics. One example of disease diffusion studies is the extensive work carried out on the spread of measles in Iceland (Cliff and Haggett 1988). This represented a particularly suitable case for analysis, being a contained island community (thus simplifying the possibilities of transmission) and one with suitably comprehensive data sources over a sufficiently long period. More recently, we have seen work on a much more extensive geographical scale devoted to the case of AIDS (Smallman-Raynor and Cliff 1990).

Perhaps the most comprehensive recent work dealing with this traditional strand of medical geography is by Cliff and Haggett (1988). This is rather mis-leadingly titled an *Atlas of disease distributions*; it is not so much a compendium of maps, more a comprehensive, 'state of the art' review of the analytical methods of medical cartography which describes some of the more sophisticated cartographical techniques. Other wide-ranging overviews in this field are provided by Learmonth (1988), while the edited collection by Akhtar (1987) contains a section that deals with such studies of tropical Africa.

Later in the book we see how this strand of medical geographical work feeds into changing perspectives on diseases and their causes. There are very close interdisciplinary links between this strand of medical geographical work and the disciplines of epidemiology and ecology, and with public health perspectives. Chapters 4 and 7 take up the discussion of changing views on the relationships between health and the social and physical environment, resulting in reappraisals of ideas of the 'epidemiological transition', and of the influence of social and living conditions on health, for example. Changing views on social and spatial inequalities in health and ill health are also examined in Chapter 4, which explores inequalities at different spatial scales and shows how studies of both individual and area attributes have been used to explore health inequalities along various dimensions such as social class, ethnicity, gender, and place of residence. In Chapter 6 we examine how information on geographical differences in health can be applied to the estimation of need for health services and resource distribution, a question which leads us to the boundary with the second strand considered below.

Strand 2: spatial patterning of service provision

The second strand was also present from a very early stage in the life of medical geography as a discipline. Here the main focus is on service provision and/or service use, particularly their spatial patterning, hence the alternative name, 'geography of health care' or (in a more restricted sense) 'geography of health services'. Three main components of the geography of health care can be distinguished. The first of these includes studies of the *structure and spatial patterning of health service facilities* like hospitals, clinics, and doctors' surgeries. Here the scale varies from national down to very local level. This field of work is analogous to the disease-mapping of ecological medical geography. A second component is provided by studies which aim to identify *patterns of inequality in supply and use of services*, in some cases comparing these with structural and spatial configurations of health service systems which are presented as being 'optimal' on certain criteria. Thirdly, there are studies which focus on *patient utilization of health services* and factors influencing the behaviour of individuals in their contact (or non-contact) with organized health services. These attempt to elucidate the factors that influence individuals' decisions to make use of health services. Once the variety of different lay and professional concepts of health are recognized, and more sophisticated theories of behaviour are incorporated, this subject area takes us into the domain of the third strand, to be considered shortly. The second and third components may be researched at levels ranging from the aggregate scale of broad population groups to the level of the individual or household; the emphasis within both is on modelling in terms of quantifying relationships between variables.

The emphasis here is on the spatial characteristics of health service systems and their utilization. It includes examination of spatial variation in the availability, quality, and use of health service facilities, while the nature of health and ill health is regarded as unproblematic in a positivist fashion. Within this strand we find numerous studies using a simple cartographic approach, examining changes in the scale and nature of various types of service provision at different levels of aggregation over time. Excellent reviews of many studies of this type carried out in a variety of countries can be found in Joseph and Phillips (1984), while the book by Haynes and Bentham (1979) offers more extended treatments of studies relating to rural accessibility; Phillips (1990) is a good starting-point for studies of low-income countries and Akhtar (1987) contains a range of studies on health service delivery in African countries.

Major themes within this strand of medical geography relate to the twin issues of accessibility and service use, with much attention given to deriving suitable quantitative measures of accessibility. Such measures are of varying degrees of complexity, but usually involve making highly simplistic assumptions about the behaviour of patients or potential patients. Most of these studies also impose a conception of accessibility constructed by the researcher, rather than deriving from the views of the patients themselves (studies that do examine patients' concepts of accessibility are considered under subsequent strands). Two general issues here are the selection of appropriate levels of spatial aggregation for analysis and the inclusion of relevant socio-economic factors affecting access. This is well illustrated by consideration of the urban poor. As Harpham *et al.* (1988) illustrate, although urban areas as a whole may be advantaged in terms of overall levels of health expenditure and availability of facilities, this does not benefit all city-dwellers equally, and radical differences in accessibility are found; those most disadvantaged are the urban poor, and in particular slum and shanty-town dwellers.

Following in the wake of the so-called 'quantitative revolution' (Burton 1963) in mainstream geography, with its emphasis on geographical systems and the application of computerized techniques such as linear programming and stochastic modelling (Cliff and Haggett 1988), it is not surprising that many studies of this type used these methods to identify 'optimum' patterns of facility location, for example Horner and Taylor (1979) and Wilson (1974). These analyses rely on the identification of specific quantifiable criteria, for example average travel time or travel cost, which are used to identify the 'optimal' configurations of facility locations and catchments, and thus to minimize the selected criteria for the population in the region studied. The concept of 'optimality' is extremely problematic, however; see for example the critiques offered by Curtis (1982), Stimson (1983), and Rosenberg (1988), in particular concerning the assumptions that are necessary in order to generate a single set of locations which is 'optimal'. One major difficulty is the selection of an appropriate criterion to use. Where use of a simple criterion like travel time or travel cost is not considered to be appropriate, then a composite criterion can be generated, but this is still problematic in terms of weighting the different components. In addition, some factors may be hard to incorporate; for example, different qualities of care may be available at different locations. There are also problems with the simplicity of approach used, for example, often assuming that catchment areas can be defined and that people do not travel outside them, whereas in reality, and for a wide variety of reasons, people do cross such boundaries, and may not visit their nearest facility. There are also practical, political, and social constraints on the implementation of optimum solutions. Finally, there is also the difficulty of dealing adequately with real-life situations: there are often multiple objectives which can generate several possible criteria to be used for the determination of 'optimality', and these criteria may conflict.

Some of these problems are explicitly recognized in some later studies (see e.g. a number of the applications in Australia, England, Italy, and the USA reviewed in Taket *et al.* 1986), where the emphasis is shifted to a simulation type of approach, providing models which can (it is hoped) inform, but not produce, planning decisions. This is based on a rather different approach, namely the use of analytical methods used to help health authorities evaluate the potential effects of different alternatives. Here there is no attempt to generate optimal locations; instead a model is used to predict the pattern of use that will arise from each feasible option under consideration, and consequences can be explored by comparing options on a number of different criteria. In situations where different criteria point to different solutions, planners and other relevant parties can then consider explicitly the trade-offs between different criteria, exposing the value judgements involved in discussion and negotiation; see for example the study by Taket (1989), which explores one such case where selection of an option based on the criterion of equity (in terms of minimizing variance in ratios of service use to need for different population groups) yielded very different results from that based on the criterion of accessibility (based on minimizing simple measures of travel distance). Although the behavioural assumptions involved in these later studies can still be criticized for their gross simplicity (in the underlying models of the determinants of service use behaviour), at least such approaches can deal with the existence of multiple (and possibly conflicting) criteria. This type of study can be seen as essentially 'applied research', which has direct and immediate application in a health service planning context. It can also be suggested that this makes such studies particularly dangerous if the criteria selected for use do not represent the most important factors influencing uptake. Less sceptically, in a wider framework of information and analysis related to health planning problems, it might be argued that this illustrates the potential usefulness and relevance of the insights and tools offered by medical geographers.

Other recent modelling approaches which examine aspects of service use include 'multi-level' modelling, used by Jones *et al.* (1991) to study ecological and individual effects in the uptake of childhood immunization. These models allow calculation of estimates of relationships at different levels of aggregation. In the case of the immunization study, effects at the level of individual service users and at the level of the treatment centres could be estimated, showing the separate importance of factors at both levels. These results suggested that policies for immunization in the area studied should aim to take into account the relevant characteristics of the clinics providing immunization as well as individual attributes of users, related to their rate of uptake. Although such a study does identify the factors *statistically* associated with high or low uptake, arguably appropriate policy responses cannot be designed without understanding the *reasons* underlying these statistical associations. Multi-level models do not provide the necessary insight for explanation; for this an alternative approach, considered under the next strand, is more appropriate and we discuss a relevant study below.

Such geographical studies of services feed into questions of resource allocation for health services. Issues of equity and efficiency are of growing importance for health service organization and use, within the context of the changing structure of health services worldwide, and in particular in the light of the growing policy emphasis on the role of market mechanisms in achieving efficient and effective allocation of resources. Chapter 6 is concerned particularly with the evaluation of successes and failures in health service provision, and examines the contribution of geographical perspectives, alongside those of disciplines such as health economics, to the development and implementation of institutionalized processes for evaluation of resource allocation and service provision in health services.

Contemporary medical geography

More recent developments in medical geography have reinforced the shift of emphasis from 'medical' geography towards the 'geography of health and health care', which is more critical of biomedical and positivist views of health.

Strand 3: the humanistic turn

In the third strand we find examples of studies in medical geography which have been influenced by humanistic approaches. These approaches are widely used in various areas of human geography, and they drew heavily on strategies developed within other disciplines such as anthropology and sociology. This strand of medical geography is exemplified by work on concepts of health and health-related behaviour. In these studies we see the replacement of highly simplistic normative behavioural assumptions, which formed the basis for much earlier work, by a focus on the nature of the motivations behind individual health-related behaviour and a concern with understanding individual decision-making. Research is most often carried out at the level of the individual, and uses more qualitative research approaches.

Studies such as these are perhaps still comparatively rare from those who classify themselves solely as medical geographers. In the UK, Cornwell (1984), Donovan (1986), and Eyles and Donovan (1986) adopted an ethnographic approach, and used extended in-depth interviews (both structured and unstructured) of small numbers of people in particular neighbourhoods or social networks to examine concepts of health and illness and their causation, perceptions of their own and others' health status, use of health services, attitudes to health and illness, and views on health services. Cornwell looked at white working-class residents of a particular neighbourhood (Bethnal Green) in east London, and Donovan at Asians and Afro-Caribbeans living in different parts of London, while Eyles and Donovan studied residents of a west midlands town in England, with a mainly working-class population.

The use of ethnographic methods is directly linked to the adoption of a humanistic perspective: 'we have employed qualitative or interpretative methods to allow people to express themselves in their own ways' (Eyles and Donovan 1986: 426). The strategy of grounded theory (Glaser and Strauss 1967) is often used, in which initial theoretical concepts are progressively refined by recourse to material obtained in the field, so that the resulting interpretation and general propositions emerge only after detailed investigation. This shift in emphasis from quantitative, statistical investigation of pre-defined hypotheses to qualitative approaches, the use of detailed case studies, and the theoretical selection of subjects rather than random sampling represents a shift in the notions of validity employed in the conduct of research, and a move away from positivism. The analysis of interview material enables the researchers to extract relevant concepts and constructs from people's explanations, from their stories, from the way they make sense of their own and significant others' lives and the incidents in them: 'the logical consistency of the relations between the identified characteristics may be seen in the fact that they are derived from complete stories about health, health care and sickness in which individuals make sense of their lives. We emphasise again that such logical consistency need not be related to "truth" but with how people perceive and act in the world' (Eyles and Donovan 1986: 417). So the concern becomes one of exploring people's own lived experience, as it relates to health, in terms and concepts derived from people's own stories, their own words. Daker-White's study (1995) of the perceptions of clients and staff of drug services in east London and Allen's (1995) study of perceived roles and responsibilities in relation to health care for children, also in east London, provide recent examples of this type of study.

While the traditional ecological approach within medical geography, discussed earlier, can be characterized as dominated by naturalist conceptions of health and illness, in this third strand of research we find a direct concern with the social construction of health and illness (e.g. this is a major focus in the studies by Cornwell, Eyles, and Donovan mentioned above). However, we do not mean to imply very sharp divides or boundaries between the strands; for example, Philo (1986), examining what he refers to as 'psychiatric geography', also identifies the seeds of this humanistic approach within studies more usually cited as exemplars of the traditional ecological approach, such as Faris and Dunham (1939) and Giggs (1973).

This third strand also includes studies which examine features of health service provision and use from the point of view of the service user, exploring consumers' notions of concepts of accessibility, appropriateness, and satisfaction, rather than imposing some predefined view of these. The results of such studies confirm the criticisms levelled at the measures of accessibility used in many of the studies discussed in the previous section, defined simply in terms of distance, travel time, travel cost, or waiting time. For example, Scarpaci (1988a) reports a study of frequent primary-care users in Santiago, Chile which revealed no association between patients' assessments of the accessibility of facilities and the length of time waiting for care. (This was with an average waiting time of 4.1 hours, so that it could not be argued that all waits were of insignificant length.) Instead, factors such as the manner of the doctor towards the patient were more associated with patients' notion of 'accessibility'. Curtis (1987a) reports a similar emphasis on rapport between doctor and patient in responses to a survey in London. The issues of measurement of accessibility and satisfaction are taken up again in Chapter 6.

With reference to the question of uptake of immunization services, already discussed above, a study by New and Senior (1991) explored qualitative aspects of parental decision-making. They showed that those who missed appointments could be distinguished from those who attended on the basis of differing attitudes to, and knowledge of, infant immunization, arising from varying personal experiences. Transport problems and time–space constraints (which were expected to affect uptake) did not account for differences found. The study further argues that once the detail of parental decision-making is explored, it becomes clear that it is mistaken to label failure to take up immunization services as 'irrational'. Explanations for such behaviour should take into account the interaction between personal experience, advice and information received, and the impact of constraints, particularly gender role constraints, on women with young children.

The terrain of this third strand is covered in detail, incorporating work from other disciplines, in Chapters 2 and 3, which examine the social construction of health, disease, and illness, and also in the sections of Chapter 6 dealing with models of access to services. In these chapters we see geographical perspectives merging at the interdisciplinary boundaries with sociology and anthropology. Besides people's 'individualistic explanations', the researchers in many of the studies considered here draw out the wider connections in the ways that individuals see work, locality, and social change as related to health and health care. These connections are illustrated by people's stories of episodes from their own lives, and enable an exploration of 'the dialectic relation between individual and society' (Eyles and Donovan 1986: 425), which takes us on to the domain of the fourth strand.

Strand 4: the structuralist/materialist/critical turn

The label we have given to this strand is meant to illustrate the range of different types of social theory that studies of this type might draw on. Alternative titles suggested by some authors are 'political ecology' (e.g. Mayer 1992) or 'medical–social geography'. In this strand we find a more recent emphasis, which initially grew out of the 'welfare' approach in human geography (Smith 1977), focused on the presumed importance of health and welfare services in improving the quality of life for people. This moved beyond descriptive analyses of 'who gets what, where, and when', discussed above under the second strand, towards a concern with explanations for the patterns found, leading to exploration of the role of socio-economic processes in the production of inequity in health and resource distribution. This area is explored explicitly for the case of Britain in Eyles's book *The geography of the national health: an essay in welfare geography* (Eyles 1987). Studies within this strand also have a particular concern with the interactions between individual agency and structural and material constraints in the shaping of people's experience of health and health services. They also examine the roles of health services in society. Research of this type may share the concerns of humanistic studies in strand 3 with individual perceptions and experience, examining the ways in which these are influenced by wider socio-structural power relations and using social theory to develop our understanding of these processes.

Part of this component are studies which include an examination of the interplay of social and political forces in shaping policy affecting health service provision and the health services themselves. Such approaches, variously referred to as having a critical, structuralist, materialist, or political economist standpoint, place their emphasis on links to the totality of society and its particular economic and social structure. The studies described in this section respond to some of the critiques from within medical geography of the perceived limitations of much of the work carried out in the name of medical geography. Rosenberg (1988) for example, argues the need to link geographical, medical, and political aspects in the study of health service delivery. He illustrates his argument by a consideration of the use of abortion services in Ontario, showing how individual access to these services was influenced by a complex interaction of socio-cultural and political economic factors. This type of approach represents a broadening of political economic perspectives on health and health services, and we find geographers drawing on perspectives from political science and social theory to respond to this concern to understand how societal structures relate to health.

Turshen (1984), in her study of the political ecology of disease in Tanzania, as well as Kloos and Zein (1993), in their edited collection of studies of Ethiopia, exemplify this approach in setting out to identify the economic, social, and political determinants of ill health. A section in the edited collection by Akhtar (1987) presents a number of studies of different African countries which explore the impact of specific development projects and policies on the health of the population. This wider focus challenges concepts of disease causation inherent in the germ theory of disease and the doctrine of specific etiology, which form the basis of the biomedical model. In doing so, it emphasizes the concerns of social medicine, based on a socio-ecological perspective on health and ill health. This type of approach in geography can contribute to the body of work discussed more comprehensively in Chapter 7, which examines public health strategies.

Examples of studies of this type focused on health service provision include Scarpaci's (1988b) detailed study of primary medical services in Chile, examining the economic and organizational structure of the delivery system and how this affects accessibility. This clearly illustrates the importance of the wider social and economic context in accounting for the spatial patterns of service use, and inequalities between different groups in the population, particularly according to income. Scarpaci extends this analysis (1991) to examine how the public and private provision of health services, particularly primary care, in Argentina, Uruguay, and Chile has been shaped by the specific ways in which these states have restructured collective consumption, and the effects this has had on the implementation of policies with aims such as decentralization. Another example of the type of work carried out in this field is provided by a paper by Mohan (1990a), which analyses state intervention in health services, and compares state policy in Britain and the United States in the 1980s to analyse the efforts of right-wing government to 'restructure' the welfare state. In both countries this has resulted in moves towards a more mixed economy of welfare. However, Mohan identifies the various tensions at work which result in a divergence between the ostensible aim of 'rolling back the state' (present in the election manifestos of both the Reagan administration in the USA and the Thatcher administration in Britain) and the increased regulatory role adopted in relation to health services. The differences in the ways that regulatory strategies have been adopted in the two countries, in particular in respect of the balance between central and local tiers of the state, illustrate the importance of moving beyond simple functionalist, particularly marxist, analyses of the state and health policy, and the importance of examining the historical and geographical context surrounding the policy under analysis. The key issues discussed by Mohan are taken up again in later chapters of the present book: efforts to promote cost containment and the internal reorganization of health care provision via deregulation and privatization are considered in Chapter 5, while concerns with decentralization and participation as policy themes are addressed in Chapter 8.

There is also a group of geographical studies, forming part of this strand, which explore the historical development of health and health services. Newson (1993) reports on part of a large-scale study concerned with demographic change in sixteenth-century Ecuador, demonstrating how the effects of 'Old World' diseases on the indigenous populations of early colonial Ecuador were mediated by different sizes of communities, their settlement patterns, forms of subsistence, social organization, and ideology, affecting both disease mortality and the ability of different indigenous communities to recover. Another example is provided by Good (1991), who examines the role of Protestant and Roman Catholic missions in pioneering Western medicine and public health in much of Africa, before Western health services began to be provided by colonial governments, viewing them as 'frontier agencies of imperialism in a multi-cultural setting – the vanguard of Western medicine' (p. 4). Chapter 9 takes up similar concerns in examining some of the historical background to the work of international organizations in health.

Strand 5: transgressing the boundaries – the cultural turn

The final strand to be considered here consists of studies informed, in various ways, by the concerns of cultural geography. We might call this the cultural turn in medical geography, representing changing views of the importance of space and place to individuals and their health. This issue is explicitly taken up by Gesler (1992), in an article which explores how an expanded meaning of the concept of 'landscape', informed by the 'new' cultural geography, is first reflected in the concerns of health-related studies carried out from within other disciplines, and secondly indicates a fruitful agenda for further exploration within medical geography. He argues that there is much to be gained in the exploration of therapeutic processes in different settings by the application of concepts drawn from cultural geography such as sense of place, landscape as text, symbolic landscapes, negotiated reality, hegemony and resistance, territoriality, legitimization, and marginalization. He integrates these concerns within the notion of the *therapeutic landscape*, understood as:

> a geographic metaphor for aiding in the understanding of how the healing process works itself out in places (or in situations, locales, settings, milieux). For example, a confrontation between a patient and physician in a treatment room is affected by the physical attributes of the room (e.g. temperature, size, colour of the walls, arrangement of the furniture), the ideas and intentions of the actors (e.g. illness and treatment beliefs, symptom description and interpretation, concealment of certain facts), and the structural forces underlying the physician – patient relationship (e.g. dominance–resistance, type of medical system, territoriality). (Gesler 1992: 743)

This is in many ways a return to some of the oldest concerns of medical geography to examine the interplay of human lives and health in particular places and spaces. However, there are important differences, such as the explicit attempt to find room for both structure and agency, and the concern with healing and the promotion of health rather than just with the causes of ill health. We also find a newly problematized 'gaze', which brings out more clearly the individual effects of relations of power and knowledge within societies. This particular blend of emphasis both on the power of human agency/individual action and on the influence and constraints of underlying structure has been labelled by some (e.g. Jackson 1989) as 'cultural materialism' – a joining of a materialist form of analysis to the concerns of cultural studies. This is one of the most recent trends to emerge in medical geography, and one that still remains to be fully exploited.

Work by Philo (1987, 1989) exemplifies some of these concerns in two rather different ways. The first is through a detailed examination of the organization of space within nineteenth-century 'lunatic asylums' (Philo 1989). Here he examines how spatial organization was explicitly considered in terms of its contribution to various objectives, not just to those of supervision, but also to serve therapeutic aims. He finds that asylums, both actual and proposed, followed a variety of different spatial organizations rather than any single institutional blueprint. Varying techniques for structuring and dividing asylums were explicitly used in different measures and for different purposes. Despite the attention to therapeutic aims – to 'reform' or 'cure' the inmates – he also finds that the practical role of such asylums was to produce and continually reproduce a population designated as 'different, deviant and dangerous' (p. 284). This finding still has resonance in the current exclusion of mental health facilities from 'respectable' (for which read 'relatively affluent') suburban communities (Dear and Taylor 1982).

A second theme in Philo's work is the examination of the importance of place and the consideration of particular locational attributes in decisions about the siting of asylums during a particular historical period (Philo 1987). He identifies the association of institutions with particular places, regions, and types of environment, and examines how these arose in the context of 'quite particular social, cultural and professional understandings of "madness" ' (p. 404). He identifies the prevailing view of the time as one requiring a 'natural', tranquil, rural setting for the achievement of any cure, noting that this represented a shift from the view common in the late eighteenth and early nineteenth centuries. A well-formed dominant consensus vision of the desirable features of the site for such institutions was formed, involving climate, soil, underlying rock type, topography, elevation, aspect, and vegetation cover. This view linked medical and hygienic dimensions to 'moral' issues, connected particularly to the possibilities for farm and garden work for patients and the aesthetic qualities of a landscape and its corresponding therapeutic effects. These considerations combined in a 'medico-moral' discourse which called for features such as extensive land availability, productive soil and rich natural vegetation, high elevation, and varied relief to produce a 'fit locality for an asylum'. The same association of elevation and health was noted earlier in Snow's writings, within the same period.

Similar themes are taken up in Bell's study (1993) which examines the links between geography and imperial emigration from Britain in the period 1880–1910. She focuses on the construction of images of South Africa's environments and British gender roles, present in different bodies of literature of the time, including medical treatises and technical guides to travellers on the care of their health. She demonstrates how 'underpinning the physiological discourse on climate was a distinctive regional medical geography in which moral judgements made about race and gender difference were given the authority of science while satisfying particular imperial goals' (p. 339). This included, in particular, arguments in support of segregation in the South African urban environment. This issue is taken up again in Chapter 9 below.

Studies showing the importance of notions of place are illustrated by the work of Kearns (1991). He studied Hokianga, one of the twelve 'special medical areas' in New Zealand, an area with a population of some 6000, over half Maori. These special medical areas are remote districts with scattered and often socio-economically deprived populations. Kearns's study focused on how the medical clinics and the hospital were experienced by users. People from the smaller and more isolated communities, in particular, used the clinics as a well-developed arena for social interaction, which suggested that the 'non-medical wellbeing of this area is positively enhanced by the particular form of health care provided' (p. 529). Kearns points out the dangers of ignoring such features when considering narrow concepts of 'efficiency' sometimes used to justify closure of public facilities.

Within the public health field, Coombes (1993) reports a detailed study of groups working on public health issues within an area of east London. She demonstrated how the agenda of public health professionals in directing work to one small neighbourhood rather than another was influenced by the very distinct views and conceptions these professionals held about the different neighbourhood: stereotypes almost, that did not always correlate well with other indicators of need, such as the actual state of housing and statistical measures of material deprivation. While the groups were successful in identifying and working within some areas of high need, there were other areas with similar levels of need that were not selected for any action. Notions of place were thus particularly important in influencing the implementation of public health policy. Other studies that take up similar concerns in practice include that of Moon (1990), which explores the use of conceptions of space and community in British health policy, arguing that they reflect a complex amalgam of sociological assumptions, providing perhaps little more than useful rhetoric for politicians.

In the field of health promotion, Woodhead (1995), taking the specific example of gay men and HIV prevention and the promotion of safer sex, is concerned to explore how culturally and spatially sensitive health promotion practices might be achieved by recognizing and examining the different ways in which 'outreach' health promotion activities need to be carried out in different kinds of space and place if they are to be successful. This, he argues, necessitates a reconceptualization of the notion of space which draws on, but questions, the categories of 'material' and 'imagined' spaces, to argue that these do not represent discrete classes of spaces, but rather 'that they are *implicated* within each other; they are *complicit*' (Woodhead 1995: 235; emphasis original). Ethical and political issues are raised, however, by an examination of radical health promotion practice which demonstrates how, despite its coding as resistant to dominant medico-moral discourse, it produces effects which are complicit with the dominant discourse. Studying services for drug-users in east London, Daker-White (1995) has demonstrated how the ways in which services are provided may act as forms of social control on drug users.

The concerns of this strand in medical geography link particularly with the exploration of lay concepts of health and ill health in Chapters 2 and 3 of this book. It contributes to the understanding of health policy development and implementation discussed in Chapters 8 and 9; and it shows

further examples of work at the interdisciplinary margins, where social
scientists from various disciplinary backgrounds exchange ideas.

Multiplying medical geographies

The foregoing account illustrates the range and diversity of medical geographical studies. It perhaps also illustrates the futility of attempting to
encapsulate the whole of the field under the rubric 'one medical geography
or two?', when the reality is much more complex. All the strands discussed
above, traditional and contemporary, continue to be present in medical
geography today, and they clearly interconnect. The text above should not
be read as providing any simple narrative of progress, but rather one of
multiplication and proliferation. The reader may prefer alternative groupings of studies, but the ones we have chosen here have illustrated how medical geography has changed, and in the process has contributed to the
changing perspectives on health and societies considered in this book.

In discussing different strands within medical geography, contrasting theoretical and methodological perspectives have been illustrated. In so far as
it is possible to distinguish clearly between the areas of work labelled 'traditional' and 'contemporary', we suggest that the two 'traditional' strands
accept disease as a naturally occurring, culture-free, and 'real' entity, where
the problems posed by questions of accurate measurement and distribution
are assumed to be technical (in the Habermasian sense) and solvable. In contrast, the other three 'contemporary' strands adopt a stance which argues, in
various ways, that notions of health, disease, and illness are problematic,
and intimately linked to power relations in society. Thus the assumption of
health professionals as invariably caring, neutral scientists is questioned,
and the different roles they fulfil in maintaining the current social order
become subjects for scrutiny.

In drawing such a distinction, however, we do not intend to imply the
superiority of one position over the other. We would, rather, argue that
each of these stances has a place and usefulness in medical geographical
research. There is much to be gained from studies based on designs that use
a judicious combination of the qualitative and the quantitative, for example.
A more extended discussion of similar arguments for the adoption of epistemological pluralism in the specific case of medical geography can be found
in Mayer (1992) and Scarpaci (1993), while, in the context of mainstream
geography, Pile and Rose (1992) illustrate the subversive and critical possibilities opened up by such stances.

In a review essay written in 1989 Mohan argued:

Local variations in health status and health care provision are certainly important but the principal concerns of medical geography as

currently practised – access to and the location and utilisation of health facilities, the use of quantitative techniques for spatial analysis in health care planning, or the extent of association between environmental factors and disease – seem largely irrelevant to an understanding of the socio-political determinants of health and access to health care. Medical geography requires radical surgery if it is to come to grips with such issues. (p. 176)

The diverse range of studies illustrated within the discussion above of the more recent work in medical geography, particularly within the third, fourth, and fifth strands, begins to answer Mohan's critique of the contemporary relevance of medical geography. They also illustrate the potential value of further exploration of the relationships between space and place in health policy and service provision, a theme explicitly developed by Kearns and Joseph (1993).

Medical geography thus represents a much more diverse field than it used to, and many researchers in this field would argue that the label 'medical geography' no longer adequately reflects this. It is perhaps more appropriate to talk instead of 'geographies of health and health services'. This diversity of perspectives and of methodological and theoretical approaches towards health, health services, and health policy is associated with changes in the academic perspective of geography, but it also reflects changes in the wider debates about health and societies going on outside as well as inside this particular discipline. The contribution of medical geography to these debates is considered alongside the interpretations offered by other disciplines in the following chapters.

|2|

Changing perspectives on the social construction of health, disease, and illness

... disease may involve a temporary or permanent impairment in the functioning of any single component, or of the relationship between [the] components making up the individual ... (Polgar 1968: 330)

the notion of the *body as machine*, disease as the consequence of *breakdown of the machine*, and the doctor's task as *repair of the machine* ... (Engel 1977: 131; emphasis added)

Health may be expressed as a degree of conformity to accepted standards of given criteria in terms of basic conditions of age, sex, community and region, within normal limits of variation. (WHO 1957: 14)

... the fact of 'health' is a cultural fact in the broadest sense of the term, which is to say at once political, economic, and social. Which is to say that it's bound up with a certain state of individual and collective consciousness. Each period has its own notion of 'normality'. (Foucault 1983: 175)

The quotations above exemplify some of the contrasting views on the nature of health, illness, and disease which will be explored in this chapter, which examines the diversity of what is meant by such seemingly innocent and self-evident terms as 'health', 'ill health', 'disease', 'sickness', 'illness', and how these interact with views on the causation of health and ill health. Such studies are an important component in understanding the basis for medical practice, as well as for an understanding of the wider relationships between the health service systems and the social context. Understanding of lay views on causation of health and ill health is important for the design of those health education and health promotion programmes aimed at making people change their behaviour in a way likely to improve health, and for understanding some of the basis for individuals' use or non-use of health services and patient 'compliance' with treatment regimes.

It has become conventional to divide discussions of concepts of health, disease, and illness into 'lay' and 'professional' views. Although such a division will be followed below, it is important to stress at the outset that such a division can be deceptive if taken to imply a unitary lay or professional discourse on the subject of health. A major emphasis here is on exploring the diversity within both lay and professional views. A distinction between the two is important in any particular context, however, since it enables an examination of the particular nature of the roles and impacts of health professionals within society.

The chapter begins by considering various 'professional' discourses on the subject, originating from health professionals and academic writers. In particular we contrast the biomedical model of Western medicine, which despite considerable contestation remains the basis for most formal health service provision in the West, and the more recently emerging (or re-emerging) socio-ecological discourse on health, which challenges the dominant position of the biomedical model. The chapter then moves on to consider research into 'lay' discourses on health and illness, reviewing selected studies of differences by gender, social class, ethnicity, and culture, and questioning how successful these studies have been in capturing the complex nature of responses to health and illness, constructed in the course of individuals' attempts to make sense of themselves and their experiences.

The chapter demonstrates how the various different discourses around health, involving notions of health, disease, symptoms, normal functioning, illness, etc., are all closely linked with the structure of the society concerned, i.e. shows that they are all socially constructed, and that furthermore these discourses are subject to constant contestation in different arenas, leading to a continual redefinition of their meanings and usage. This is further illustrated in the following chapter, which contains three case studies of contested definitions of health and disease, involving different lay and professional views, and their implications. These deal with: women and mental illness, in particular the 'disease' of hysteria and the problem of 'nerves'; the construction and deconstruction of homosexuality as a mental illness; and finally the proliferation of discourses on HIV and AIDS.

The mechanistic discourse of the biomedical model

The world view that has shaped modern Western science and technology, including Western medical science, commonly referred to as the mechanistic or Cartesian paradigm, emerged in the sixteenth century. This was a radical departure from the thinking of antiquity. The entire universe, including the bodies of all living creatures, was viewed through the metaphor of a huge mechanical system functioning like a clock with great precision according to mathematical laws. Understanding of these laws was held to be attainable through the application of the scientific method to component parts.

The key characteristics of this mechanistic paradigm were the separation into component parts and the emphasis on the detailed study of each, i.e. a reductionist approach. Discoveries such as the circulation of the blood (Harvey) and the physics of the functioning of the eye (Kepler) supported the development of this mechanistic view of the human body (Rhodes 1985). This mechanistic view of the world began to have an impact on Western medical practice during the late eighteenth and early nineteenth centuries. In this period, experiments by Pasteur, Koch, and others demonstrated that some diseases could be produced by the introduction of single specific factors, virulent micro-organisms, into animals. This led to a corresponding search for the 'magic bullet' – the specific substance required to counteract the disease-causing factor and cure the disease; the use of vaccination provides an example of the successful application of this doctrine. A detailed study of the turning-point in Western medical practice is provided in Foucault's examination of the birth of modern clinical medical practice in France (Foucault 1973), which describes, *inter alia*, the emergence of this mechanistic discourse.

Within biomedical discourse, disease is a 'temporary or permanent impairment in the functioning of any single component, or of the relationship between [the] components making up the individual' (Polgar 1968: 330) or, more succinctly, 'the consequence of *breakdown of the machine*, and the doctor's task as *repair of the machine*' (Engel 1977: 131; emphasis added). Ill health, manifested by various signs and symptoms, results from pathological processes in the biochemical functions of the body. The specific constellation of signs and symptoms enables the identification of the specific disease responsible, whose occurrence results from the action of some specific pathogen.

This conventional rendition of biomedicine tends to obscure the fact that we are talking largely about male scientists and physicians in Western societies. Other views, some of which offer direct contestation of the biomedical view, have always coexisted, in particular in the perspectives and paradigms used by traditional midwives, herbalists, and other practitioners (see e.g. Donnison 1977; Cooter 1988; Ehrenreich and English 1974, dealing with female practitioners). There are also important differences in non-Western societies, where the basis of medical practice involves radically different ontological assumptions (see e.g. Bannerman *et al.* 1983 for a general coverage of different systems of medicine (often inappropriately labelled 'traditional' in a pejorative sense); Kleinman *et al.* 1975, Kaptchuk 1983, and Aakster 1986 for a discussion of the basis of Chinese medicine; Clifford 1984 and Donden 1986 for Tibetan Buddhist medicine; Kakar 1984 for a discussion of Indian systems, Obeyesekere 1977 for Ayurvedic medicine; Lambo 1964, Ayoade *et al.* 1978, Janzen 1979, and Janzen and Prins 1981 for an examination of concepts of health in African cultures). We return to some of these different views in Chapter 5.

Within biomedical discourse we can identify four major assumptions (Mishler 1981). The first of these is the notion of disease as *deviation* from *'normal'* biological functioning; with an eye to critiques of the biomedical model that will be considered later, we should note the difficulty of establishing the limits to the range of variation that is normal. The second assumption is the doctrine of specific etiology, the notion that each disease is caused through a *specific* pathogenic agent, micro-organism, or disease vector. Third is the assumption that diseases are generic: that any case of a particular disease will have the same disease symptoms and processes in different times and locations. The emphasis here is on physico-chemical data, and observations made by doctors (what Foucault calls the physician's 'gaze') rather than the patient's perception. Finally we can identify the implicit assumptions of the scientific neutrality and rationality of medicine. Within the discourse of biomedicine there is thus an emphasis on ill health due to diagnosed disease rather than on health, and correspondingly, the direction of attention and intervention towards *cure* rather than *prevention*, i.e. a concentration on the treatment of ill health once it has arisen, rather than on its prevention, or on the promotion of good health. The normative definition of health that lies within the biomedical model is well expressed in the definition put forward by a WHO study group on measurement of levels of health: 'Health may be expressed as a degree of conformity to accepted standards of given criteria in terms of basic conditions of age, sex, community and region, within normal limits of variation. It is a relative concept' (WHO 1957: 14). We may note particularly its emphasis on objective measurement, and also the roughly equivalent definition: 'health is an absence of disease'.

Within the discourse of biomedicine it is usual to make a distinction between *disease* and *illness*. The distinction is well expressed by Field (1976: 334): 'Disease . . . refers to a medical conception of pathological abnormality which is indicated by a set of signs and symptoms. Illness . . . refers primarily to an individual's experience of ill-health and is indicated by the person's feelings of pain, discomfort and the like.' So while patients experience 'illnesses', physicians diagnose and treat 'diseases'; illness is subjective, disease is objective. It is thus possible to have illness without disease and disease without illness. Illness is 'allowed' to be culturally specific and dependent on lay conceptions of normality which may or may not have a relationship to biomedical definitions. Illness is allowed to have moral, psychological, and social as well as physical dimensions. Studies of lay conceptions of health (which are considered later) illustrate the differences that can exist in what is regarded as illness between different cultures and between different social groupings within a single society. In the light of this distinction between illness and disease, the definition of health in the biomedical model as 'absence of disease' is (deliberately) different from 'absence of illness', emphasizing the fact that the biomedical definition is rooted in *professional* views, in the judgement of the medical 'expert'.

Contesting the biomedical model

The concentration on the individual and a focus on component parts of the body, together with the separation of the mind and body and the emphasis on measurable physiological or chemical data, has meant that social, cultural, economic, and even environmental factors in the causes of ill health are often neglected (Doyal 1979; Elling 1982; Navarro 1986; Tesh 1988), and further, that in a medical consultation emphasis is placed on diagnosis by tests. A further critique of the biomedical model argues that, as disease and illness are both socially determined, the diagnosis of disease is not objective.

The biomedical model has proved particularly unsatisfactory for dealing with psychiatric and mental illness. This is exemplified by the classic experiment of Temerlin (1968) in the field of psychiatry, in which three groups of psychiatrists and clinical psychologists were shown a videotaped interview with an actor who had been trained to give a convincing account of 'normal' behaviour. Before viewing, the first group was allowed to overhear a high-prestige figure comment that the 'patient' was 'a very interesting man: he looked neurotic but actually was quite psychotic'; the second group overheard: 'I think this is a very rare person, a perfectly healthy man', while the third group received no suggestions. Following the viewing, individuals in each group made their diagnosis. In the first group, only 8 out of 95 professionals pronounced the patient to be 'normal'; in the second group there was unanimity: 20 out of 20 pronounced the patient 'normal'; in the third group, 12 out of 21 diagnosed 'normal'. The results clearly illustrate the plasticity of the diagnosis and its inconsistent nature, casting doubts on the 'objectivity' of the process of diagnosis. Studies of diagnostic agreement in the case of schizophrenia (reviewed by Clare 1980) demonstrate considerable variation, both between and within countries; the majority of studies were carried out in the United States and/or Britain. More recent work (van Os *et al.* 1993), which compared psychiatrists working in Britain and France, found differences in diagnostic criteria and treatment methods between these two countries. A series of studies by Rosenhan (1973) reinforce the same points, and also illustrate how the diagnostic process is crucially dependent on the interpretation of behaviour in the context of the particular social environment in which it is observed. In the context of psychiatric hospitals, the 'normal' behaviour of Rosenhan's pseudopatients (sane people who had gained admission to psychiatric hospitals under false pretences, but after admission behaved 'normally') was interpreted so as to be consistent with their diagnosis at admission, as were details of their life histories. Similarly, the 'normal' behaviour of other patients, such as requests for information and overtures for social interaction, were reacted to by staff in hospitals in a way that they would not have been in social situations *outside* the hospital setting: with rudeness, with completely unconnected responses, or even with physical abuse. Other studies have demonstrated that the process of medical diagnosis can be highly situation-dependent. Besides pointing to the inadequacies of the assumptions underlying the biomedical model, such studies provide illustration of the extreme stickiness of diagnostic labels and the phenomenon of stigmatization of those diagnosed as having particular diseases (Goffman 1963).

Studies that have demonstrated the influence of racism in fashioning the concepts of mental health and mental disorder in the biomedical model and much of the resulting practice of Western psychology and psychiatry (see e.g. Gilman 1985; Dalal 1988; Fernando 1991; Alladin 1992) provide further evidence of biomedicine's failure to live up to its own ideals. This is also seen in the construction of 'diseases' peculiar to the oppressed group, so, for example, in the historical context of slavery there are the examples of 'drapetomania' and 'dysaethesia aethiopis' used by the physician Cartwright in the nineteenth century (Cartwright 1851), and more recently 'West Indian psychosis' (Mercer 1986) or 'cannabis psychosis' (Littlewood and Lipsedge 1989). Western psychiatry is based on concepts and beliefs prevalent in Western cultures such as materialism, the separation of mind and body, and about the nature of forces that operate or do not operate in the world; it is thus ethnocentric, not universal. The Present State Examination, an interview schedule common in Anglo-American psychiatry, bases its line of questioning on such ethnocentric concepts (Fabrega 1989). The anthropological literature (briefly reviewed in von Thurn *et al.* 1993) provides a wealth of evidence that mental health is a culturally constructed domain. These examples, as well as the example of gender, sexuality, and mental illness considered in the next chapter, all serve to put the neutrality and rationality of biomedical practice into question, since the use of such diagnoses is seen to be instrumental in maintaining existing power relations within particular societies. Another stark example of this is the case of 'psychiatric' treatment in the former USSR (Bloch and Reddaway 1984), where, in particular, a diagnostic category of 'reformist delusions' served a clear political and social control function (Stone and Faberman 1981).

The (re-)emerging socio-ecological discourse on health

Despite the criticisms outlined above, the biomedical model is a useful conceptual tool for the analysis of diagnostic behaviour, and it remains the

basis for most Western hospital medical treatment. But more recently its
dominance has been increasingly contested, both within and without
Western medical practice. There is an increasing tendency for individual
health to be regarded in a more holistic manner, with an emphasis on the
treatment of the whole person and not just a component subsystem. There
are also the beginnings of a move towards expansion of the medical model
to include explicitly social, economic, and cultural factors and their influence.

This 'new' development represents a resurgence of some of the earlier
concepts and frameworks offered for thinking about health that existed
prior to the emergence of the biomedical model; see for example Rosen's
(1979) discussion of the evolution of social medicine and its historical
antecedents. This contestation of the dominance of the biomedical model
can be seen particularly in the different definitions of health currently in use,
examined later in this chapter. A number of different factors can be identified in accounting for this shift. The first of these is the role of non-medical
health practitioners such as nurses, health visitors, and midwives, whose
professional practice encompasses different perspectives on health and ill
health (see e.g. CETHV 1977; Fitzpatrick and Whall 1983; Aggleton and
Chambers 1986; George 1990). There are links here with the development
of feminist research (Ehrenreich and English 1973) into the impact of medical practice in constructing restrictive roles for women. The growth of
alternative health and self-help movements within Western societies (Katz
and Bender 1976; Hatch and Kickbusch 1983) has also been important, in
particular in use of paradigms from non-Western medical systems of
thought. This has been reinforced by shifts within the WHO towards recognition of the value of non-allopathic systems of medicine in use in other cultures (WHO 1978; Vuori 1982), which work within paradigms which
include a much wider social and cultural context, discussed in more detail in
Chapter 5.

The socio-ecological perspective is reflected in the definition of health
offered in the WHO constitution: 'Health is a state of complete physical,
mental and social well-being and not merely the absence of disease or infirmity' (WHO 1988: 1). This emphasizes health as a positive quality,
extremely wide in scope, and presents health as a general value, ideal, or
slogan. Many critiques of this definition have been offered (see e.g. Noack's
1987 review), first on the basis that health is conceptualized as an ideal goal
that can be approached but never achieved. Also, the notion of well-being is
as ambiguous as the notion of health, with associated problems of measurement of 'complete'. The definition has also been criticized as being too wide.
Another line of argument is that health is not a state but a task: health is a
means to an end, such as the fulfilment of role obligations, rather than the
end itself. These last criticisms come largely out of schools of thought that
conceptualize health as a process, activity, or potential, as exemplified in
Seedhouse's definition of health as the 'foundation for achievement' (1986:

3), which moves away from relating health to 'normality' towards a rather more individualistic definition. A related definition is that offered by Duhl (1976: 33): health is 'the ability to command events (internal and external) that effect [*sic*] our life. The issue is to have choice and the freedom of movement', conveying clearly the idea of health as a resource or potential possessed by the individual. This idea is also expressed by another definition emanating from within WHO, where health is seen as 'the extent to which an individual or group is able to realize aspirations and satisfy needs, and to change or cope with the environment. Health is therefore seen as a *resource for everyday life*, not the objective of living; it is a positive concept encompassing social and personal resources as well as physical capacities' (WHO/EURO 1984: 653–4; emphasis added). Notions similar to the second part of this WHO definition are echoed in the following quotations from Dubos, which visualize health in terms of *adaptation* to changes in nature and society: 'a physical and mental state fairly free of discomfort and pain, which permits the person concerned to function as effectively and as long as possible in the environment where chance or choice have placed him [*sic*]' (Dubos 1965: 351); 'the ability of the individual to function in a manner acceptable to himself [*sic*] and to the group of which he [*sic*] is part' (Dubos 1960: 206). In all of these we find a strong emphasis on functioning, with health related to ability to perform 'appropriately'; this raises the question of whose values are to be used to judge appropriateness. It also carries with it the notion that health is tied to the specific social position occupied in society by a person or group of people.

In contrast to the normative definitions of health implicit in the biomedical model, the socio-ecological model uses definitions of health representing some combination of the *perceptual* (e.g. the WHO constitution definition quoted earlier), which is individualistic, based on the person's perspective and views, and the *functional/adaptational*, examining the ability to perform customary tasks and role, often together with the (implicit) assumption that this will be strongly dependent on a person's particular place in society.

Early studies of 'lay' perceptions of health: desperately seeking simplicity?

Many of the earlier studies of lay concepts were concerned with identifying whether differences existed between different broad groups in the population defined by age, sex, social, and cultural group, in the hope, amongst other things, of illuminating different patterns in service use and compliance. Thus studies were often interested in examining the concepts of health implicit in people's definitions of themselves as 'healthy' or 'ill', and in perceptions of the causes of ill health. In particular, many studies were concerned to explore whether these causes were seen as primarily external or internal and whether they were avoidable or not; the findings have obvious consequences in terms of individual health-related behaviour.

Details of many of these early studies can be found in Fitzpatrick (1984), Currer and Stacey (1986), Calnan (1987), and Stainton-Rogers (1991), while more detailed analysis of the historical development of lay concepts is provided by Herzlich and Pierret (1987) and Unschuld (1987). In the remainder of this section, some illustrations are given of the type and range of studies, together with their drawbacks, focusing on examples drawn from a limited number of studies carried out in high-income countries, where the dominant system of health service provision is based on the biomedical model; most of these are highly localized studies carried out in specific communities.

Studies of lay views of health illustrate the differences that can exist in concepts and expectations of health within the same society. The study carried out by Blaxter and Patterson (1982) focused on working-class Aberdeen women. Their findings emphasized a functional definition of health: women reported that they were ill only when they could not perform certain functions, associated with their usual activities, in terms of waged work, housework, childcare, etc. So, for example, many of the women reported their health as satisfactory despite disability and discomfort, for example:

> After I was sterilised I had a lot of cystitis, and backache, because of the fibroids. Then when I had the hysterectomy I had bother wi' my waterworks because my bladder lived a life of its own and I had to have a repair . . . *Healthwise I would say I'm OK.* I did hurt my shoulder – I mean, this is nothing to do with health but I actually now have a disability, I get a gratuity payment every six months . . . I wear a collar and take Valium . . . then, just the headaches – but I'm not really off work a lot with it. (Blaxter and Patterson 1982: 29; emphasis added)

We might note the very low expectations of health implied here, and the important potential implications in terms of behaviour in using health services. Doctors or other health professionals might well view some of the problems reported in Blaxter and Patterson's study as soluble, or at least alleviable, in terms of some relief or pain control being possible, but the respondents tended to accept them stoically. The study found very few positive conceptions of health, and with regard to causation of illness, the respondents emphasized external factors, outside their control, and heredity. Pill and Stott (1982), studying the wives of skilled manual workers in south Wales aged 30 to 35 with children living on a single housing estate, found that 40 per cent of the women employed concepts of cause involving behavioural choice and individual responsibility (i.e. what might be referred to as their lifestyles). This was thought to be linked to their higher level of education (compared to the women in the previous study) and greater feeling of control over their own lives. When results were analysed further by level of education of women and housing tenure (home-owner or tenant in public housing), further differences emerged that supported this; those with higher education were more likely to stress links between health and individual behaviour, and this also applied for home-owners compared to public-housing tenants. As in the Blaxter and Patterson study, the definitions of health employed tended to be expressed in terms of capacity to function and ability to cope, i.e. a functional definition. Some identified positive conceptions of health, associated with being 'cheerful', 'enthusiastic', and 'effervescent', and more positive conceptions were found than in the Blaxter study; this has been linked by some to the different social class covered.

Herzlich (1973), studying middle-class Parisians, found three basic definitions of health: 'health-in-a-vacuum' or the absence of illness; having reserves of health, various strengths, or capacities to maintain a state of good health (akin to Noack's notion of health potential); a personal sense of equilibrium or balance in life, which is regarded by Herzlich (p. 58) as the 'superior' form of health, representing 'the realization in experience of the possibilities of the reserve of health, contrasted with the negative and inferior form regarded as health-in-a-vacuum'. This final view is very reminiscent of views of health as well-being, a holistic notion of health found in the WHO constitution definition. Similar results were also found in a study by Williams (1983), based on open-ended interviews with 70 informants aged 60 and over in Aberdeen (including both working- and middle-class respondents), and the Health and Lifestyle Study carried out across the UK (Blaxter 1990). Herzlich's study also distinguished three different views of illness: as *destructive*, imposing retreat, causing desocialization, inactivity, dependence, social exclusion, characterized by denial and then passivity (going on till you drop!); as *liberator*, providing the opportunity to withdraw and find oneself in pleasant retreat, a valuable experience freeing the individual from everyday commitments; as *occupation*, involving active participation by an individual, with professionals, in achieving recovery, and requiring the individual to fight and attempt to control illness.

A later French study (d'Houtaud and Field 1984) used a self-selected sample drawn from people having a regular health check at a centre for preventive medicine in France, and carried out a systematic content analysis of the ways people talked about health. Conceptions of health were characterized by positivity in many of the phrases used, and in discussing causality the sample emphasized the possibility of preventing ill health; it could be argued that neither of these features is surprising, given how the study participants were obtained. This study also analysed results by socio-occupational categories, finding that the higher-status groups had more holistic, wider concepts of health (nearest to the WHO constitution or health balance/health potential idea). Health was positively conceived, and carried with it the idea of personal responsibility for health, and of personal ability to control or influence it (i.e. definitions were positive, personalized, and expressive), while the lower-status groups had definitions more based on biomedical model, negatively conceived, with a lack of perceived responsibility or control (i.e. definitions were negative, socialized, and instrumental). Similar findings are reported by Calnan (1987) from studies in England.

Studies such as Blaxter (1990) have also found age differences in expectations of health, so that the older respondents considered a much higher level of pain and discomfort to be a 'normal' expectation for their age. There are, however, problems with interpreting some study results on this point, owing to a lack of clarity about how the different questions used were interpreted by individuals. In addition, results from cross-sectional studies do not allow us to separate effects of changes over the life-course of individuals from effects due to differences between generations.

For the variety of reasons reviewed above, we therefore need to be very wary of concluding that we can predict an individual's beliefs in this area from a knowledge of factors such as class, culture, sex, or that these beliefs will change along well-defined paths throughout the lifespan. Such an interpretation is suggested, for example, by Stacey (1988: 152):

> the beliefs of lay people have their own logic, a logic which can be seen when the believers are located in their social context. . . . Certainly a good deal of evidence suggests that aspects at least of lay concepts vary from one social class to another in ways that appear to relate to the material difference between the classes . . .

or by Calnan (1987: 178):

> the working class women in particular saw health as being the absence of illness or the absence of serious illness . . . the middle class tended to integrate mental and physical health into their concepts of health, whereas working class women tended to see health predominantly in physical terms.

Such statements can be misleading if they are read in too deterministic a fashion. There is far more complexity to be found. Some of the early studies point to the important notion that such beliefs will not necessarily follow a well-defined course as the individual passes through life, identifying the fact that beliefs vary according to the context in which they are elicited. For example, Linn *et al.* (1982) asked American terminal cancer patients and patients suffering from other chronic (but non-fatal) diseases about the causes of cancer. They found that both groups cited smoking and work as causes, but cancer sufferers were much more likely to emphasize 'God's will' or inherited factors, whereas the other group stressed more controllable factors such as diet. We can interpret this as pointing to the need for the individual to make personal sense of why he or she in particular has been affected, in a way that does not impute blame to themselves. The implication of this is that perceptions will be influenced by the context, and should be viewed as dynamic rather than static belief frameworks, a point taken up in the next section, where we look at some of the more recent work in this area. Another illustration of plasticity in beliefs is provided by research into lay recognition of the category 'mental illness', which has found that this is strongly influenced by clues about a person's association with different professionals: 'guilt by association'. A study by Phillips (1967) combined different behavioural descriptions with information about contact with different types of person: clergyman, physician, psychiatrist, admission in mental hospital, no contact. The (lay, white, married) women interviewed categorized the case description of normal behaviour plus admission to mental hospital as 'severely mentally ill', while the case description of schizophrenic behaviour plus no help-seeking was categorized as normal, again demonstrating the phenomenon of stigmatization (Goffman 1963).

Yet another warning against a too-simple categorization of beliefs is provided by Helman's studies of general practitioners (1978), which illustrate the interaction of lay and professional beliefs. Helman found that general practitioners in England actively use lay or folk models and the associated treatments in relation to conditions such as colds, fevers, and chills (see also the discussions in Chapter 5). Herzlich and Pierret, over a much wider historical period, further demonstrate how an 'interplay and exchange' (1986: 75) between medical and lay views can be observed rather than any simple dependency.

Finally, there is also the concern that some of the differences found may be methodological artefacts. The identification of public and private accounts (Cornwell 1984), which will be discussed in the next section, lends support to this argument, as do the results of studies such as Blaxter (1990) and van Dalen *et al.* (1994), which find less clear-cut differences between different social groups. Some researchers (Calnan 1987; Blaxter 1990) have argued that findings reflect a greater facility in middle-class groups to express more complex multidimensional definitions when faced with abstract questions in an interview setting (and often with a 'middle-class' academic interviewer). There is also the possibility that differences found are related to the nature of the questions asked, in particular the difference between 'health in the abstract' and 'health for oneself' (Blaxter 1990; van Dalen *et al.* 1994).

Enriching the picture: more recent approaches to 'lay' discourses

The preceding discussion of some of the early studies of lay beliefs has identified the need to pay attention to the potentially complex and dynamic nature of such beliefs. The methods used in many of the early studies were not well suited to this task, consisting of structured or semi-structured interviews or questionnaires. In this section, in order to look at these issues, we examine some more recent studies, based on qualitative, particularly ethnographic, approaches.

Much more complex distinctions are reported in Cornwell's study of a group of white working-class residents in Bethnal Green, a district of east London (Cornwell 1984). The first interesting aspect for us is the distinction she finds between public and private accounts. Public accounts are those which accord with what the speaker considers will be found acceptable, and which reproduce the culturally normative pattern. These tend to exclude anything that might be considered unacceptable or not respectable; in other words, they are censored. On the other hand, private accounts are those in which the response described only what the speaker and people personally known to the speaker would think and do. These spring directly from personal experience and the thoughts and feelings accompanying it, are uncensored, and represent what individuals 'really think'. Cornwell stresses that she only obtained access to private accounts after building up a considerable rapport over repeated meetings with her study participants. It is thus questionable whether many of the earlier studies discussed above, which were based on single questionnaires or interviews, managed to get beyond public accounts.

The concepts of health and illness found in the public accounts were dominated by the moral aspect: whether the illness was 'legitimate' or represented 'malingering'. Respondents were concerned not to stray outside what they conceived of as doctor's beliefs (i.e. there was a tendency to conform to the biomedical model). In contrast, private accounts were more practical and pragmatic, and not so bound by perceptions of medical models. In examining beliefs about causation, the public accounts were again dominated by the medical model: 'they say', 'they know', 'I've heard', linked to a concern for discovering who was to blame or responsible for the illness. When it was not the ill person's fault, sympathy and compassion were indicated; if it was considered to be the ill person's fault, then no sympathy was merited. Private theories, on the other hand, related the person's own (or close other's) explanation. These were less concerned with responsibility, and were more personalistic and diverse, describing a causal process in which there was movement between factors internal to the person and external factors in their environment, so that illness was not portrayed as something that could be separated from the person or their life circumstances. Similar evidence of public and private accounts was found in Donovan's study of beliefs about health and illness amongst Afro-Caribbeans and Asians living in east London (Donovan 1986). Her study also found a preponderance of functional conceptions of health amongst the Afro-Caribbean group, while the Asians espoused more positive conceptions. She cautions, however, against hasty interpretation, pointing to the differences in socio-economic status between the groups, the Asians in her study being drawn from higher-status groups than the Afro-Caribbean.

Crawford (1984) reports research into concepts of health based on a sample of American adults, which, although he attempted to achieve as much variation in social class, 'race', sex, and age as possible, was dominated by white middle-class females aged under 40. He identifies two contrasting notions of health: health as self-control and health as release, linking these to various cultural themes, both historically and contemporaneously. He warns, however, against simplistic interpretation, arguing that such cultural meanings are not only shared or given, they are also fragmented and contested, and represent sites of resistance:

> the opposition of control and release . . . is only one dimension of a highly differentiated and complex health discourse . . . even though I have emphasised . . . the hegemonic meanings that can be found in these clearly individualistic conceptions of health, it is also possible to see liberatory aspects in both modalities. Even if repressive cultural mandates are unavoidably internalized, the categories of self-control and release have no inherent ideological content. They can be and are appropriated for ends that are not wholly conformist. (Crawford 1984: 96)

Most recently, working in England, Stainton-Rogers has focused on exploring the diversity of accounts for explaining health and illness, criticizing much of the earlier work for assuming 'that people's understandings are sufficiently artless, lawful and common property that they can be expressed along some simple dimension, or be encapsulated within a small number of pre-ordained categories' (Stainton-Rogers 1991: 3). Her studies first of all focused on explanations of health and illness generated in response to questions about individuals' beliefs about what affected their current state of health, their capacity to become healthier in the future, whether or not they become ill, and how quickly they will recover when ill. Her analysis yielded five factor complexes (see Table 2.1) which point to the inadequacy of a simple categorization of beliefs about illness causation into external or internal. The first of these complexes was characterized by perception of a wide range of factors influencing current state of health and future illness. The factors were both internal (and to do with state of mind as well as bodily factors) and external. Respondents reported a sense of control over these factors, and they did not see a strong role for medical services. The second complex was dominated by factors internal to the person, and most strongly linked to the body, rather than the mind. In terms of recovery, a combination of individual actions and medical treatments was seen as effective; little role was played by fate or chance. The third complex was dominated by external factors in the causation of current health and future development of illness; but within this account, luck was distinguished as well as

Table 2.1 Explaining health and illness: five different factor complexes

Factor complex	Strongest influences on			
	My current state of health	My capacity to become healthier in the future	Whether or not I become ill	When I am ill, how quickly I will recover
1	My state of mind My emotions Whether I feel 'on top' of my life My overall lifestyle 'Taking good care of myself' My working environment The circumstances in my home life The current circumstances at work Particular events in life at the time	Promoting a positive attitude Seeking out things that make me happy Tackling unresolved inner conflicts Taking charge of my own life Changing my day-to-day behaviour Giving up unhealthy habits Improvements in work circumstances	My body's natural defences My state of mind becoming negative Working in a poor environment Stressful conditions at work Stressful, nasty, or unsettling events Inbuilt weaknesses	None
2	My body's natural defences Actively taking action to be healthy	Changing to a more healthy lifestyle Giving up unhealthy habits	Behaving in stupid ways Adopting an unhealthy lifestyle	Taking responsibility for myself Looking after myself Being careful about my behaviour Making my lifestyle more healthy Giving up unhealthy habits Circumstances conducive to recovery Treatments My body's own natural defences

3	Good or bad luck Simple probability Exposure to infectious organisms My age	Changing to a more healthy lifestyle Giving up unhealthy habits Good or bad luck My age Exposure to infectious organisms What happens in the future Exposure to substances	My body's natural defences weakened Adopting an unhealthy lifestyle Bad luck Exposure to infectious organisms Working in a poor environment Exposure to harmful chemicals Other people's stupid actions Virulence of infective organism My age	Quality of medical treatment Virulence of the disease Taking drugs or medicines Treatments 'Alternative' therapies Seeking medical advice soon enough My age
4	Simple probability Exposure to infectious organisms	Simple probability Seeking out preventive services	My body's natural defences weakened Simple probability Effects of poor medical treatment Exposure to infectious organisms Virulence of infective organism	Simple probability Virulence of the disease
5	'Inner forces' of my psyche The culture within which I live God or some other supernatural power	Improvements in family relationships God's power or influence Some other supernatural influence	God's will Other supernatural influences A curse or ill-wishing	Care from my family and friends Intervention of a spiritual healer Prayers said for me God's will Some other supernatural power

Source: Stainton-Rogers (1991: tables 7.3–7.7). The strongest influences shown are those which scored +6 or more on a scale that ran from 0 to 7.

probability. The effectiveness of medical treatments to control the course of an illness and speed recovery was identified, but there was little emphasis on self-help, or the respondent's own actions, except for seeking help. The fourth complex also saw causes of health and ill health as external. Within this complex, however, medical treatment was not found to rate highly as an influence on recovery, which was felt to be associated instead with probability and the virulence of the disease caught. Thus health was dominated by chance, with very little room for effects of individual or medical action. The fifth complex might be characterized as a religious or personalistic view, dominated by external religious or supernatural powers in combination with cultural and kinship factors; this is close to the view offered within personalistic medical systems.

Stainton-Rogers also explored the broader cultural contexts and explanatory frameworks within which people construct these explanations, resulting in eight different 'accounts' (see Table 2.2) which were used as the basis for people's explanations about health and illness:

Table 2.2 Cultural accounts of health and ill health (Stainton-Rogers 1991)

Account	Features of account
Body as machine	illness naturally occurring and real belief in efficacy of modern Western biomedicine operating within modernist world-view of science
Body under siege	individual under threat from germs, disease, stress individual on own relatively powerless to resist threat
Inequality of access	belief in efficacy of modern Western biomedicine inequalities can be remedied by access to suitable services old-style marxist critique
Cultural critique	base in 'dominance' sociological world-view of exploitation and oppression knowledge socially constituted and ideologically mediated
Health promotion	health as fundamental human right recognizes collective and personal responsibility for health stresses importance of healthy lifestyle message of self-empowerment
Robust individualism	individual's right to 'satisfying life' individual's right to choose way of life
God's power	health a product of right living, spiritual well-being, God's care recovery attained by intercession
Willpower	individual pre-eminently in control moral responsibility for use of 'will' to maintain health

Source: Stainton-Rogers (1991).

people have access to and utilize a range of alternative accounts to 'make sense of' health and illness . . . this [is] a more plausible understanding . . . than claiming that there are specific, enduring personality traits, psychological mechanisms or social forces that constrain people to think in particular ways. (Stainton-Rogers 1991: 226)

The accounts she identifies are not all mutually compatible; there are contradictions between them. However, the choice of contradictory accounts at different times and in different situations should not be viewed as evidence of irrationality (or even illness). She argues that these accounts are best regarded as complementary or 'sympatric' (a term borrowed from ecological biology, where it is used as a descriptor for species which compete within an ecosystem for survival but which at any particular point in time will appear as coexisting). We can also draw a parallel here with theoretical physics, where, for example, (contradictory) theories of light as wave and particle phenomena are required to explain the properties of light in different circumstances.

In this view, people's use of these accounts, or texts, is a dynamic process, and will be intimately connected with the particular context in which the account is elicited or used. Here we are reminded (and Stainton-Rogers makes the same connection) of Jocelyn Cornwell's private accounts, where she found these to be closely connected to context and particularity of circumstance. As Cornwell emphasizes (1984: 145): 'it is not enough to know that health is interpreted as "functional ability" or "capacity to work" without knowing something about the nature of the work people do and of their relation to it.' Thus we are again reminded of the importance of understanding the context in which the concepts are used, and their social construction. These need to be viewed as complex phenomena, not in any simplistic or deterministic manner. This more complex interpretation may help us to understand apparently 'contradictory' or 'illogical' attitudes or behaviour. This is emphasized very well by Ntozake Shange, discussing responses to her choreopoem 'The love space demands':

A White critic in New York was outraged about the way the play went from 'Crack Annie' giving her daughter away into a joyful sequence. Yeah, that's what Black people have had to do. We have had to have our children taken away and then we've had to go pick fucking cotton, this is correct. Life does not stop, we do not stop, we may be sad but we keep going. I thought this critic's response was a most condescending one to put to people who have had to rear families in slavery and apartheid. Yes we still have to have romance in the face of adversity, this is a fact. *This, to me, is not sickness on our part, it's absolute viable stamina, emotional stamina.* (Shange 1992: 19; emphasis added)

Here Shange illustrates how the interpretation of a particular sequence of behaviour as a sign of 'sickness' relies on an inappropriate vantage-point, one that does not take account of the context in which the behaviour occurs. The reverse interpretation, that the sequence of behaviour represents a sign of emotional stamina or health, becomes apparent once context is taken into account. Nancy Scheper-Hughes, in her study of the women and children of a Brazilian shanty-town (1992), also emphasizes the importance of context, for example, in describing the effect of a high expectancy of child death (consequent on poor socio-economic circumstances) on maternal thinking and practice. Similarly, Anderson *et al.* (1991), studying immigrant Chinese and Anglo-Canadian women living with diabetes, describes how views about illness and styles of managing it were closely linked to the specific details of the women's material circumstances, rather than being interpretable in terms of 'ethnic' belief systems.

Rather than a fixed and somewhat simplistic view of different types of people favouring different and unvarying accounts, a view that sees 'health beliefs' as fixed and singular essences, we see the much richer picture of people continually re-making sense of their experience and that of others close to them, with different accounts forthcoming according to the particular circumstances. To give some specific examples, first from Stainton-Rogers, in the situation where you have just been bitten by a rabid dog, the account used (in the West at least) is likely to be the 'body as machine' (find a doctor fast) perhaps mixed with the 'god's power' (pray hard). Secondly, consider a quote from an undergraduate student in response to an assignment on exploring their own views about health and illness: 'Health is a combination of many things, to a great degree it is a state of mind (unless you have been hit by a car). It is a lot to do with your feelings.' In the case of being hit by a car, one is less likely to have recourse to internal accounts favouring state of mind and more likely to call for the ambulance. In contrast, in the presence of symptoms such as insomnia or stomach trouble, a much wider range of accounts is likely to be drawn on.

Conclusions: diversity and difference – undoing the 'lay'/'professional' divide

Looking at historical changes in assumptions about nature of disease and illness as reflected in Western medical practice, we can trace several transformations or stages. In the eighteenth century and before, there were different rival systems or schools of thought, rooted in Greek humoral pathology. Explanations of health and illness were given in terms embracing the total mental and physical disposition of patient, a 'person-centred' approach. During the nineteenth century biomedicine came to growing dominance, with a focus on specific etiology and on internal pathology. This corresponded to the growth of hospital medicine: the 'birth of the clinic'

(Foucault 1973), the start of the medical 'gaze'. During the twentieth century the growth of 'laboratory medicine' followed, with increasing use of increasingly more sophisticated laboratory tests and diagnostic procedures. However, this century has also seen increased contestation of the biomedical model, through many different oppositional discourses, with a consequent movement back to multi-causal, holistic frameworks, contesting professional dominance within health services and outside. The formal health services within Western societies have become more accepting of alternative medicine, now usually referred to as 'complementary' medicine (an important linguistic shift); for example, the British Medical Association, the heart of the medical establishment in the UK, issued a guide to good practice using complementary medicine in 1993 (British Medical Association 1993), while surveys throughout the Western world show considerable, and often increasing, acceptance and use of complementary therapies, in the USA (Nelson *et al.* 1990; Eisenberg *et al.* 1993), in Germany (Payer 1988), in Belgium, Italy, and France (Abbott 1992). We return to this in Chapter 5.

This chapter has examined how concepts of health and illness are socially constructed and contested, illustrating the tremendous range of different professional and lay concepts, together with their dynamic and contingent nature. Such concepts cannot be regarded as neutral; they are intimately linked to social relations and power structures. At the extremes of physical robustness and well-being or of disabling disease or mental disorder, there may be near-universal agreement about who is in good health and who is not. Between these extreme states, judgement has been found to be conditional upon many factors, including: age, sex, family status, occupation, ethnicity, culture, religion, class position, and geographical and temporal location of the person; the activity of the individual, interaction with, relationships to, and dependence upon others; individual and societal attitudes and beliefs regarding specific symptoms and illness. Perceptions of health and illness also vary over time (between generations) and in space (between different communities in different parts of the world). We have noted some international differences, but most of the examples here demonstrate significant differences within societies. This diversity poses particular challenges for the interpretation of results from surveys; see for example the discussions by von Thurn *et al.* (1993) of the implications for question design for the measurement of mental health in the US National Health Interview Survey.

There has been a considerable shift towards the adoption of a socio-ecological paradigm of health, marked by specific struggles to contest the previously dominant professional definitions of health and illness, further illustrated in the three examples considered in the next chapter, and resulting in a move away from exclusive dependence on the biomedical model of health. This theme is taken up again and examined in health policy terms in the discussions of the new public health movement in Chapter 7, and of

'Health for All' and health promotion in Chapter 9, both of which illustrate that this process is a continuing one.

There has also been increasing recognition of the dynamic nature of individuals' responses in accounting for, and making sense of, their own experiences of health and ill health; as Kleinman expressed it, talking about people with chronic illness:

> The illness narrative is a story the patient tells . . . to give coherence to the distinctive events and long term course of suffering . . . Over the long course of chronic disorder, these model texts shape and create experience. The personal narrative does not merely reflect illness experience, but rather it contributes to the experience of symptoms and suffering. (1988: 49)

These narratives are thus crucially dependent on the details of the particular context concerned, and the dangers of trying to capture these within a simple classificatory system, which implicitly privileges a particular subject position (usually that occupied by the male white heterosexual Western health professional), should be recognized.

|3|

Contesting concepts of health

The previous chapter, in its examination of lay and professional discourses about health, provided ample illustration of the diversity of discourses around health, involving notions of health, disease, symptoms, normal functioning, illness, etc., and further illustrated how they are closely linked with the structure of the society concerned, i.e. that they are all socially constructed. Such discourses are subject to constant contestation in many different arenas, leading to continual redefinition of their meanings and usage. Continuing with this theme of examining the dynamic and situated nature of notions of health and ill health, this chapter discusses three different examples in order to illustrate further the ways in which different concepts of health and ill health have been conceptualized, contested, and changed. The first of these, and historically the most distant, deals with 'hysteria', particularly in the period from the eighteenth century onwards in America and Europe, describing the ways in which medical notions of the disease and the causal factors underlying it were contested by the feminists of the time, and how these interactions have been more recently reinterpreted by feminists. The second example, also taken from the field of mental health, deals with the construction and deconstruction of homosexuality as a mental disorder in America. The third example deals with HIV and AIDS, discourses about which provide the sites for many different struggles, many still continuing, both within and between countries.

All the examples have in common the use of a medical diagnosis of ill health in a way that reinforces some aspect(s) of the social status quo at the time concerned, in the sense of contributing to the maintenance of existing relations of power, combined with the use of moral or cultural censure and sanctions against marginalized and/or disempowered groups in society. This is not to be read, however, as simple repression of one group by another using medical diagnosis as a source of legitimation; in each of the examples the active struggles and resistance of the disempowered can also be identified.

Women and mental illness: hysteria and the problem of 'nerves'

The material for this first case study is based particularly on four works produced by feminist researchers on historical material relating mainly to the United Kingdom and the United States, and particularly to the period that covers the late nineteenth and early twentieth centuries. The first of these is Vieda Skultans' book (1979), which includes a detailed study of ideas about the relationship of notions of female vulnerability to mental illness in England over the period 1580–1890, and explores how these were related to the social circumstances that governed women's lives. Elaine Showalter's study (1991) examines themes and images of female insanity through three periods of English psychiatry covering the period 1830–1980, showing how cultural ideas about 'proper' feminine behaviour shaped the definition and treatment of 'female insanity'. Carroll Smith-Rosenberg's paper (1972) focuses on the role of hysteria within sex roles and role conflict in nineteenth-century America. Lastly, Barbara Ehrenreich and Deirdre English's work (1973) on women and medicine in the late nineteenth and early twentieth centuries in America is particularly notable for its careful consideration of how notions of health and ill health within medical and scientific discourse were strongly class-related. Three of these works consider much besides hysteria, which we shall concentrate on here to provide a reasonably focused subject.

Contrasting views on hysteria and its causes

Grant suffrage to women, and you will have to build insane asylums in every country, and establish a divorce court in every town. Women are too nervous and hysterical to enter into politics. (Massachusetts legislator, nineteenth century, quoted in Ehrenreich and English 1973: 27)

In regard to the possible effect on health and physical vigour of women students, it was feared that the opening of new facilities for study and intellectual improvement would result in the creation of a new race of puny, sedentary and unfeminine students, and would destroy the grace and charm of social life, and would disqualify women for their true vocation, the nurturance of the coming race and the governance of well-ordered, healthy and happy homes.

Over stimulation of the female brain causes stunted growth, nervousness, headaches and neuralgias, difficult childbirth, hysteria, inflammation of the brain and insanity. The female character is likewise altered by education; the educated woman becomes cultured, but is unsympathetic; learned, but not self denying (Clouston 1906: quoted in Skultans 1979: 94)

[Freud, in response to Dora's disgust at an uninvited kiss on the lips from her father's friend] I should without question consider a person hysterical in whom an occasion for sexual excitement elicited feelings that were preponderantly or excessively unpleasurable; and I should do so whether or no the person were capable of producing somatic symptoms. (Freud 1905: 28)

Contemporary feminist analysis of the cult of chronic invalidism of affluent women in the late nineteenth and early twentieth centuries may be represented by the following:

Constantly considering their nerves, urged to consider them by well-intentioned but short-sighted advisors, they pretty soon become nothing but a bundle of nerves. (Mary Putnam Jacobi 1895, quoted in Ehrenreich and English 1973: 23)

I think, finally, it is in the increased attention paid to women, and especially in their new function as lucrative patients, scarcely imagined a hundred years ago, that we find explanation for much of the ill-health among women, freshly discovered today. (Mary Putnam Jacobi 1895, quoted in Ehrenreich and English 1973: 30)

[American men] have bred a race of women weak enough to be handed about like invalids; or mentally weak enough to pretend they are – and to like it. (Charlotte Perkins Gilman, quoted in Ehrenreich and English 1973: 23)

Mental illness . . . for women [is] often a form of logical resistance to a 'kind and benevolent enemy' they are not permitted to openly fight. In a sick society, women who have difficulty fitting in are not ill but demonstrating a healthy positive response. (Charlotte Perkins Gilman 1892, quoted in Ussher 1991: 307)

The illness of hysteria has a long history, traced by Veith (1965) from ancient Egypt through to this century. It was seen as a disease which affected females and was associated with the 'peregrinations of a discontented womb' (Veith 1965: ix), taking its name from the Greek word for womb. Although for centuries hysteria had been the 'quintessential female malady', between 1870 and the First World War in England, America, France, and Germany it assumed a particularly central role in psychiatric discourse (Showalter 1991), and its incidence (along with other female nervous disorders of anorexia nervosa and neurasthenia) 'became epidemic' (Showalter 1991: 18). This can also be seen as part of what Foucault (1976a: 146–7) refers to as the 'hysterization of women': 'which involved a thorough medicalization of their bodies and their sex, was carried out in the name of the responsibility they owed the health of their children, the solidity of the family institution, and the safeguarding of society', with a purpose both disciplinary and regulative. Showalter argues that it is no coincidence that, over the same period, middle-class women were beginning to organize on behalf of higher education, professional entrance, and political rights. Hysteria was seen as an extension of women's 'natural frailty', and doctors directly linked this to women stepping outside the prescribed, limited social roles assigned to them. It is important to emphasize here that the ideas of women's natural roles were strongly structured along class and 'race' lines, so that, for example, while the women of the upper and middle classes were perceived as naturally sickly and frail, poor, working-class, and black women were perceived as being robust and fit for hard work. Ehrenreich and English (1973) provide one of the clearest analyses of the different stereotypes that operated and how these were buttressed by the legitimation offered by medical opinion, concluding:

> beneath all this ran two ancient strands of sexist ideology: contempt for women as weak and defective, and fear of women as dangerous and polluting. Here we see the two separated, and applied to wealthy and poor females respectively. Upper- and upper-middle-class women were 'sick'; working class women were 'sickening' (p. 18)

The basic explanatory model offered by Darwinian psychiatry (Skultans 1979; Showalter 1991) was one of faulty heredity exacerbated by biological and social crises of puberty, with a role played by sexual frustration, rather than intellectual frustration. Smith-Rosenberg draws an important distinction, in terms of medical explanations, between what might be seen as doctors' explanations of *immediate* causes ('physicians saw hysteria as caused either by the indolent, vapid and unconstructive life of the fashionable middle and upper class woman, or by the ignorant, exhausting and sensual life of the lower or working class woman' (p. 657) and their explanation (in accordance with the biomedical model) of an underlying organic etiology. Hysterics were found to express 'unnatural' desires for privacy and independence, and hysteria was more likely to appear in young women who were 'especially rebellious', more independent and assertive than 'normal' women, and in 'unconventional' women such as writers and artists. Darwinian psychiatrists were quick to point to the dangerous 'psychological and physical consequences of feminist rebellion' (p. 145), resulting in some instances in radical women's committal to lunatic asylums. Showalter quotes the case of Edith Lanchester, kidnapped by her father and brothers and committed, with the supposed cause of her 'insanity' being 'over-education', and the doctor in the case arguing that her 'opposition to conventional matrimony made her unfit to take care of herself' (p. 146). Lanchester was released through the intervention of the Commissioners in Lunacy, but the doctor was supported by the *Lancet* and *British Medical Journal* (except in the matter of his involvement in kidnapping).

The period of Freudian psychiatry represented a shift in the underlying model of causality, escaping the crude physical essentialism of Darwinian psychiatry. Showalter argues that the initial promise in the psychiatrist's willingness to listen to women's words and feelings later became lost as the meaning of hysteria was interpreted in terms of psychological, specifically sexual, disturbance, thus replacing a crude physical essentialism by a more sophisticated/complex psychic essentialism. Perhaps the classic example is Freud's account of one of his patients, 'Dora', a 'hysteric'. Freud ignores the social circumstances of Dora's life (and in particular the unwanted sexual advances made to her by a friend of her father's) and instead dictates the meaning of her hysteria in terms of masturbatory fantasies, incestuous desires for her father, and possible homosexual or bisexual wishes. The case was later to be one of the key steps in Freud's formulation of the Oedipus complex. This has been seen as a clear example of the substitution of a psychological analysis for the acknowledgement of a real oppression. This analysis was certainly not found satisfactory by the patient, who terminated her sessions with Freud. Some see proto-feminism in Dora and other hysterics (Cixous and Clément 1975; Gallop 1982; Rose 1986), differing, however, in the extent to which they see hysteria as a productive tactic, so that while Cixous sees Dora's hysteria as a powerful form of rebellion, to Clément the hysteric's deviance and rebellion are programmed and delimited by the social order. They all agree, however, in viewing the role of the psychiatrist in such cases as one of reconciling women with their socially ordained roles.

Ehrenreich and English (1973) see part of the rise in the incidence of hysteria as women attempting to subvert the sick role and use it to their own advantage, for example, as a form of birth and sex control, and a strategy to gain a limited measure of power within the family. They see this form of revolt as of limited efficacy, remaining as it did individualized, and leading to increasingly repressive and brutal 'treatment' methods (Ehrenreich and English 1973; Masson 1986; Ussher 1991), which amounted to direct methods of social control. They also point out the direct financial interests that doctors had in the maintenance of a high incidence of such illness within affluent families able to pay for the doctor's services. Thus the medical view of women's frailty served two purposes: it helped disqualify women as healers (both those seeking formal training in the medical profession and those working as lay healers or midwives), and it made them highly qualified as patients. Showalter also considers that, in many cases, the supposed mental pathology was in fact suppressed rebellion, and concludes that hysteria represented a mode of protest for women deprived of other social or intellectual outlets or expressive options. Smith-Rosenberg offers a similar analysis, arguing that hysteria was itself a product of role conflict, and often represented a role choice by women as an alternative to life in rigid family roles, and was used as a way of redefining or restructuring their position within the family. Her analysis is particularly interesting as she finds evidence of the occurrence of hysteria across *all* classes, although it was most frequent amongst women aged 15–45 of the urban middle and upper middle classes.

The example chosen here is quite far removed in history, and hysteria is no longer a 'fashionable' medical diagnosis, categories such as depression, anorexia, and neurosis having replaced it (Ussher 1991). Other historical examples could equally have been chosen – for example, the psychiatrization of crime and the birth of forensic psychiatry (Foucault 1978), or debates about the role of psychiatry in the maintenance of social order (Ingleby 1983). The next section presents a more recent historical example, also taken from the field of mental health, on the construction and deconstruction of homosexuality as a mental illness. Psychiatry and psychoanalysis today still remain fields where concepts of health and illness are hotly contested along gender lines. Ussher (1991) provides an overview, which is, however, unfortunately heavily limited to the white Western world, and does not deal adequately with questions of 'race', culture, class, and sexuality, and the critiques levelled against psychiatry from black, working-class, lesbian, and gay standpoints. Unfortunately, Juliet Mitchell's conclusion that 'there seems overwhelming justification of the charge that the many different psychotherapeutic practices . . . have done much to re-adapt discontented women to a conservative feminine status quo, to an inferiorised psychology and to a contentment with serving and servicing men and children' (Mitchell 1974: 299) still remains true today, and is applicable to other groups as well as women. Different case studies could have been presented, of relationships between mental illness and class, ethnicity, or culture, focusing on resistance by 'patients' and other groups, or on contrasting the concepts of Western psychiatric and psychological medicine and their underlying world views with those of other medical systems (see e.g. the section in the December 1970 issue of the *American Journal of Psychiatry* on racism in American psychiatry; Black Health Workers and Patients Group 1983 (UK); Mercer 1986 (UK); Fernando 1988, 1991; Littlewood and Lipsedge 1989 (US and UK) on the case of schizophrenia in particular; Torkington 1991 (UK); and Alladin 1992 (UK)). There are increasing numbers of oppositional therapeutic practices being formulated, demonstrating the success of different groups in achieving a shift in psychiatric and psychological theory and practice (see e.g. Ernst and Goodison 1981; Deleuze and Guattari 1984; Baruch and Serrano 1988; Alladin 1992; Greene 1992; Watson and Williams 1992; Comas-Díaz and Greene 1994;).

The construction and deconstruction of homosexuality as a mental illness

It has been apparent to anthropologists and sexologists since at least the nineteenth century that homosexual behaviour has existed in a variety of different cultures, and that it is an ineradicable part of human sexual possibilities. But what has been equally apparent are the range of different responses towards homosexuality. (Weeks 1989: 96–7)

The second case study examines changes in responses to homosexuality within Western medicine, and investigates how, over the period from the late nineteenth century to the present day, the construction and deconstruction of homosexuality as a mental illness occurred. In Western countries, before the rise of sexology in the late nineteenth and early twentieth centuries, passionate 'friendships' between women aroused little comment (Faderman 1980). As Kitzinger (1987) points out, the development of the 'disease' theory of lesbianism can be seen as an attempt to suppress women's developing political analysis of gender and sexuality. Psychoanalysts like Abraham and sexologists like Krafft-Ebing explicitly linked the (pathological) lesbian with the feminist. Szasz (1971) identifies the concept of homosexuality as heresy as being prevalent in the days of the witch-hunts; as Weeks (1989) points out, throughout the Christian West, there was prohibition of particular sexual acts, especially buggery (not just homosexual buggery), as a 'sin against nature'. McIntosh (1968) identifies the late seventeenth century as the time of emergence of a specific male homosexual role in England; a specialized, despised, and punished role which 'keeps the bulk of society pure in rather the same way that the similar treatment of some kinds of criminal helps keep the rest of society law abiding' (p. 184). McIntosh points to the close links between medicine and the law in the emergence of the category of the 'homosexual', a connection also highlighted by Foucault (1976a).

As late as 1871, in the UK at least, concepts of homosexuality were extremely undeveloped both in the Metropolitan Police and in high medical and legal circles (Weeks 1989), but in the latter part of the nineteenth century there was the emergence of 'new conceptualizations of homosexuality', with elements in this development being changing legal and medical attitudes. The most commonly quoted European writers on homosexuality at the time were the German medical doctor Casper and the French legal writer Tardieu, both of whom concerned themselves with the need to define the new type of 'degenerates' and to assess whether they could be held legally responsible for their acts. Increasingly the 'problem' came to be seen in terms of sickness or mental illness rather than sin; as Weeks identifies it, the medicalization of homosexuality took place:

In the great classificatory zeal that produced the complex definitions and the new sexual types of the late nineteenth century (and in which Ellis was the main British participant) we can discern the supplanting of the old, undifferentiated, moral categories of sin, debauchery and excess, by the new medical and psychological categories of degeneracy, mental illness and disease. The vast majority of the late nineteenth-century pioneers of sex research were concerned, like Krafft-Ebing, with the variations from the norm. (Weeks 1989: 144)

Foucault (1976a: 43) credits an 1870 article by Westphal as giving birth to the 'psychological, psychiatric, medical category of homosexuality' in place of aberrant, sinful behaviours or acts.

Contrasting views on homosexuality and its 'causes'

From the standpoint of the psychiatrist . . . homosexuality . . . constitutes evidence of immature sexuality and either arrested psychological development or regression. Whatever it be called by the public, there is no question in the minds of psychiatrists regarding the abnormality of such behaviour. (Menninger, in introduction to the authorized American edition of the Wolfenden Report, Great Britain Committee on Homosexual Offences and Prostitution 1963: 7)

Homosexuality is to a very large extent an acquired abnormality and propagates itself as a morally contagious disease. (Moore 1945: 57)

Psychoanalytical experience teaches us that the unconscious reason for female homosexuality is to be found in an unsolved oral-masochistic conflict of the pre-Oedipal child with the mother. (Bergler 1951: 326)

Heterosexuals may or may not recognise the homosexual as a human and political equal; however, the homosexual may choose whether or not to engage in prohibited sexual conduct. In short, the homosexual makes a choice – a deviant one – and society retaliates by declaring that he is 'mentally sick' and hence incapable of making a 'real' choice! Were he able to choose 'freely' – 'normally' – he would choose, like everyone else, to be heterosexual. This is the logic behind much of psychiatric rhetoric. The patient's behaviour is the product of irresistible compulsions and impulses; the psychiatrist's, of free decisions. The cognitive structure of this explanation conceals the fact that its imagery only serves to degrade the patient as insane, and to exalt the psychiatrist as sane. (Szasz 1971: 244)

Most doctors now accept that there is no rational basis for regarding homosexuality itself as an illness. A homosexual lifestyle is compatible with all the criteria of health except possibly fertility – and voluntary infertility is not regarded as an illness. (Bancroft 1988: 308)

Sexuality is increasingly recognized as a strong human force that makes a positive contribution to health, when it is allowed expression in the context of caring, supportive, mutually consensual relationships. Member States should endorse the view that, in accordance with fundamental human rights, consenting adults can decide how to lead a healthy sexual life. Important differences in cultural values and traditions continue to exist between countries and in populations, but the rights of individuals to self-determination in choice of sexual orientation must also be considered. The challenge for health promotion is to support positive expressions of sexuality in a manner sensitive to cultural values. This means taking action to support the rights of all adults to form sexual relationships with consenting partners of their own choice, and to promote respect and tolerance for individuals' decisions in this matter. (WHO/EURO 1993a: 77–8)

As Foucault reminds us, this should not be interpreted in simplistic and purely negative terms:

> there is no question that the appearance in nineteenth-century psychiatry, jurisprudence, and literature of a whole series of discourses on the species and subspecies of homosexuality, inversion, pederasty, and 'psychic hermaphrodism' made possible a strong advance of social controls into this area of 'perversity'; but it also made possible the formation of a 'reverse' discourse: homosexuality began to speak on its own behalf, to demand that its legitimacy or 'naturality' be acknowledged, using the same categories by which it was medically disqualified. There is not, on the one side, a discourse of power, and opposite it, another discourse that runs counter to it. Discourses are tactical elements or blocks operating in the field of force relations; there can exist different and even contradictory discourses within the same strategy; they can, on the contrary, circulate without changing their form from one strategy to another opposing strategy. (Foucault 1976a: 101–2)

In 1952 the American Psychiatric Association (APA) issued its first official listing of mental disorders, the Diagnostic and Statistical Manual of Mental Disorders or DSM-I (American Psychiatric Association 1952). This had evolved from the efforts of a working group brought together by the US Public Health Service to design a nosological scheme adequate to the needs of modern psychiatry. Homosexuality, together with 'other sexual deviations', was included among the sociopathic personality disturbances: 'these disorders were characterized by the absence of subjectively experienced distress or anxiety despite the presence of profound pathology. Thus it was possible to include homosexuality in the nosology despite the apparent lack of discomfort or dis-ease on the part of some homosexuals' (Bayer 1981: 39). As Bayer goes on to point out, this explicitly acknowledges the centrality of dominant social values in defining such conditions; DSM-I asserted that individuals so diagnosed were 'ill primarily in terms of society and of conformity with the prevailing cultural milieu' (American Psychiatric

Association 1952: 34). Within the revised nomenclature issued in 1968, DSM-II (American Psychiatric Association 1968), homosexuality was removed from the category of sociopathic personality disturbances and listed (together with the 'other sexual deviations': fetishism, paedophilia, transvestism, exhibitionism, voyeurism, sadism and masochism) among the 'other non-psychotic mental disorders' (American Psychiatric Association 1968: 44).

The 'pathological' model of homosexuality remained predominant until at least the 1970s; for example, according to Morin (1977), as much as 70 per cent of pre-1974 psychological research on homosexuality was devoted to the three questions: 'are homosexuals sick?'; 'how can it be diagnosed?'; 'what causes it?'. As Bayer (1981) points out, discussing America during the first half of this century, many homosexuals (of those willing to express themselves publicly) welcomed the psychiatric effort to 'wrest control of the social definition of their lives from moral and religious authorities. Better sick than criminal, better the focus of therapeutic concern than the target of the brutal law' (Bayer 1981: 9). To which we might add 'better sick than damned'; as Nugent and Gramick (1989) point out, this medicalization of homosexuality inclined some schools of thought within the Protestant, Catholic, and Jewish religions towards a more accepting attitude towards homosexuality. But Bayer goes on to add: 'by the late 1960s, however, homosexual activists had discarded whatever lingering gratitude remained toward their former protectors and in a mood of militancy rose up to challenge what they considered the unwarranted, burdensome, and humiliating domination of psychiatry' (p. 9). This very active resistance against the dominant psychiatric classification of homosexuality, which resulted in the removal of homosexuality from the Diagnostic and Statistical Manual by the American Psychiatric Association in 1973, provides a very clear example of something particularly unprecedented, namely the active face-to-face negotiation of the boundaries of disease and illness between the medical establishment and those deemed by that establishment to be 'sick'. Bayer has provided us with a detailed study of this episode in the history of American psychiatry, which is drawn on below to illustrate some of the factors at play.

It is important to acknowledge that, although eclipsed by the pathological view, dissenting views on the nature of homosexuality have been present throughout the period under discussion. In the early twentieth century for example, reformers like Havelock Ellis and Magnus Hirschfeld offered a view of homosexuality as a normal variant of human sexuality. In the 1940s and 1950s a number of studies were carried out which provided supporting evidence, and which were to be crucial in contesting the definition of homosexuality as a mental disorder. The first of these was Kinsey's study of the sexual behaviour of white American males (Kinsey *et al.* 1948). This study found that 37 per cent of the population had had physical contact to the point of orgasm with other men at some time between adolescence and old age; it also found a continuum ranging from no homosexual behaviour to exclusively homosexual behaviour, with 10 per cent reporting more or less exclusively homosexual behaviour for at least three years between the ages of 11 and 55. In presenting their results, Kinsey *et al.* rejected the conventional dichotomy between normal and abnormal, assuming instead that differences were a matter of degree. Given the frequencies they found – which they acknowledged were higher than they had expected to find (p. 625) – they argued that it was inconceivable that the homosexual reaction could be an indication of psychopathology, arguing instead that, far from heterosexuality representing a biological directive, our 'mammalian heritage' made possible both heterosexual and homosexual responses, and that the pattern of sexuality chosen by individuals represented but one example of the 'mysteries of human choice' (p. 661). They concluded that the dominance of heterosexuality was due to restrictive cultural norms, and that rather than 'helping' homosexuals to conform (which was enforcing the cultural hegemony of heterosexuality), assisting human beings to accept their diverse sexual orientations was a goal of significantly greater merit (p. 660). Bayer (1981: 199) notes also that their data and theoretical assumptions were incompatible with those who argued that redirection of sexual orientation was impossible, something that was to serve as a central feature of homophile ideology at certain times.

The second study, published in 1951, was a cross-cultural analysis of 77 cultures, together with an investigation of behaviour of non-human primates and other animals (Ford and Beach 1951). In 49 of these societies, male homosexual activity of some variety was considered normal and socially sanctioned for some members of the community. In some cases homosexual behaviour was considered appropriate for all men at one stage of life, in others, exclusive homosexuality for some men, fulfilling special social functions, was not only accepted but valued. No societies were found where male homosexuality represented the dominant form of sexual activity for adults. In the remaining societies in the study, homosexual behaviour was considered unacceptable, with explicit social pressure applied against it; however, indications of homosexual activity were still found. From their review of studies of sexual behaviour in animals, Ford and Beach found evidence of sexual activity between monkeys of the same sex, which they argued could *not* be interpreted as efforts to assert dominance (owing to the existence of signs of erotic arousal and satisfaction), nor as substitutive behaviour occasioned by absence of females, since concurrent homosexual and heterosexual activity was observed. Their conclusion was that in virtually all animal species there was an inherent biological tendency for 'inversion of sexual behaviour', as they termed it (Ford and Beach 1951: 143). Overall, therefore, they concluded that the homosexual responsiveness found in humans reflected our 'fundamental mammalian heritage' (p. 259), refuting the argument that homosexuality was pathological because it violated a biological directive. Like Kinsey, they argued that cultural experience was the determinant of the predominance of heterosexuality:

> men and women who are totally lacking in any conscious homosexual leanings are as much a product of cultural conditioning as are exclusive homosexuals who find heterosexual relations distasteful and unsatisfactory. Both extremes represent movement away from the original indeterminate condition which includes the capacity for both forms of sexual expression. (p. 259)

A third set of key studies are those carried out by the psychologist Evelyn Hooker into groups of homosexuals in the community, who did not fit the image of the tortured and disturbed homosexual. These studies were in marked contrast to those of many psychologists and psychiatrists, whose research focused on clinical populations or on settings such as prisons and barracks. One of these (Hooker 1957) compared a group of male homosexuals, drawn from names provided by two Californian homosexual rights groups, with a group of male heterosexuals matched by age, IQ, and educational level. The use of Rorschach and other projective tests categorized two-thirds of both homosexuals and heterosexuals as of average or better adjustment. The judges of the test results were unable to distinguish between the homosexuals and the heterosexuals in the matched pairs. Hooker concluded that, while a homosexual orientation might be viewed as constituting a 'social' maladjustment, it did not invariably affect the psychological well-being of the individual (p. 30). Hooker's further work examined family relations, since much of the orthodox literature argued that homosexuality was a pathological development rooted in a disturbed family background. She concluded that, for most homosexuals, disturbed family relations were neither necessary nor sufficient as determinants of their psychosexual development. She argued, like Kinsey, that the causes of homosexuality involved many different factors, including 'biological, cultural, psychodynamic, structural and situational' (Hooker 1968: 225). She also argued (Hooker 1955, 1957, 1965) that traits such as 'fear of intimacy', 'obsessive concern with homosexuality', and 'withdrawal and passivity', all of which formed part of the evidence cited for 'homosexual pathology' were, when they existed, traceable to the pressures on homosexuals from the heterosexual world; the 'disturbed' behavioural patterns were 'ego defensive', linked to the victimization individuals had experienced.

These studies provided increased stimulus to the homophile movement, which was becoming increasingly organized and open over the 1950s and 1960s in America to contest the dominant view of homosexuality as pathology, adopting instead an affirmative stance towards homosexuality (Bayer 1981). Bayer notes the important influence of the experience of the civil rights struggle on the homophile movement's adoption of an affirmative stance, explicitly recognized in the resolution adopted by the 1968 North American Conference of Homophile Organizations (NACHO): 'BECAUSE the Negro community has approached similar problems and goals with some success by the adoption of the motto or slogan: *Black is Beautiful* RESOLVED: that it be hereby adopted as a slogan or motto for NACHO that GAY IS GOOD' (Bayer 1981: 90–1). Over the same period as the struggle to remove homosexuality from the DSM-II was taking place, the American Psychiatric Association was also beginning to acknowledge the reality of racism and the need to counter its effects within psychiatry (see e.g. the special section on racism in the *American Journal of Psychiatry* in December 1970 (Fernando 1991)). The homophile movement also adopted the view that the therapeutic posture was morally wrong, and that 'those who offered assistance to *voluntary* patients who expressed profound distress over their sexual orientation, did so as the agents of society and should be attacked' (Bayer 1981: 85; emphasis in original). The movement found new allies in the psychiatric profession, those who were critical of the profession and who pointed to its 'illegitimate power' – namely the tendency, not only in the sphere of homosexuality, to pass off moral judgements as ostensibly value-neutral and scientific (see Szasz 1971, 1974).

Initially the tactics adopted were those of education and reason, but some argued, drawing on the experience of the civil rights movement of the 1960s, that militant action in the courts and on the picket lines was the only effective tool available to those seeking social change: 'we would be foolish not to recognize what the Negro rights movement has shown us is sadly so. Mere persuasion, information and education are not going to gain for us in actual practice the rights and equality which are ours in principle' (Kameny 1965, quoted in Bayer 1981: 83). Bayer links this change in tactics also to the climate created by the contemporaneous struggles of many different marginalized groups within American society. There followed the use of a new mode of struggle: picket lines began to appear at the lectures of those who defended the orthodox psychiatric position on homosexuality; in 1968 the convention of the American Medical Association in San Francisco was leafleted by activists with demands that those who opposed the pathological view of homosexuality be represented at further conventions, and that representation from homophile groups also be invited; there were similar protests at Columbia University's College of Physicians and Surgeons (Bayer 1981: 92).

Alongside increasing action in other fora (gay pride demonstrations, pickets and demonstrations against the Catholic Church and the mass media), there was continued action against psychiatrists who gave public lectures on the 'disease of homosexuality', and protests against university departments that used articles and books considered unacceptable in their portrayal of homosexuality. There was soon a shift from demonstration to disruption (mirroring the change in tactics adopted in anti-war groups' actions); for example, in October 1970 the Gay Liberation Front interrupted a session showing a film on aversive conditioning techniques for elimination of homosexual behaviour at the second annual Behaviour Modification conference in Los Angeles, demanding reconstitution of the session in small discussion groups to address issues such as 'homosexuality as an alternative life style' (Bayer 1981: 99). With a view to the demand for the deletion of homosexuality from the American Psychiatric Association's (APA) official Diagnostic and Statistical Manual of Psychiatric Disorders, gay groups in alliance with feminists conducted a systematic effort to disrupt the 1970 annual meeting of the APA, which systematically refused to let homosexuals appear on the official programme. The panels on transsexualism and homosexuality (which included papers by psychiatrists such as Bieber, whose 1962 book on homosexuality represented a typical statement of the view of homosexuality as a pathology requiring clinical intervention) and on issues on sexuality (which included a paper on aversive conditioning techniques) were successfully disrupted. In response to this action, the following year saw the first panel discussion by homosexuals at an APA convention, following which the gay activists made their first requests to present their demands for the deletion of homosexuality from the APA's DSM-II to members of the Association's Committee on Nomenclature; this request, however, came to nothing at that stage.

The following year there was a fully institutionalized gay presence at the APA convention in Dallas 1972; tactics moved away from disruption towards an emphasis on trying to open dialogue. A flyer produced by Kameny stressed both scientific and social arguments: that psychiatrists had acted 'unscientifically' in labelling homosexuality as a disorder and that the social consequences for gay women and men of being so stigmatized had been disastrous:

> We are trying to open dialogue with the psychiatric profession. . . . In past years it has been necessary, on occasion, to resort to strong measures against a resisting profession in order to achieve such discussion of *our* problems *with* us instead of merely about us. We sincerely hope that resolution, constructive discussion and dialogue, followed by meaningful reform of psychiatry will soon proceed. . . . Psychiatry in the past – and continuingly – has been *the* major single obstacle in our society to the advancement of homosexuals and to the achievement of our full rights, our full happiness and our basic human dignity. Psychiatry *can* become our major ally. ('Gay, Proud and Healthy', quoted in Bayer 1981: 108; emphasis in original)

It was, however, as the result of yet another action, at the Association for the Advancement of Behaviour Therapy conference in October 1972 by the New York Gay Activist Alliance (GAA), that Robert Spitzer, a member of APA's Committee on Nomenclature, first came into contact with homosexuals demanding revision of psychiatry's attitude towards homosexuality. According to Bayer (1981: 116), Spitzer was impressed by their passion and arguments, and he agreed to arrange for a formal presentation of their views to a full meeting of the committee, and to sponsor a panel at the APA's 1973 convention on the question of whether homosexuality ought to be included in the Association's official listing of psychiatric disorders. Charles Silverstein of the Institute for Human Identity (a homosexual and bisexual counselling centre) was chosen to prepare the GAA's statement for presentation at the Nomenclature meeting. This presentation, given in February 1973, focused first on reviewing the research findings of psychologists, psychiatrists, and social scientists to demonstrate that the classification of homosexuality was inconsistent with a scientific perspective, using Hooker, Kinsey, Ford, and Beach, as well as articles by Marmor, Green, and Hoffman in the 1972 special issue of the *International Journal of Psychiatry*. Secondly, it considered the documentation on the use of diagnostic labelling to buttress society's discriminatory practices. Thirdly, it pointed to the psychological damage to homosexuals caused by the labelling of homosexual preference as pathological. Bayer reports that much of the data presented was new to the committee members, and they found it very interesting. In spite of considerable opposition from a hastily formed *ad hoc* committee against the deletion of homosexuality from DSM-II, under the leadership of Irving Bieber and Charles Socarides (Bayer 1981: 122), this initial presentation led fairly rapidly to the preparation of a proposal by Spitzer for the deletion of homosexuality as a mental disorder from DSM-II, which was passed by the Board of Trustees of the APA in December 1973 (Bayer 1981: 134).

Following this, the 1975 edition of the *Comprehensive Textbook of Psychiatry* saw the replacement of Bieber's essay on homosexuality (which had appeared in the 1967 edition) by a contribution by Marmor: 'in the final analysis, the psychiatric categorization of the homosexual outcome as psychopathological is fundamentally a reflection of society's disapproval of that outcome, and psychiatrists are unwittingly acting as agents of social control in so labelling it' (Marmor 1975: 1517). In August 1979 Surgeon-General Julius B Richmond ruled that government physicians would no longer consider homosexuality a mental disease or defect, and that the Immigration Service officers should cease referring suspected homosexuals to the Public Health Service for examination, the required practice during the previous twenty-seven years. However, in a particularly curious reversal pointed out by Bayer (1981: 193–4), the Justice Department required a reinstatement of the exclusionary policy, on the basis that Congress did not empower the Surgeon-General to disagree with Congress's passing of the Immigration and Nationality Act in 1952, which considered homosexuality as a disease.

Pathological notions are still found in professional texts of psychoanalysis and clinical psychology (see e.g. the review provided in Kitzinger 1987), and homosexuality still has not entirely escaped medicalization; see, for example, the recent debate about male homosexuality in terms of the possible existence of a 'gay gene' (Baron 1993; Hamer *et al.* 1993). It is, however, now becoming more common to find implicit or explicit acceptance of the existence of healthy homosexuality within professional texts; see for example the quotes from Bancroft, written in an editorial for the *British Medical Journal*, and from WHO shown on pp. 52–3.

Discourses around HIV/AIDS: an epidemic of signification[1]

> AIDS does not exist apart from the practices that conceptualize it, represent it, and respond to it. We know AIDS only in and through those practices. This assertion does not contest the existence of viruses, antibodies, infections, or transmission routes. Least of all does it contest the reality of illness, suffering, and death. What it *does* contest is the notion that there is an underlying reality of AIDS, upon which are constructed the representations, or the culture, or the politics of AIDS. If we recognize that AIDS exists only in and through these constructions, then hopefully we can also recognise the imperative to know them, analyze them, and wrest control of them. (Crimp 1987: 3)

The final example we consider in this chapter is the case of discourses about HIV/AIDS. What we shall be concerned with here is to analyse the different ways AIDS has been presented, particularly in terms of discussions about its origin, and the social consequences these have had. Before exploring these issues, however, in order to provide a basis for the later analysis, we summarize the current state of knowledge about HIV and AIDS, and then move on to identify some of the commoner terms which are often used inaccurately, creating misleading impressions and lending support to the myths we will consider below. Table 3.1 defines some key concepts in terms of current knowledge about HIV and AIDS.

AIDS (acquired immune deficiency syndrome), as the name implies, is a syndrome – or collection of signs and symptoms – arising from a number of conditions resulting from an impairment in the body's ability to fight disease, which involves a suppression of the immune system, and a weakening of the body's defences against a variety of infections, viruses, and malignancies. AIDS is diagnosed according to the case definition produced by the US Centre for Disease Control (CDC); where laboratory confirmation of the presence of HIV is not available, this is based on the presence of one or more of a list of opportunistic infections, such as thrush in the throat or Kaposi's sarcoma, in a patient under 60 years old. Where HIV-positive status is established, the presence of a wider range of opportunistic infections, such as recurrent septicaemia, is used to diagnose AIDS (Miller and Rockwell

1 The phrase is taken from an article by Paula Treichler (1987).

Table 3.1 Key concepts to do with HIV and AIDS

AIDS (acquired immune deficiency syndrome)	a syndrome that involves the presence of a number of conditions resulting from an impairment in the body's ability to fight disease, which involves a suppression of the immune system, and a weakening of the body's defences against a variety of infections, viruses, and malignancies
antibody	a blood cell that recognizes and targets a specific substance
antigen	a substance that stimulates production of antibodies
HIV (human immuno-deficiency virus)	the virus that is thought to be the sole or foremost causative agent of AIDS
HIV test	the most commonly used test detects the presence of antibodies to HIV in the blood
immune system	the body's system of defence, in which specialized cells and proteins in the blood and other body fluids work together to eliminate disease-producing agents and other toxic substances
opportunistic infections	infections caused by organisms that do not normally cause infections in people with healthy immune systems
seroconversion	initial development of antibodies specific to a particular antigen, e.g. the development of antibodies to HIV

Source: Panos (1990); Banzhaf et al. (1992).

1988). A virus called HIV (human immuno-deficiency virus) is considered to be the sole or foremost causative agent of AIDS (for a recent analysis of the debates that still rage over the cause of AIDS, see Fujimura and Chou 1994). Although HIV has been isolated from a number of bodily fluids and tissues, the virus has only been found to be transmissible between people in certain ways, and is not easily communicable though normal day-to-day contact. It can be transmitted when the blood, menstrual fluid, semen, or vaginal secretions of someone who is infected with HIV enters another person's body through unprotected sex, needle re-use, injury, unscreened transfusion of blood or blood products, or from mother to foetus through the placenta. Transmission rates vary both between and within these different routes. HIV does *not* spread from mosquito bites, toilet seats, swimming pools, telephones, sharing cups or cutlery, shaking hands, coughing, hugging, or kissing. There have been no proven cases of transmission through tears or saliva. Once infected by HIV, seroconversion (the initial development of antibodies to HIV) occurs after a period between a few weeks and eighteen months. The current HIV tests work by detecting the presence of antibodies to HIV in the blood, and thus will only show a positive result after seroconversion has occurred. The average period between infection with HIV and the onset of AIDS can be extremely variable – up to at least twenty years according to some estimates.

Misleading messages

Many of the terms and phrases used in discussion of HIV and AIDS, both in the medical literature and in the media at large, act to foster stigmatizing myths and/or to convey inaccurate information. Table 3.2 lists some of the misleading terms/phrases which are still commonly found, together with the reasons why they should be avoided. The terms in Table 3.2 provide some

Table 3.2 Misleading messages

Term(s) to avoid	Why?	Use instead
AIDS carrier AIDS positive carrying AIDS	This confuses the two distinct phases of being infected with HIV and having AIDS. People 'have' AIDS, they don't 'carry' it.	*Either* HIV antibody positive, person with HIV, *or* person with AIDS, *as appropriate*
AIDS test	There is no test for AIDS; this is a diagnosis made according to clinical symptoms. The most commonly used test detects antibodies to HIV.	HIV antibody test
AIDS virus	Can easily cause confusion between HIV and AIDS.	HIV
catching AIDS	It is not possible to catch AIDS. HIV is what you can be infected with. It is even misleading to say 'catch HIV', since this tends to suggest transmission is similar to colds or flu, it is not.	contract HIV become HIV positive
AIDS sufferer	Having AIDS does not mean being ill all the time. Someone with AIDS may continue to work and live a normal life after diagnosis. 'Sufferer' is therefore inappropriate.	person with AIDS person living with AIDS
AIDS victim	Suggests helplessness, so inappropriate.	person with AIDS person living with AIDS
AIDS plague	Suggests a contagious disease (one contracted by casual contact).	AIDS epidemic
innocent (AIDS) victim	Suggests that others with AIDS are guilty.	
high-risk group	There are risky behaviours, not risk groups. Membership of a particular group does not put the individual at greater risk, but what she or he does, regardless of group membership, may do.	

Source: Grover (1988); ACT UP London; Panos (1990); Banzhaf *et al.* (1992).

of the basic building-blocks that contribute to the various myths that continue to circulate. They continue to appear regularly in the media, and even within health service and medical literature. In the examples we consider below, we shall find that these terms are repeatedly misused. A second strand in the analysis below is provided by the use (or abuse) of metaphor (Sontag 1979, 1989), as Altman (1986: 194) argues: 'AIDS, it is clear, touches a whole set of fears and prejudices that go very deep. The link with sexuality and blood makes AIDS particularly susceptible to metaphorical use.'

Treichler, in her 1987 paper from which the title of this section is taken, lists thirty-eight ways in which AIDS has been conceptualized, coming chiefly from printed sources in the USA, and explores in detail some of the homophobic conceptualizations. She argues compellingly:

> we cannot effectively analyze AIDS or develop intelligent social policy
> if we dismiss such conceptions as irrational myths and homophobic
> fantasies that deliberately ignore the 'real scientific facts'. Rather they
> are part of the necessary work people do in attempting to understand
> – however imperfectly – the complex, puzzling and quite terrifying
> phenomenon of AIDS. . . . This work is as necessary and often as dif-
> ficult and imperfect for physicians and scientists as it is for 'the rest of
> us' (Treichler 1988: 34)

Homophobia is not the only factor at work; sexism and racism are also implicated in the shaping of discourses on HIV and AIDS. These are illustrated below by exploring three of the many myths about HIV/AIDS: 'the gay plague', 'the African problem', 'the white man's disease'. The examples serve to illustrate the interplay of heterosexism, sexism, and racism operating to produce a diverse set of often contradictory readings about the nature, cause, and modes of transmission of HIV and AIDS. These frequently result in distancing large sections of the population from the problem, making it the concern, and the fault, of some undesirable, marginalized, 'other' group. Throughout, the role of the media has been crucial in shaping these myths, confusing and confounding the messages originating from health service sources, in terms of information concerned with prevention. The separation of the three myths in the discussion below should not be taken to imply that they are sequential; it should instead be recognized that they are often present simultaneously and interact in particular ways.

'The gay plague'

The myth of AIDS as the 'gay plague' appears perhaps as the first myth in the history of HIV and AIDS, following the discovery of unusual occurrences of Kaposi's sarcoma and pneumocystis carini among homosexual men in America in the early 1980s. Altman (1986) illustrates how the American media responded by establishing a homosexual character for AIDS; this was reinforced by the use of the term 'GRID' (gay-related immune deficiency) in some medical circles (Altman 1986: 33), in contrast to the Centre for Disease Control, who used the neutral term 'Kaposi's sarcoma and related opportunistic infections'. This perception was not altered by the discovery of cases in other groups: drug-users, Haitians, and haemophiliacs (mid-1982), in the female partners of those in the so-called 'risk groups' (beginning 1983), and then in Africa in both sexes (from 1982 on, but not reported in the American media until much later). The media reaction was not to change its reporting language, but instead to make the report more 'newsworthy' when a non-homosexual was affected (Altman 1986: 17), leading to the multiply contradictory report in the British press quoted by Altman (1986: 20) of the death of a drug-free white heterosexual grandmother on Long Island under the headline: 'Gay Bug Kills Gran'.

The headline provides an example of another theme within the 'gay plague' myth, that of fear of contagion. This, combined with an emphasis on 'risk groups', resulted in a multitude of debates about restricting the activities of certain groups, not to mention quarantine. So that, for example, in 1983 the *New York Post*, in two articles entitled 'AIDS Disease: It's Nature Striking Back' (24 May, p. 25), argues for restrictions on employment of homosexuals in food-handling, donating blood, and childcare. The headline also illustrates another subtext of the 'gay plague' myth, that of deserved retribution. In other variants, retribution is visited by divine agency rather than a reified nature.

The very real consequences for people living with HIV or AIDS were refusals of contact and direct discrimination through loss of jobs, discrimination, and prejudice in the health service sector (Altman 1986; Gilman 1987; Panos 1990), extending in some instances to all members of the so-called risk group, so that, for example, Altman details discrimination against Haitians, and a devastating effect on their tourist industry (see also Farmer 1992). The mechanisms at work were recognized by CDC (1983): 'the classification of certain groups as being more closely associated with the disease has been misconstrued by some as to mean these groups are likely to transmit the disease through non-intimate interactions. This view is not justified by available data. Nonetheless, it has been used unfairly as a basis for social and economic discrimination.'

In France, Herzlich and Pierret (1989) studied the coverage of AIDS in the French press over the period 1982–6, illustrating first its early identification as 'homosexual pneumonia', 'homosexual cancer', 'gay cancer', or 'homosexual syndrome'. Labelling of this type was contemporary with the recognition that it was not a cancer, and that homosexuals were not the only people affected. They also carefully illustrate how AIDS was constructed as a cultural and moral phenomenon, on the basis of concepts borrowed from epidemiology. Terms such as 'risk factor', 'risk category', 'risk group', 'risk individual', 'risk population', with shifts from risk factor to cause and back again, were used to justify anticipation and apprehension about 'contamination' or contagion, constructing a discourse about the dangerous 'other', in this case the homosexual. In relation to transmission of HIV, there are risky behaviours, not risk groups. Membership of a particular group does not put the individual at greater risk, but what she or he does, regardless of group membership, may do. As Altman argues, 'the equation of AIDS with gay men carried with it a strain of blaming gays for the introduction and spread of the disease, and the concomitant idea that others who fell sick were somehow "innocent victims"' (1986: 24–5).

The 'gay plague' myth and its various subtexts have been the subject of vigorous contestation on both sides of the Atlantic. Crimp (1987) and Gever (1987) review some of the initiatives taken by 'cultural activists' to produce highly visible public information about AIDS in the USA, in response to inactivity by the health and welfare services and misleading information in the media. They also produced other material in a variety of media: film, video, music tapes, all offering critiques of official and media responses to AIDS. Gever (1987) analyses in detail one such piece of work, Stuart Marshall's 'Bright Eyes', a video produced for Channel 4 in the UK, demonstrating how it is constructed to offer political and historical analysis of the current discourses on AIDS, and specifically to counteract the homophobia that permeated the dominant media's representation of AIDS. The association of AIDS particularly, or uniquely, with gay men and lesbians still continues to circulate, however.

'The African problem'

illustrating the classic script for plague, AIDS is thought to have started in the 'dark continent', then spread to Haiti, then to the United States and to Europe . . . It is understood as a tropical disease: another infestation from the so-called Third World. (Sontag 1989: 51–2)

Although the myth of the 'gay plague' was the first to arise in connection with AIDS, and was vigorously contested from the start, other equally oppressive myths soon appeared, fuelled partially by the work carried out to locate the origins of HIV. As Patton puts it, 'the de-gaying of AIDS discourse was simultaneous with a heightening of class and race anxieties around AIDS' (Patton 1989: 115). Chirimuuta and Chirimuuta (1989), Watney (1990), Williamson (1989), Treichler (1989), and Patton (1990) have provided particularly detailed analyses of how the various myths and subtexts connected with the construction of AIDS as 'the African problem' and the notion of a very particular 'African AIDS' circulate and are maintained in both scientific and lay discourses. These discourses about AIDS in Africa have been shown to be influenced by the presence of racial stereotypes, moralistic reasoning, and xenophobic policies (see also Miller and Rockwell 1988; Dada 1991; Packard and Epstein 1991; Schoepf 1991).

Watney (1990) identifies some specific aspects of the notion of 'African AIDS'. First, it speaks of a special affinity between a virus and a continent, and connectedly, it regards 'Africa' as the source of the HIV infection not only in the sense of origin but also in the sense of cause. It also reads the modes of transmission of HIV as signs of a generalized and homogeneous 'primitiveness', both sexual and medical, and supports this by identifying alleged 'misreporting' of HIV and AIDS statistics as further 'evidence' of 'backwardness' and 'unreliability'. Finally it equates black Africans and Western gay man as wilful 'perverts' who are equally threatening to 'family values'. The use of a term such as 'African AIDS' serves to camouflage the very different characteristics of the AIDS epidemic in different countries in Africa, as well as creating the notion of a singular unitary 'Africa', neglecting the cultural, social, economic, and ethnic diversity of the continent.

Williamson identifies many of the narratives about AIDS as seeking a source of contamination in Africa. For example, from Randy Shilts's *And the Band Played On,* she quotes: 'Many [with AIDS] were ailing among the uncounted sick of primitive Africa. Slowly and almost imperceptibly, the killer was awakening'; and she notes particularly the use of 'primitive' and the implicit animation of the HIV virus as a monster (Williamson 1989: 73). This serves to animate AIDS as a subject with intentionality, also implicit in other familiar phrases – the 'killer disease', which 'claims its victims' as if they belonged to it: 'AIDS can choose you' (Williamson 1989: 74). Watney (1990) and Williamson (1989) demonstrate how debates on origins often equate a source of something with its cause (extended into imputing blame or fault; see earlier discussions on lay discourses in Chapter 2). Gilman (1987) makes a similar point, discussing the identification of AIDS as an African or Haitian disease in America, noting also that, for the French and the former Soviet Union, AIDS was viewed as an American disease.

The theory of a simian origin for AIDS (in the African green monkey), together with a hypothesized species crossover occurring in Africa, was propounded in 1985, to be definitively rejected some three years later at the beginning of 1988 (Sabatier 1988; Chirimuuta and Chirimuuta 1989). Despite that, the scientific literature continues to circulate the myth. Chirimuuta and Chirimuuta quote the example of an article entitled 'The origins of the AIDS virus' in the October 1988 issue of the *Scientific American,* by Essex and Kanter, two of the researchers involved in research with the green monkey who earlier in the year had acknowledged that earlier research was flawed and the presence of HIV in laboratory monkeys had been due to laboratory contaminants, invalidating the conclusions on origin. The October 1988 article fails to mention this and features the green monkey prominently. The article fails to mention the literature discussing alternative hypotheses such as a Euro-American origin (Katner and Pankey 1987), accidental or deliberate laboratory creation (Segal and Segal 1987; Seale 1988), or other genetically similar viruses in sheep and cattle (Gonda *et al.* 1985). This example can be taken further; the following year, 1989, the article appeared again, reprinted in a book, entitled *The Science of AIDS*, of articles taken from the *Scientific American*. It is the only article in the book that discusses the origins of HIV, and the introduction provided to the book's contents on the cover mobilizes the full force of myths about the factual nature of 'science' and the book's contents:

> today, when scare tactics and controversy seem to dominate the coverage of AIDS, this new work presents a level-headed, scientific look at this emotionally-charged medical and social phenomenon. Assembled from the pages of *Scientific American*, the articles contained in this Reader address what *is* known about AIDS; there is not a more authoritative account of the situation. . . . *The Science of AIDS* offers what is most needed to understand the causes and effects of this deadly and controversial affliction: clear, factual, and current information from the people most involved in controlling and attempting to cure AIDS' (italics as in original).

Intermingled with the simian-origin theory was the notion that AIDS represented an 'old African disease', something that had been present in the continent for a long time; this was supported only by speculation on the basis of extremely unreliable studies of seropositivity carried out on small numbers of old blood samples. Chirimuuta and Chirimuuta (1989) provide a careful analysis of the racist notions of 'exotic' cultural, and specifically sexual, practices (unsupported by evidence) which were held responsible for the species barrier being crossed. They also examine how the interpretation of many of the early studies carried out in Africa is usually based on implicit assumptions of HIV transmission from Africans to Europeans, with little consideration of the reverse possibilities. In the face of this perpetuation of the myths of 'African AIDS' in sections of the scientific literature, it is perhaps not surprising that it remains one of the themes constantly referred to in discussions of AIDS and Africa in popular magazines and newspapers, as illustrated by the articles analysed by Watney (1990) and Williamson (1989).

A particular subtext here is provided by interaction with some of the sexist myths of women as sources of contagion, and of female sexuality as dangerous. Williamson (1989) gives the example of a particularly pernicious piece, 'Happy Hookers of Nairobi', which appeared in the UK *Guardian* in 1987. The operation of sexism and racism in combination can clearly be seen: 'the best time to observe the Nairobi hooker is at dusk when the tropical sun dips beneath the Rift Valley and silhouettes the thorn trees against the African skyline. It is then that the hooker preens itself and emerges to stalk its prey . . . white men looking for fun and with money to burn' (*Guardian*, quoted in Williamson 1989: 77). The style of reportage draws on the familiar conventional tropes of the wildlife documentary to construct the prostitutes as 'beasts' displaying a dangerous animal sexuality, while their 'prey' are freed of any responsibility, this notion being reinforced by the use of the term 'hooker', placing the prostitute in the position of initiator of the encounter, rather than her customer. Chirimuuta and Chirimuuta (1989) examine how the nature of the epidemiological studies carried out and their reporting in the scientific media work to support such myths, commenting: 'it is quite extraordinary that these researchers can perceive the widespread sexual exploitation of poor African women by their former colonial masters only as a threat posed by African sexuality to the health of Europe' (1989: 161).

As with the 'gay plague' myth discussed above, the construction of 'African AIDS' brought in its wake discrimination against Africans, including discussions about screening programmes in many countries aimed particularly at immigrants, migrant workers, and students, institutionalized in some cases into restrictions on entry (Waite 1988), as well as direct violence and attacks against Africans studying or working abroad (Chirimuuta and Chirimuuta 1989).

'The white man's disease'

Both within and outside the 'gay world' in America and the UK, AIDS is often seen as largely a white disease. The myth is succinctly summarized by Madhubuti in the opening paragraph of his article which explores the potential impact of AIDS on the black population: 'like most misinformed, confirmed heterosexuals, I was convinced that AIDS was a white middle-class homosexual disease that, at worse, would only touch Black homosexuals. I also felt that the AIDS reports coming out of Africa were exaggerated' (Madhubuti 1990: 51). Some, for example Gayle (1987), Rumsey (1990), and Ports and Banzhaf (1990), see links to a lack of acknowledgement or open discussion of homosexuality within black communities. As Altman points out, this can be seen as 'a reflection of a larger reality – namely, the tendency of both the gay and the non-gay press to conceive of homosexuality in largely male, middle class and, above all, white terms' (Altman 1986: 100), compounded by racism within homosexual communities and the health service system (Dada 1990).

The myth persists despite the official reported statistics on AIDS cases, showing, for example, a high proportion of AIDS cases in America amongst the black population. Gayle (1987) suggested several contributing factors: that the data is not believed; that it is viewed as a conspiracy by the white majority, particularly to '(once again) blame a major catastrophe on blacks'; or that individuals rationalize away their personal vulnerability by asserting that the disease only threatens 'others' who are members of 'risk groups' unfamiliar to themselves. Again here, the use of the term 'risk groups' rather than 'risk behaviour' operates to support such distancing. Gayle also notes the influence of the media, with television documentaries and drama portraying AIDS as affecting the 'white, usually male and homosexual citizen', while newspapers and magazines rarely confront the issue of black people and AIDS, particularly in visual terms.

The circulation of this myth has obvious consequences in terms of its potential effects on the allocation and targeting of resources, as well as in terms of its interference with preventive programmes. Altman (1986: 100) gives the example of how, when the organization AID-Atlanta lobbied for city money, this was one of the myths it had to dispel, while Gayle (1987), Rumsey (1990), and Ports and Banzhaf (1990) identify the lack of relevant resources to address the needs of black communities for AIDS education. A more recent example of the perpetuation of the myth is given in the following quote from an interview with Earvin 'Magic' Johnson, the basketball player, shortly after he was diagnosed as HIV-positive and decided to go public in a bid to combat ignorance about HIV and AIDS:

> I particularly hadn't paid attention to the figures showing that AIDS is a huge problem in the black community. I didn't know that half of the Americans currently suffering from the disease are either black or Hispanic. Like most other blacks, I was denying that AIDS was spreading through our community like wildfire while we ignored the flames. To me, AIDS was someone else's disease. It was a disease for gays and drug users. Not for someone like me. (Johnson 1991)

We might note particularly the invisibility of black homosexuals within the statement.

Metaphor in health and ill health

'Official' AIDS information participates actively in the ideological foreground of all Western societies, seemingly validating social values and boundaries with the full authority of 'science' and excluding

whole population groups from what Stuart Hall has described as the
'imaginary community of the nation'. (Watney 1989: 18)

The widespread resistance to acknowledging the long-established fact
of heterosexual transmission is not simply an example of 'ignorance'
or 'misinformation': it stems directly from the ideological construction
of AIDS as emblematic of otherness. Indeed the complex history of
AIDS related legislation and official AIDS publicity demonstrates time
and time again that the epidemic has been used to articulate values and
beliefs that have nothing to do with AIDS. In effect health education
has been recruited to the prior purposes of political and ideological
struggle. for such purposes it has seemed far more important to establish the idea that homosexuality is an intrinsic wrong, than to communicate the relatively simple information that explains how and why
different people are at different degrees of risk from HIV. (Watney
1989: 19)

So one way and another the AIDS discourse of our society is structured and coded precisely to fend off transgression, or what Julia
Kristeva has called 'the weight of meaninglessness', the *abject* which
cannot be contemplated except as something to be ejected from the
self. (Williamson 1989: 70)

The ways in which discourses about AIDS have reproduced a circling set of
myths about the fault of (multiple) others, identified in the extracts above, is
captured succinctly in Fig. 3.1, reproduced from a Chilean AIDS education
magazine. Williamson identifies the function of the (AIDS) monster as a
projection of unacceptable parts of the self, of society, in this case of 'unacceptable' sexualities (Williamson 1989: 77). She notes also the dynamic of
alternation 'between fascination with, and denial of, otherness and difference, the simultaneous wish to know and the wish not to know, all play
their part in people's perceptions of AIDS' (p. 79).

A particularly sustained attack on the use of metaphor in the case of cancer and other physical disease is provided by Sontag (1979), who concludes
that metaphor should be stripped from all talk of disease. Her conclusion
remains problematic in its implicitly idealized notion of the existence of a
language somehow uncontaminated by any use of metaphor, and is one that
is arguably impossible to fulfil. Instead, to reflect back to the quotation from
Crimp which opened this section, what is needed is first a recognition that
no disease exists independently of the practices that conceptualize it, represent it, and respond to it, and secondly, the fostering of strands of cultural
analysis and cultural activism which act to know, analyse, and wrest control
of such practices (for examples of such practices of resistance, see Crimp
1988; Boffin and Gupta 1990; King 1993). This is emphatically not to deny
the reality of pain, suffering, and disease, but to act to seek to ensure that
this is not, deliberately or inadvertently, worsened by any epidemic of signification which may attend the situation.

Fig 3.1 Assigning blame for AIDS; myths about the fault of
(multiple) others.

Conclusions: recognizing and working with diversity and difference

As with the examples taken from the field of mental illness and psychiatry,
in place of the case of HIV and AIDS considered above, other examples
could have been chosen from the field of physical health. For example,
Martin (1987) reports an anthropological study carried out in Baltimore
which compares the views of Western medical science with those of ordinary American women about the different ways that women's reproductive

processes are perceived, examining in detail views about menstruation, menopause, and childbirth. She finds that the current medical imaging of reproductive processes is based around metaphors of production, typically with negative descriptions of menstruation (failed reproduction) and menopause (end of productivity). Childbirth is presented in terms of: baby as product, woman as labourer, uterus as machine, doctors as management team. With regard to menstruation, she finds that middle-class women typically bring forth the dominant medical metaphors about menstruation, while there are more resisting or alternative discourses among working-class women. An earlier study by Prudence Rains (1971), carried out in America in the late 1960s, examined the ways in which two groups of young pregnant women responded to therapeutic constructions of their experience in two different institutional settings. The young middle-class white women, in an expensive private residential facility, accepted and internalized a therapeutic perspective which regarded their pregnancies as not simple 'mistakes', but as unconsciously motivated by latent emotional problems, for example a refusal of parental authority or a demand for parental love, rewriting their physical condition in psychiatric terms. In contrast, the second group, of young black women in a non-residential municipal facility, resisted the terms of the psychiatric discourse offered to them, sometimes directly challenging it, sometimes subverting it indirectly by humour, choosing to accept only those aspects of the services offered to them that suited their own definition of their needs.

The examples considered in this chapter have illustrated further how concepts of health and illness are socially constructed and contested, reinforcing the conclusions of Chapter 2 that such concepts cannot be regarded as neutral, since they are intimately linked to social relations and power structures in societies. Recently there has been a considerable shift towards the adoption of a socio-ecological paradigm of health, marked by specific struggles to contest the previously dominant professional definitions of health and illness (such as those examined in some of the case studies above), and resulting in a move away from exclusive dependence on the biomedical model of health. This theme is taken up again and examined in health policy terms in the discussions of the new public health movement in Chapter 7, and of 'Health for All' and health promotion in Chapter 9, both of which illustrate that this process is a continuing one. The interaction between different medical systems, with increasingly widespread acceptance of medical pluralism (Janzen and Feierman 1979) furnishes other examples illustrating the themes of this chapter. As we shall see in Chapter 5, the formal health sector within Western societies has become more accepting of alternative medicine, now more usually referred to as 'complementary medicine' (an important linguistic shift).

The importance of this discussion of views about 'health', 'illness', and 'disease' is not to try to find the most satisfactory definition of a set of terms, but to illustrate the multiple and complex ways in which these terms are used discursively. They are carried into and shape every interaction that takes place within health service settings, whether in the name of treatment, cure, care, prevention, or even health promotion. Fox (1993) analyses how the organization of health services and the caring relationship, as currently constituted, frequently operates to reinforce existing relations of power between the health professional and individual in support of the status quo, but also how there is resistance, and thus scope for recasting these relationships to support the empowerment of the individuals involved. Although we shall not explore this area in detail, we shall take up the theme again in Chapter 8, where we explore this and other Foucauldian studies in the context of health policy development and implementation. It is also relevant to the discussions of Chapter 10 on the interactions between health, health services, and wider aspects of the social context.

|4|

Social and spatial inequalities in health

Health variations: differences and inequities

This chapter considers the evidence for differences in health within and between different societies. We place particular emphasis in the following discussion on what can be learned from differences between populations in different places. Health differences can be distinguished at all geographical scales, from global differences between groups of countries to local differences between small areas. Geographical variation in health status is an important barometer of health inequalities because indicators of population health for geographical areas are widely available, and are often more comprehensive than data on the health of particular social groups. Spatial differences in population health are often suggestive of contextual or environmental influences on health. Studies which compare the geographical patterning of social groups with spatial variations of health indicators also provide indirect evidence of social inequalities in health (although such interpretations need careful qualification, as discussed below).

This chapter concentrates particularly on what can be learned about health inequalities among populations associated with differences in their social and economic profile. The causative processes which give rise to these social inequalities are apparently quite complex, and are likely to be associated with differences in opportunities for the basic requisites for health. The World Health Organization, for example, describes the types of resource which are relevant to health inequalities:

> the right kind of food in sufficient quantities, safe drinking water, good sanitation, and universal, free primary education. . . . the chance to choose healthy lifestyles . . . adequate housing and employment opportunities . . . and good health care. (WHO 1988: 140–1)

Much of the global and local inequality in health discussed in this chapter is not inevitable; it could be reduced if world resources were distributed more equitably, so that basic human needs could be satisfied more effectively.

Measuring inequalities in health

In this chapter we are concerned to examine inequalities which are discernible for groups of people, especially those in different geographical areas; so we will consider differences between average conditions of aggregated populations. This is very different from assessing the health of an individual person. One important aspect of aggregated analysis is that it requires the use of standardized definitions of health and illness. Chapters 2 and 3 showed that health can be defined in many different ways, and that between individuals and social groups there are varying perceptions of what constitutes health and illness. The previous chapters also demonstrated how the dominant groups in society often impose their definitions of health and illness on other, less powerful groups. It is likely that most of the measures of health status used in this chapter reflect the influence of 'dominant' views about what constitutes health and ill health.

Another feature of aggregated analysis is that it tends to oversimplify the variety of human experience. Chapters 2 and 3 stressed the idiosyncratic dimensions of health perception, and showed that studies of the health and illness experience of individuals give us very valuable information on how health differences affect different people in different contexts. In particular, these studies are useful because they show us that while there are some rather consistent elements of experience between individuals there is also considerable diversity. This means that we should be very careful in making assumptions about an individual's health on the basis of other characteristics, such as socio-economic position or family composition. Most importantly for the following discussion, it is important to avoid the dangers of the ecological fallacy (see Chapters 1 and 6). The risks of the ecological fallacy become greater as the scale of analysis becomes wider, so that, for example, international comparisons at the level of whole countries always conceal a considerable amount of regional difference in health, and even regional statistics will obscure the very local patterns observable at the scale of neighbourhoods or individual people.

Bearing in mind these caveats, aggregated analysis does have certain features which are complementary to the in-depth analyses favoured by a more individualistic approach. In particular, it shows how far health differences seem to be generalized for populations as a whole, and gives us a clearer impression of the scale of the burden of morbidity and mortality in societies. Perhaps it is partly for this reason that, in terms of the 'politics' of public health debates, statistical information which is applicable to large groups of people sometimes carries more weight in arguments over allocation of society's resources than do detailed accounts of the experiences of individuals. The justification for this preference for statistical as opposed to qualitative evidence is, of course, spurious, reflecting the particular emphasis in biomedicine (and in Western societies in general) on the supposed objectivity of positivist evidence and its ability to estimate statistically the generalizability of experimental findings. Furthermore, it is not always the case that the weight of numbers speaks most loudly in political terms. In some cases, public reaction to individual cases of health deprivation has done as much to spur politicians to action as the statistical evidence. For example, recent concern over a case in which a young man with a mental illness was mauled by a wild animal in a London zoo while undergoing 'care' in the community produced shock waves throughout the National Health Service concerning the quality of mental health care.

In making aggregate comparisons of health status, we are constrained by the information which can be compared systematically between populations. The main sources of data are statistical information on health, illness, and death. Unfortunately, the emphasis tends to be on the two latter aspects of health status, since information available on good health and healthy development is very limited. In fact, the most comprehensive and systematic data collected around the world relates to death, rather than health or illness during life. We therefore begin the discussion with a consideration of variations in mortality, and go on to consider some of the more limited information which is available on variations in morbidity and more positive aspects of good health.

Geographical analyses (in spite of their imperfections) are very revealing of the diversity in health of different populations and of the whole range of factors which can impinge on health. It is to some extent possible to 'explain' the health differences between areas in terms of differences in living conditions, in so far as we are able to measure them. However, area measures of living conditions have their limitations, since they tend to summarize complex differences in socio-economic and physical conditions in terms of rather crude indicators of national or individual income, occupational status, housing conditions, and access to health services. The examples discussed below show that, even when several indicators are used in the analysis, these variables can only 'explain' some of the variation in health between areas. The geography of health inequalities therefore teaches us a good deal about the complexity of health differences, as well as the limitations of the tools which we have at present to measure them.

There is a vast literature on geographical health inequalities. Much of this has been reviewed elsewhere (e.g. Jones and Moon 1987; Meade *et al.* 1988; Townsend *et al.* 1988), and it is not intended to cover the same ground here. The following discussion is intended, rather, to provide illustrations of some of the more recent work on health differentials, and it focuses on two issues in particular: first, we consider international health inequalities and the debate over the epidemiological transition; second, we examine social inequalities in health at the infranational level and the evidence for a geographical effect in these differences.

International variations in mortality and the epidemiological transition

Most countries have formal medical and legal systems for recording deaths, and when aggregated together these registrations produce statistical information which is relatively easy to compare and to interpret. There are also internationally agreed classification systems for recording the cause of death (ICD 1977). This is not to say that the reporting systems are always identical between countries, but at least there is some scope to compare the methods used and make informed assessment of the comparability of data. There are examples of detailed research on some of the disparities in mortality data between countries. Kaminski *et al.* (1986) compare information on mortality among young people in the countries of the EC, and discuss the differences in procedures used to classify and record deaths in childhood. When there are significant differences between countries as close as the member states of the European Union, it is clear that there will be even greater disparities between countries with more widely varying resources for information gathering. For example, Curto de Casas (1993) reports that there are no death certificates for 30 per cent of deaths in Guatemala and 50 per cent of deaths in Ecuador, and that there are other difficulties with mortality data from a number of other countries in Latin America. Thus, while mortality data is the most widely available information for comparison, and is reliable enough to give a reasonable indication of health differences, we should not over-stress its consistency or reliability.

It is common for information on deaths to be converted to various forms of mortality indicator to enable comparisons. These include mortality rates and ratios (standardized to allow for the demographic structure of the population), and measures of life expectancy and of years of life lost. Each of these indicators has different properties, and gives a different perspective on mortality differences.

Crude mortality rates show the numbers of deaths occurring in a given population over a given period. Measures expressed as *rates* are often employed to compare the *burden* of mortality because they indicate numbers of deaths in relation to the population size. They allow us to compare mortality between populations while controlling for variation in the absolute size of the populations. *Standardized mortality rates* are calculated to control also for differences associated with the age and sex composition of the population, and are therefore usually more useful as measures of difference in population health status. *Standardized mortality ratios* are calculated as the ratio of *observed* deaths to *'expected'* deaths in the study population. Observed deaths are the number of deaths actually occurring in the study population over a given time period. 'Expected' deaths are the hypothetical number which would have occurred if the pattern of death in each age and sex group corresponded to that occurring in a 'reference' population. Usually the information for the reference population corresponds to national or international averages. Where the standardized mortality ratio is 100, the mortality observed in the study area corresponds exactly to the expected number. Ratios of less than 100 show that the study area has fewer observed deaths than the number 'expected' (i.e. relatively low levels of mortality), while populations with ratios above 100 are interpreted as having relatively high mortality. Standardized mortality ratios are designed to control for age and sex composition of areas. They facilitate comparison between areas because they vary around a 'reference' value of 100, but they do not directly show the burden of mortality in terms of numbers of deaths. (For detailed explanation of the method of calculating standardized mortality rates and ratios, see e.g. Lilienfeld and Lilienfeld 1980; Jones and Moon 1987.)

National data on *life expectancy at birth* is a statistical indicator which shows the average length of life for a baby born in a country at a particular point in time, given the rates of death at different ages which apply for men and women in the country at the time of the birth. Analyses published by the World Bank (1993) provide a good illustration of global health inequality in life expectancy for groups of countries classified into demographic groups according to their level of development. Fig. 4.1 shows that in 1990 there were wide differences in life expectancy between countries of sub-Saharan Africa (with average life expectancy of 52 years) and the countries with established market economies (EME), where the average was 76 years. The chart also shows differences in the relative improvement in life expectancy achieved since 1975. The countries with the greatest improvements are not always those with the highest levels of life expectancy. For example, there

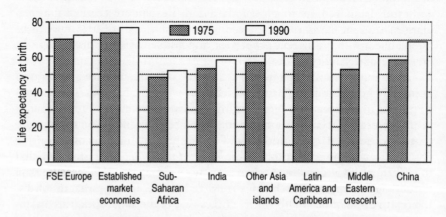

Demographic regions in increasing order of improvement in life expectancy, 1975–90

Fig. 4.1 Life expectancy at birth for demographic regions, 1975 and 1990. (*Source:* World Bank 1993.)

has been most significant improvement over this period in the countries of the Middle Eastern crescent and Latin America, which are undergoing rapid economic development and an associated demographic transition, as initially high levels of fertility and mortality are significantly declining. EME countries which have already undergone the demographic transition showed less impressive gains in life expectancy. The lowest rate of increase in life expectancy was in the Former Socialist Economies (FSE) of Eastern Europe and the former USSR. Also notable is the situation in sub-Saharan Africa, where life expectancy is comparatively low and the improvements have been much less than in the other low- and middle-income countries mentioned.

Another perspective on health, which attempts to represent morbidity as well as mortality, is presented in the form of loss of *Disability Adjusted Life Years* (World Bank 1993). This is a measure which uses mortality data to estimate loss of healthy life for the populations of different world regions in 1990. For each recorded death, the number of years of life lost was estimated as the difference between actual age at death and the life expectancy at birth which would have been typical in a country with low mortality rates. The loss of healthy life due to disability was estimated using information on the incidence of diseases, derived from community surveys or expert opinions. For each type of disease the typical duration of the illness was combined with a weighting to reflect the likely severity of the disease. Death and disability losses of healthy life at each age were then combined to give the total numbers of years of health life lost due to death or disability. A discounting process was used to give more 'value' to healthy life at young ages than at older ages. The resulting indicator of Disability Adjusted Life Years (DALYs) was compared for different countries. The indicator is obviously debatable in terms of its accuracy and validity. The basis for classifying the causes of death and disability was the Ninth Revision of the International Classification of Diseases, which would cover causes of death comprehensively, but would omit about 5 per cent of possible causes of disability. The incidence of disability due to different causes had to be estimated approximately in many cases, using information from community surveys or expert opinion. The method also depends, to quite a significant extent, on estimations, and on the expert judgements involved in the calculation. However, it is an interesting attempt to express the full impact of disease on the populations of different parts of the world. Fig. 4.2 presents the loss of DALYs in relation to the population size of the world regions in 1990. It shows that the contrasts between countries are even more striking on this measure than the differences in life expectancy considered above. In sub-Saharan Africa, the relative loss of healthy life is five times that in the EME countries on this measure.

The World Bank Report also compares the world regions in terms of the type of diseases which contribute to the loss of DALYs. In Fig. 4.3 it can be seen that non-communicable diseases (such as cardiovascular diseases, cancers, and nutritional deficiencies) play a major role in the EME and the FSE groups of countries, and in China. However, in countries such as sub-Saharan Africa, India, and the Middle Eastern crescent, the largest loss of healthy life is due to communicable diseases such as malaria and respiratory infections. This type of analysis places greatest emphasis on loss of healthy life at a younger age, so that death due to injury, especially from accidents (which is more common in younger age groups), also contributes significantly to the total loss of DALYs.

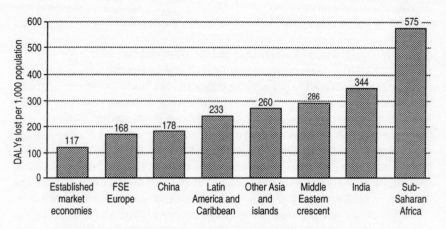

Fig. 4.2 DALYs lost per 1,000 population in demographic regions of the world, 1990. (*Source:* World Bank 1993.)

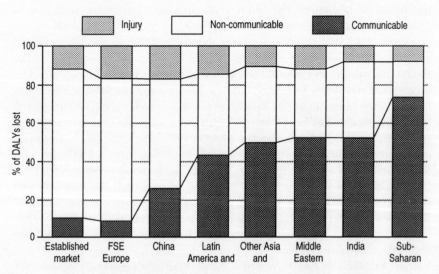

Fig. 4.3 Distribution of DALYs lost, by cause, in demographic regions of the world, 1990. (*Source:* World Bank 1993.)

These global disparities in life expectancy and the causes of life lost have often been interpreted in terms of combined demographic and epidemiological processes which have been summarized by the demographic and the epidemiological transition models. These models are based mainly on the historical experience of EME countries. The models show that, before the transition takes place, countries are typified by high levels of mortality (mainly due to infectious diseases, which are poorly controlled), associated with high levels of fertility and a relatively youthful population (which is particularly susceptible to communicable, as opposed to non-communicable, diseases). As societies begin to control the level of mortality from infectious diseases, especially among children, the mortality rate falls and population numbers begin to rise. It is postulated that fertility then begins to decline, and gradually the rate of population growth begins to slow down as mortality and fertility become more balanced at a lower level. As fertility and childhood mortality decline, the non-communicable diseases, which are primarily degenerative diseases more common in old age, begin to become more important.

These changes are summarized in Table 4.1, which summarizes the changes postulated by Omran's model of the epidemiological transition from dominance of communicable diseases to dominance of non-communicable diseases (first proposed in Omran 1971). This model emphasizes links between epidemiological change and social and economic development. These processes summarized in the model vary over time and space (Omran 1983), so that countries experience change in their epidemiological profile in different ways. Jones and Moon (1992) and Picheral (1989) are among those who have reviewed recent work in this field and comment on the importance of geographical and historical context for the epidemiological transition model, which makes it a fertile area for research by medical geographers as well as those in other disciplines.

Omran (1983) suggests that, while the classic/Western model applies to countries like Western Europe and North America, other variants of the model are needed to capture the pattern of change in other countries. An accelerated transition in Japan, Eastern Europe, and the former USSR has largely taken place since 1945. In contrast, in countries such as Mexico the transition process has been delayed because the social and health care infrastructure has failed to sustain the process of change after early improvements in mortality. In other developing countries, such as Jamaica and Singapore, social and medical changes have, on the other hand, made possible a transitional variant of the delayed model, in which Western levels of mortality are finally being attained.

Powles (1992) points out that the relative importance of non-communicable diseases varies among countries which have undergone the transition and are controlling infectious diseases effectively. In North-West and Eastern Europe and North America, the gains in adult mortality due to control of conditions such as tuberculosis have been partly offset by an 'epidemic' of non-communicable diseases such as heart disease and cancers.

Table 4.1 Summary of the epidemiological transition model

Profile	Stages of transition				
	Pestilence and famine	Early phase of receding pandemics	Late phase of receding pandemics	Degenerative and man-made diseases	Possible stage of reduction in degenerative diseases
Demographic	Cyclical growth. Mortality dominates. High and variable fertility. Population growth constrained. Predominantly young population.	Mortality high but declining and less variable. Net population growth. Proportion of elderly starts to increase. Some migration to towns.	Mortality slowly but progressively declines. Fertility begins to decline. Strong rural–urban migration.	Mortality rapidly declines, then begins to stabilize. Fertility declines to low, fluctuating level. Population ages. Urban residence increases, but urban pressures lead to counter-urbanization trends.	Population continues to age with low variable fertility. Counter-urban migration continues as well as expansion of urban zones of influence.
Economic	Subsistence economy. Predominantly agrarian, based on manual labour. Labour efficiency reduced by illness.	Improvements in agriculture and landuse and early development of transport, communications, industry.	Sustained economic growth and industrialization.	Initially high economic growth, then tapering off. Production shifts to the service sector. Public welfare and leisure grows.	Post-industrial economy with service sector very dominant. Concern over sustainability of economies. Globalization of capitalist modes of production. Developed countries become less economically competitive than developing countries in manufacturing.
Social	Traditional, rigid, hierarchical society. Strong family structures. Subordinate role for women. Low living standards. Poor, unreliable food supply, especially affecting women and children.	Upper/middle classes begin to adopt rising expectations. Strong family structures persist. Low but improving living standards. Some improvement in food supply.	Rising expectations throughout society. Extended family begins to weaken in urban areas. Public hygiene and sanitation improve, but still problematic in city slums. Improved agriculture. Better food supply.	Rational–purposive life styles expected. Tendency towards greater bureaucracy and depersonalization. Nuclear families become the norm. Women increasingly emancipated and better educated.	High living standards threatened by over-population, urbanization, reduction in economic growth. Social inequalities remain marked in spite of emphasis on equal opportunities for women and minorities.
Health	Life expectancy low (<20 yrs). Child and infant mortality still high. Mortality highest in urban areas. Main cause of death is infectious epidemics, exacerbated by malnutrition. No biomedical care systems. People rely on traditional medicine.	Life expectancy rises. Infant and child mortality still high. Urban mortality still high. Main causes of death: endemic parasitic and deficiency diseases and epidemics. Industrial disease increases. Some improvement to personal hygiene.	Life expectancy rises to 40–50. Infant, child, and maternal mortality improve. Pandemics recede. Infectious diseases still important but non-communicable diseases increase. Medicine and environmental health become better established.	Life expectancy reaches 70+. Childhood and infant mortality decline to low levels. Heart disease, stroke, cancer are leading causes of death. Most infectious diseases controlled. Problems of chronic illness grow. Medicine highly developed.	Life expectancy is further extended towards hypothetical biological limit to life. Mortality from some degenerative diseases reduced (especially heart disease). Compression of morbidity may result. Other diseases may increase (e.g. Alzheimer's)

Source: Omran (1971, 1982, 1983).

In contrast, countries such as Japan and Southern European nations have not so far seen such a steep rise in non-communicable diseases, and the same is true for some other parts of the Far East.

More recent developments also suggest that the epidemiological transition is not completed in the EME of Western Europe and North America. Recently, in the case of heart disease, rising mortality began to be checked. The United States was one of the countries where this trend became established particularly early, and Fig. 4. 4 shows the declining trend since the 1950s. Other high-income countries have seen a similar improvement (see Chapter 6); this has contributed significantly to the recent improvements in life expectancy in such countries. It is possible that advanced capitalist countries are seeing a further stage in the transition, which we might think of as a post-industrial situation of reduction of degenerative diseases. This is a speculative part of the transition model, shown in Table 4.1 as a 'possible' next stage.

In contrast to the increases in life expectancy enjoyed in the West, according to several authors (Boys *et al.* 1991; Chruscz *et al.* 1991), countries of Eastern Europe and the former USSR have experienced a continuing increase in cardiovascular and cancer mortality which is leading to an East/West health divide. There has been some debate over the precise nature of the changes in mortality in countries such as the former USSR (Anderson and

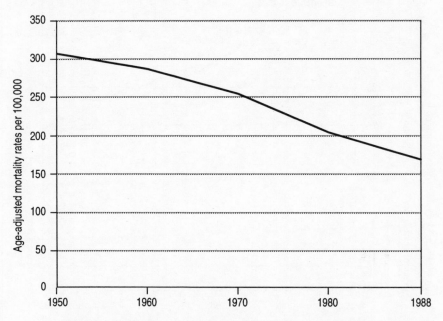

Fig. 4.4 Declining age-adjusted death rates due to heart disease for the US population, 1950–88. (*Source:* US Department of Health and Human Services 1991: table 27.)

Silver 1989; Blum and Monnier 1989). However, there does seem to be some agreement that during the late 1960s and 1970s there was a stagnation in improvement in mortality rates and that, at least for middle-aged men, mortality rates actually worsened. The explanation for these adverse trends is likely to be complex. It may partly relate to the poor conditions during the Second World War, which weakened generations reaching middle age in the 1970s. However, there is also some evidence that the trend may be due to a worsening of contemporary living conditions during the late 1960s and 1970s. Although the situation may have improved slightly during the 1980s, the trends in mortality in Eastern European countries have not shown the same gains as in Western Europe. For example, Fig. 4.5 shows trends in age-standardized mortality under 65 years for three Eastern European countries, compared with Western European and North American trends. The causes

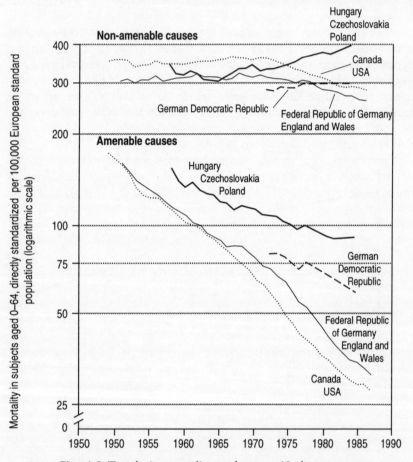

Fig. 4.5 Trends in mortality under age 65 due to causes amenable and non-amenable to medical intervention, for selected countries. (*Source:* Boys *et al.* 1991.)

of mortality have been divided into those amenable and those non-amenable to medical intervention (see Chapter 6 for an explanation). The trends for categories of mortality are unfavourable for the Eastern European countries compared with the other groups of countries.

Furthermore, there seems to be the possibility of a 'regression' in epidemiological change in some countries as certain infectious diseases gain in importance. The AIDS epidemic has produced considerable uncertainty about the likely pattern of future epidemiological development. It is likely to have a significant effect on mortality rates over the next few years, especially in countries such as Africa, where it appears to be most prevalent and particularly difficult to treat and to control (Smallman-Raynor and Cliff 1990; Smallman-Raynor *et al.* 1992). Even in countries such as the USA and Western Europe, the epidemic presents an important threat to health. Gould (1993) discusses the suggestion that the AIDS epidemic may weaken herd immunity to infectious diseases such as tuberculosis, and might result in a resurgence in communicable disease mortality of various types, although they had previously been controlled in high-income countries, according to the epidemiological transition model. Bobadilla *et al.* (1989) also point out that, in middle-income countries, the limited resources available for health care may result in neglect of the 'left-over ills' (infectious diseases and malnutrition, residual from an earlier phase in the demographic transition) as these countries struggle to cope with the increasing problems of degenerative diseases and 'new' conditions like AIDS.

The epidemiological transition model postulates an association between national economic development and health, and most of the evidence points strongly to links between the socio-economic situation in the country and the level of mortality, so that Curto de Casas (1993) summarizes international differences in the Latin American context in terms of a wealth model of health, a poverty model, and an attenuated poverty model for countries whose situation is intermediate (see Table 4.2).

Although poverty is an important factor in the health status of nations, total national income is not the only important factor for health status. For example, Table 4.2 shows Jamaica and Cuba in a relatively favourable position, though these are not very rich countries, and this is thought to reflect the priority given by their governments to health and health care, especially in primary health care measures. Bobadilla *et al.* (1991) also propose a 'protracted-polarized' model of the transition for some middle-income countries. The features of this model include extreme inequalities of wealth distribution within countries, reflecting very uneven progress towards the health changes associated with socio-economic development in the classic epidemiological transition model.

The pattern of wealth distribution within countries is related to overall health status even among high-income nations. Wilkinson (1992) shows that in Western Europe and North America the rate of improvement in life expectancy is linked with the degree of economic inequality in the country.

Table 4.2 Poverty and wealth models of health in North and South America

	Poverty model	Wealth model
Life expectancy	less than 60 years	over 70 years
% of children dying aged under 5	>34%	<20%
% of deaths over 65 years	<40%	>50%
% of deaths due to heart/cerebrovascular/ cancer diseases	<40%	>50%
Infant mortality rate	>30/1000	<15/1000
Examples: in Americas	Nicaragua Honduras Haiti Peru	Jamaica Cuba USA Canada

Source: Curto de Casas (1993).

Thus, the United Kingdom, where income shares became more unequal during the 1970s, showed a rather small increase in life expectancy compared with France, Italy, or Canada, where income distribution was becoming more equal over the same period. Wilkinson suggests that this may be partly because countries with a more equal income distribution are likely to have relatively generous social welfare provision and public services to protect the health of poorer people, but he also argues that the evidence suggests that more equal resource distribution in society is likely to result in lower average mortality for *all* groups in the population.

The debate over the epidemiological transition model therefore gives an impression of a dynamic situation which will continue to develop in ways which are not easy to predict. Certainly, the trends in mortality change in countries vary, even among those which might be expected to be similar according to the classic transition model. Differences in national mortality seem to be linked not only to the overall wealth of nations but also to the ways in which these resources are distributed in the population and in the extent to which they are targeted towards improving health, particularly through primary health and public health interventions. While the epidemiological transition model provides a useful general framework within which to consider national differences in health, it needs to be interpreted with care in each national situation.

Variation in health at the infranational scale

Within nations there is as much variation in health status as exists between countries. A large body of work in Britain (Townsend *et al.* 1989) shows persistent infranational differences in a number of aspects of health, morbidity, and mortality. Similar evidence exists for other European countries (Fox 1989; *Social Science and Medicine* 1990) and in developing countries (*Social Science and Medicine* 1993). Much of the evidence relates to mortality data, and some examples are discussed below. However, morbidity and 'positive' health indicators are also often available to compare health within countries, and we consider some illustrations in this section.

Countries vary considerably in the ways they record information on morbidity in the population. Although these differences limit international comparison, the data collected often allows study of the health inequalities within countries and it can be instructive to contrast the different approaches used (Aiach and Curtis 1990). Some infectious diseases must be formally notified to health authorities, for example, and several countries also have registers for cases of other serious diseases, particularly cancers. Finland has particularly good cancer registers, and has been able to demonstrate inequalities in morbidity as well as mortality. This is important, since the factors which result in development of the disease are not necessarily identical to those which govern mortality. Survival may, for example, be influenced by variation in the effectiveness of detection of disease and in treatment once the disease is diagnosed. Table 4.3, for example, compares the incidence of breast cancer among white and black women in the United States, standardizing for age. In both 1980 and 1987, the incidence of cases was higher among white women, although both groups had seen an increase in incidence over the period. However, the five-year survival rates for those contracting the disease in 1981 were better for white than for black women, and the mortality rates in 1987 were also slightly better for the white group. Differences in survival were attributed partly to the fact that the disease tended to be detected at a later stage among black women, which suggests less effective screening. Allowing for this difference, there were still dispari-

Table 4.3 Breast cancer among black and white females in the United States: age-adjusted rates of incidence, survival, and mortality

	White women	Black women
Incidence rates in 1980[a]	86.8	73.4
Incidence rate in 1987[a]	116.5	90.2
Five-year survival rates, 1981–7[b]	78.2	63.1
Mortality rates 1987[c]	22.8	26.5

[a] Number of new cases per 100,000 population.
[b] Ratio of the observed survival rate for the patient group to the expected survival rate over the same period for persons in the general population with matching age and 'race'. Represents the probability of surviving the effects of cancer.
[c] Deaths per 100,000 population.
Source: US Department of Health and Human Services (1991: tables 31, 50, 51).

ties in survival which also suggest that, for a given stage of development of the disease, treatment was less effective for black women (US Department of Health and Human Services 1991).

As well as information on diagnosed illnesses, there has been a considerable expansion in the availability of data on the effects of illness on people's ability to carry out daily activities of living. This represents a different perspective on illness, since similar types of disability may result from more than one type of disease. By focusing on disabilities or limitations imposed on people by illness, epidemiologists are also placing more emphasis on the human impact of illness and the level of dependency which may result from disabling illness. (See the discussion of life expectancy without disability below.) This type of analysis has been particularly relevant to elderly populations, in which relatively large proportions of people will have chronic disabling illnesses with related needs for social care as well as medical services. Of course, disability can also occur in younger age groups and, as we see the growing impact of AIDS in younger populations, we shall need to be able to adjust measures of disability originally developed in the context of elderly populations to make them relevant for younger groups of chronically ill people.

Some countries also collect information on 'positive' dimensions of good health such as self-perceived good health (Fig. 4.6) or measures with a positive dimension such as height and weight or lung function. The Health

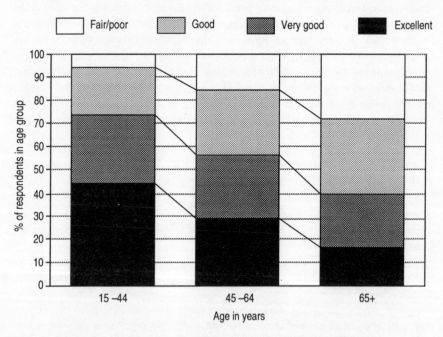

Fig. 4.6 Perceived health reported in the US Health Interview Survey, 1989. (*Source:* US Department of Health and Human Services 1991.)

Interview Survey in the USA and the Health and Lifestyle Survey in Britain (Cox *et al.* 1987, 1993; Blaxter 1990) are two examples of surveys collecting data relevant to 'positive' aspects of health which show variations in the pattern of good health and fitness between regions and between social groups. Some examples of measures of health and fitness are also discussed in Chapter 6.

The social dimensions of health inequality

Just as for international differences, the infranational pattern of health typically follows the differences in wealth, advantage, or social position. Various theories have been discussed by medical sociologists which postulate alternative processes by which more deprived groups in the population experience poorer health.

Much of this discussion focuses on the health differences between individuals in different social classes. Social classes are often defined in terms of occupation, although conceptually they represent broader notions of social relations and divisions of interest and allegiance in societies. There is also a growing body of evidence linking health inequalities with employment status and particularly with unemployment, which represents an especially marginalized socio-economic position. The explanations for these health inequalities are complex. They involve, on one hand, the damage to health resulting from exposure to material and social disadvantage and, on the other, processes of health-related selection, by which the fittest individuals are able to compete most successfully for the most prestigious and privileged jobs. There has been a lively debate over the relative importance of these two processes (e.g. Illsley 1986, 1987; Wilkinson 1986a, 1986b, 1987). More recent discussion has tended to acknowledge the possibility of joint or additive effects for material deprivation and health selection (West 1991), and to suggest that health selection may be to some extent an indirect process operative through factors such as childhood deprivation, educational attainment, and height. These factors can all have an influence on life chances and employment opportunities, particularly given intentional or unintentional discrimination in employers' recruitment procedures.

It has also been demonstrated that some of the social-class inequalities in health might be 'artifactual', resulting from statistical effects of classification methods used, rather than from 'real' social inequalities. It is certainly true that the method of social-class classification is likely to affect the impression of health inequality gained from different studies. However, the evidence suggests that artifactual effects do not account for all the inequalitities observed. Much of the data, including longitudinal analyses, demonstrates that social class is not the only socio-economic variable associated with health. Links between health and other characteristics reflecting social position, such as working conditions, level of education, housing tenure and quality, income, and ownership of consumer goods show associations which are often even more marked than social-class inequalities (Fox *et al.* 1985; Lundberg 1991; Blane *et al.* 1993).

These studies seem to mark a move towards a wider perspective on health difference, which looks beyond theories of class-based social relations, towards a better understanding of the complex processes through which social disadvantage and health are linked. In order to further this understanding, studies drawing on economics, public health, and anthropology are seeking to discover how social factors influence health at the micro-level of individual households (Berman *et al.* 1994). There is also beginning to be more concern with the factors which may mediate the impact on health of occupational circumstances. For example, Hammarstrom (1994) suggests that differences in social support, attitudes and beliefs, and sense of empowerment will influence the extent to which the health of individuals is affected by unemployment.

In addition to the inequalities in health which are evident in relation to occupational class and employment status, there are also important differences according to characteristics such as gender. Miles (1991) reviews the evidence of different health experience and mortality for men and women. In some low-income countries, female life expectancy is less than that of males, and this is thought to be associated with the hard working conditions for women in these countries and with the relatively high levels of maternal mortality resulting from poor living conditions and poor antenatal and obstetric care. Fauveau *et al.* (1991) have examined excess female mortality in the population aged under 5 in rural Bangladesh, and found that the risk of death is 1.8 times greater for girls than for boys. Severe malnutrition and diarrhoeal disease contributed most to the differential. Freeman and Maine (1993: 148) comment: 'A shamefully large number of girls and women die each year because of the unique risks inherent in being female in a world where females are second class citizens.'

In higher-income countries, gender differences in mortality tend to operate in the opposite direction and female life expectancy is typically longer than for males. Factors contributing to these differences include the fact that levels of accidental or violent death are significantly lower for younger adult women than for men. For women the onset of increased risk of death due to heart disease associated with increasing age is later than for men, and historically women have had lower levels of lung cancer mortality. (Cancers, especially cancer of the breast, are nevertheless an important cause of death for younger women.)

However, while women in developed countries typically live longer than men, they do not seem to experience the same advantages when it comes to morbidity. Research has frequently shown that women are more likely than men to suffer from minor physical disorders and mental health problems.

Research shows that this gender difference is independent of social class. A proportion of the gender differences in health may be explained by gynaecological problems, but this certainly does not explain all of the differential. Mental health problems seem to be related to the heavy dual workload of many women inside and outside the home, as well as social pressures and issues of identity (Gisbers Van Wijk *et al.* 1992; Arber and Ginn 1993; Popay *et al.* 1993; Walters 1993; see also Chapter 3). Sweeting (1995) reports that higher morbidity rates for females emerge in early adolescence, and suggests that this may be partly linked to the lowering of young women's psychological well-being during adolescence. The stress of multiple life roles undertaken by women in high-income societies may also be linked to physical illness. Elliott (1995) considers the evidence of this link with respect to coronary heart disease, and concludes that more detailed socio-ecological research is necessary to test for these associations. Graham (1994) demonstrates associations between smoking behaviour and stressful factors such as caring responsibilities and restricted access to material resources among women.

The different experience of morbidity between men and women is illustrated by national survey data collected in some countries on disabilities in the population, which has been used to calculate life expectancy without disability. When this indicator is compared for men and women, in each of the countries shown in Table 4.4, the relative advantage for women is less than that which is indicated by total life expectancy.

Lahelma and Arber (1994) have also examined the degree of social-class disparity in self-reported health apparent among men and women in four countries with contrasting welfare systems and different levels of female participation in the workforce. Their analysis showed that in three of these countries, the social-class disparities tended to be greater for men than for women, although in Sweden the reverse was true. The detail of variation in

Table 4. 4 Life expectancy without disability

Country	Total life expectancy		Life expectancy without disability	
	Males	Females	Males	Females
USA (1985)	70.1	77.6	55.5	60.4
Canada (1986)	73.0	79.8	61.3	64.9
Netherlands (1981–5)	72.8	79.5	58.8	57.3
Great Britain (1985)	71.7	77.5	63.6	66.5
France (1982)	70.7	78.9	61.9	67.2

Source: Robine *et al.* (1992).
Note: the life expectancies without disability were calculated using different data and methods in each country, and are not directly comparable.

the links between health and social position between men and women in the four countries suggested that social context, sex, and social class interact in quite a complex fashion with perceived health status.

Several studies have also shown differences in health experience according to 'race' or ethnic group. For the British population this evidence has been reviewed, for example, by Donovan (1984), Cruikshank and Beevers (1989), Ahmed *et al.* (1989), Poldenak (1989), and Ahmad (1993a). Differences in mortality, morbidity, and survival between ethnic groups in the United States are also documented by US Department of Health and Human Services (1991) (see e.g. Table 4.3, above).

Much of the work in this area in Europe, including Britain, is based on migrant populations which are distinguished from the majority population by their foreign place of birth (e.g. Balarajan *et al.* 1984; Marmot *et al.*1984; Bennegadi and Bourdillon 1990; Mechali 1990). These studies show variations in age- and sex-standardized rates of mortality associated with place of birth. Not only are migrant populations distinct from the host population in terms of their health experience, but also they differ significantly from each other.

A fundamental problem in this area of research is that of categorizing populations (McKenzie Crowcroft 1994; Senior and Bophal 1994). Using country of origin as an indicator of ethnicity is inadequate for a number of reasons. First, it may not reflect ethnic variations between different groups from a particular country. Balarajan *et al.* (1984) have shown, for example, that among immigrant populations from the Indian subcontinent, Punjabi, Gujarati, Muslim, and south Indian groups displayed different patterns of mortality. A second difficulty is that identification of minority populations on the basis of place of birth will fail to distinguish people born into ethnic minority families after immigration. In Britain, the 1991 population census introduced a question on self-perceived ethnicity which is intended to overcome this difficulty by asking people to classify themselves according to a list of alternative categories (see Table 4.5a), and data collection in the National Health Service has moved towards use of ethnic monitoring using categories which are compatible with the census. However, even this type of classification is problematic in terms of its links with health, since it mixes ideas of racial, national, political, religious, and cultural groupings, and also because one person's self-classification is not always consistent from one survey to another (Sheldon and Parker 1992). In addition, Raleigh and Balarajan (1994) point out that under-enumeration of certain minority ethnic groups in the 1991 census has been relatively significant, which may make it an unreliable denominator for calculation of rates and ratios.

In the United States, since 1976, Federal Government statistics have classified populations according to four categories (American Indian or Alaskan native; Asian or Pacific Islander; Black; White). Some of these groups are subdivided in health-related statistics, and in some states information is also available on Hispanic populations (Table 4.5b). As in Britain, there has

Table 4.5 Classifications by 'race'/ethnicity in Britain and the United States

(a) 'Ethnic groups' used to classify the British population in the 1991 census (as described in the question on the census form)
White
Black-Caribbean
Black-African
Black- Other
Indian
Pakistani
Bangladeshi
Chinese
Other

(b) Classification of 'race'/ethnicity in US Federal Government statistics and by Department of Health and Human Services
American Indian/Alaskan native
Asian or Pacific Islander (sometimes subdivided as Chinese, Japanese, Filipino)
Black
White
Hispanic (for some states only; sometimes subdivided as: Mexican American, Puerto Rican, Cuban, Central and South American)

been a lively debate over the usefulness of these classifications, especially with respect to the problem of identifying ethnicity of various groups classed as 'Hispanic' (Hayes-Bautista and Chapa 1987; Yankauer 1987; Giminez 1989).

These issues of classification make it more difficult to establish the factors which underlie racial and ethnic health disparities. Sheldon and Parker (1992) point out that the observed differences in health, associated with the ethnic classifications used, can be misinterpreted as indicating that ethnicity *causes* health difference, when in fact the explanations may be complex and not always a direct result of 'race' or cultural practices. Links between health and ethnicity may operate partly through the social and political position of populations (often collectively identified as 'black') which suffer particularly serious racial prejudice and inequality of opportunity in terms of living conditions and health care (Donovan 1984). Undue emphasis on certain illnesses which seem to be linked to 'race' can distract attention from the broader patterns of illness which are associated with poor living conditions, and would therefore be likely to affect any ethnic group living in similar conditions of socio-economic deprivation. There is a risk of blaming ethnic minorities for their own poor health, because illness is seen as resulting from culturally determined individual behaviour and not from wider constraints of the host society. Kurtz (1993) and Navarro (1990) both comment, for example, that the apparently poorer health status of black and ethnic-minority children compared with white children in both Britain and the United States needs to be understood in the context of their poorer average living conditions and access to health services. The question of ethnic differences in health is closely bound up with problems of racism at the individual and the institutional level (Ahmad 1993b; Stubbs 1993). Furthermore, there may be a tendency to overlook instances where there are health advantages enjoyed by ethnic minority populations, perhaps associated with healthy aspects of culturally specific lifestyles. All of these considerations need to be borne in mind when we consider data on racial or ethnic health differences, such as the information in Fig. 4.7 on years of life lost before the age of 65 due to different causes for the black and white male population in the United States. The figure shows that black males are relatively disadvantaged in terms of a number of conditions, although the extent of the disparities varies considerably from one cause of death to another. The racial classification used here represents a number of different aspects of social and economic disadvantage which particularly affect the black population in the United States, and it reflects factors operating at the level of society, as well as individual patterns of behaviour or susceptibility to disease.

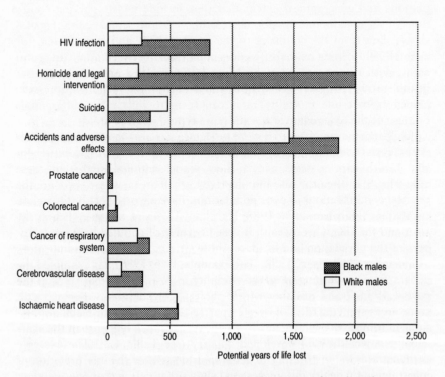

Fig. 4.7 Years of potential life lost before age 65 for selected causes of death for black and white males in the USA, 1988. (*Source:* US Department of Health and Human Services 1991: table 26.)

Thus various dimensions of social differentiation, including class, gender, and ethnicity, can be seen to be interrelated in their association with health differences for different groups in the population. Most of these differences reflect the differential life chances and risk exposures of these subgroups, although there may also be an element of biological variation. Simple comparisons along a single dimension of difference, such as social class, material deprivation, sex, or ethnicity, is likely to be misleading if we are seeking a proper understanding of the complex of factors which produces differences in health between social and economic groups in societies.

The spatial expression of health inequalities

These debates over the social inequalities in health are important for our understanding of regional health differences within, as well as between, countries. We have already noted that some countries have such large populations and geographical extent that they include widely varying conditions of socio-economic development, living conditions, and health. Fig. 4.8 shows data from the countries of the former USSR and from China, for example, illustrating major differences in mortality between urban and rural areas, with rural areas generally being less developed and having poorer health status. However, as shown by the data for the USSR, the impression gained from crude mortality rates can seem to inflate the differentials because of the older average age structure typical of rural areas.

High-income countries of a more moderate size also show striking differences, even though they have been able to establish reasonably comprehensive health care systems across their whole national space. In these countries, it is often the case that deprived urban areas, where there are the greatest concentrations of poor populations, have a worse health status than rural areas (Harpham *et al.* 1988; Harpham 1994a). Countries such as the USA and the UK, where wealth is very unevenly distributed between members of the population and in space, show striking differences in indicators of mortality (Wilkinson, 1992). For example, Pyle (1990) has examined the geographical patterning of infant mortality in North Carolina, USA, in the period 1983–7 and has shown that there is considerable variation, with some areas showing rates of over 21 per 1,000 live births for black populations in some areas, which, as shown above, are much higher than the average for a wealthy and developed country (see Table 4.2). The example demonstrates the importance of geographical scale for the interpretation of information on health differences, and shows that national or regional data can conceal considerable local disparities. The areas of highest infant mortality were in poor and isolated rural areas and in deprived metropolitan areas (see Fig. 4.9).

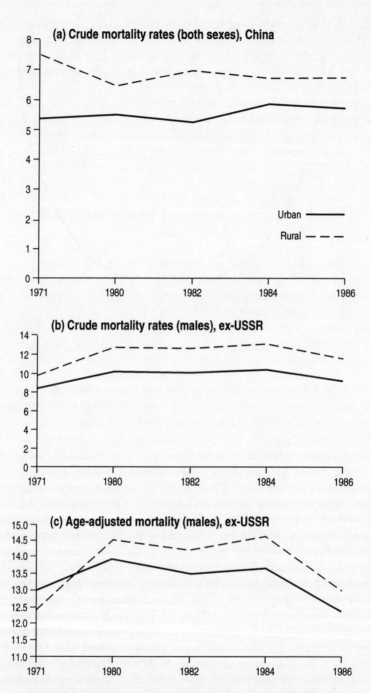

Fig. 4.8 Mortality in urban and rural areas in the former USSR and China. (*Sources*: Mezentzeva and Rimachevskaya 1990; Fang 1993.)

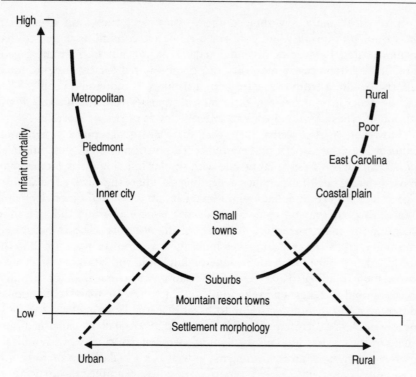

Fig. 4.9 Geographical variation in infant mortality in North Carolina. (*Source:* Pyle 1990: 444. Reproduced with kind permission of the author.)

Mortality rates in the adult population also vary markedly within countries. Typically the differences are most evident for middle-aged men and for mortality under the age of 65. The amount of variation and its geographical pattern varies according to the disease and the socio-demographic group being studied; it may also vary over time, although studies in several countries have shown a good deal of consistency in the general pattern of geographical difference over time. Britton (1990) shows for England and Wales the pattern of mortality for counties which has shown quite persistent disparities over the decades since 1951. Mortality rates are higher in the north and west of the country and in the metropolitan areas, especially those urban areas which are socio-economically deprived. At the more local level there is even stronger geographical correlation between mortality and variation in measures of social conditions. Townsend and colleagues (1988) demonstrated this clearly in the north of England, using an indicator of material deprivation which combined statistics on car ownership, housing tenure, unemployment, and overcrowding at the scale of electoral wards. This indicator showed a very close association with health differences at ward level.

Ecological analysis within countries, therefore, shows an association between high mortality and poor socio-economic conditions similar to that demonstrated by studies of individuals. A significant part of these geographical differences in mortality can be accounted for by *compositional* effects, resulting from the aggregated attributes of individual people. That is, the geographical differences in mortality arise from the uneven distribution of types of individual with differing levels of risk of mortality.

However, it also seems likely that there are geographical, *contextual* influences on the association between social position and health (Duncan *et al.* 1993). That is to say that people with similar individual risk factors may have differing health experience depending on where they live. For example, Fox *et al.* (1984) found that longitudinal data on individuals in England and Wales showed that differences in social-class composition did not fully account for the geographical differences in mortality. The standardized mortality ratios for men in professional jobs were lower than for those in unskilled manual jobs in all regions of England, but mortality of professional men in the north and west of the country was worse than for the same occupational social group in the south and east. In their analysis of regional differences in rates of death due to heart disease, Townsend *et al.* (1988) note a similar effect. Shaper (1984) showed that social class, climatic difference, and water hardness all showed independent associations with mortality rates, suggesting an environmental as well as a social dimension to health inequality. Eames *et al.* (1992) report work which examined the association between mortality rates and socio-economic indicators at the level of wards in England, and compared the relationships found in different regions of the country. One of the social indicators used in this study was the one proposed by Townsend *et al.* (1988) (see above). In some parts of the country there were steep gradients in mortality between more and less deprived areas, while in other regions the association was weaker.

Studies have also been made of geographical patterning of morbidity. Curtis (1990) found an association between an individual's propensity to report illness and the type of area where they were living in parts of London, but Jessop (1992) was not able to demonstrate the same differentiation in East Anglia. Blaxter (1990) showed that social-class differences in health reported in the Health and Lifestyle Survey were variable from one region of Britain to another. At a very local level, Phillimore and Morris (1991) compared areas of Sunderland and Middlesbrough, two areas of North East England which shared a generally similar, relatively deprived, socio-economic profile, and found that their mortality experience was different, with higher mortality in Middlesbrough. One possible explanation was that the proportion of heads of households who were unskilled or semi-skilled manual workers was slightly higher in Middlesbrough, so that poverty might have been in some way more severe in Middlesbrough (although other indicators of poverty were almost identical for the two areas). Another explanation might concern differences in the physical environment, especially air pollution resulting from the petrochemical industry around Middlesbrough. A different small-area study by Phillimore and Reading (1992) showed that the statistical association between mortality and socio-economic conditions for small areas seemed to be weaker in rural, as opposed to metropolitan, areas. However, the weaker association between deprivation and health for rural wards might be due to the smaller population size of rural wards. The correlation coefficients between deprivation and health were found to be stronger in rural areas when wards were aggregated up into larger units.

In order to investigate the interrelationship of place and socio-economic factors, methods need to be used which capture the complexity involved. Evidence for a complex relationship between geographical context and mortality also comes from research on regional variations in mortality from other countries. Casselli and Edigi (1988) have used principal components analysis to show how mortality due to different conditions varies regionally in France; they recognize different geographical contexts which can be distinguished in terms of mortality profile as well as socio-economic cultural and environmental factors. Other researchers have used multi-level modelling techniques which test more effectively for the relative importance of individual (compositional) and area (contextual) effects on local health inequalities. Most of the studies to date employing this methodology (Von Korff *et al.* 1992; Duncan *et al.* 1993; Gould 1994; Shouls *et al.* 1994) have concentrated on self-reported health and health-related behaviour. These studies suggest that, while most of the differences in health are due to individual attributes (such as social class, sex, ethnicity, or personal living conditions), there is some variation which seems to be attributable to the characteristics of the area of residence.

These examples of geographical studies of mortality all suggest that the relationship between mortality and indicators of living conditions can vary between different geographical settings. This might be because the indicators used to measure geographical differences in living conditions have different meanings in different types of area. Another interpretation may be that different aspects of living conditions combine to influence health in different ways, depending on the geographical context. Theories of association between social conditions and health should therefore be adapted to take account of local geographies. If policy measures to improve health and reduce health inequalities are to be effective, they should also be sensitive to the possibility of contextual, as well as compositional, effects.

Health inequalities in space and time

While the small-area studies considered above have generally considered inequality in mortality as a rather static phenomenon, the geographical pattern of mortality is in fact dynamic. This is well illustrated by studies of epidemics of various diseases which take a space–time perspective on health difference. Some of the classic studies of this type have investigated the spread of infectious diseases, such as influenza, cholera, and measles, using historical data (Pyle 1969, 1980; Cliff *et al.* 1981). These studies have demonstrated the influence of geographical as well as temporal dimensions of human interaction for the spread of diseases. They show, for example, that contagion spreads from one area to another by a process of expansion from original centres of infection to other areas in close proximity. The rate of spread is also partly influenced by the relatively high levels of human interaction between major urban centres, which occurs even when such large cities are distant from each other. This high level of interaction between the main cities in the urban hierarchy provides many opportunities for communication and contagion between the main cities in the urban hierarchy.

This understanding of the significance of hierarchy as well as proximity has enabled geographers to go beyond descriptive models of past epidemics to produce predictive models (see Chapter 1). Today, one of the most urgent applications of such models relates to the spread of the HIV and of the AIDS syndrome. Gould (1993) argues that geographers' understanding of the factors which determine rates of human interaction in contemporary networks of world cities and city systems can produce useful analyses predicting the national pattern of contagion in the population. Data presented by Gould (1993), Smallman-Raynor *et al.* (1992), Shannon *et al.* (1991), and Loytenen (1991) show how uneven the rate of spread of the virus is likely to be, with high levels of infection in the United States, Britain, and Finland, especially in deprived urban areas.

Another aspect of the development of epidemics which has begun to attract attention is the uneven rate at which epidemics decline in time and space. Although the term 'epidemic' is normally applied to infectious diseases, it has also been used to describe the dramatic rise in mortality due to heart disease which has typified many developed countries. As noted above, in some of the most affluent countries we are now beginning to see the epidemic of coronary heart disease (CHD) mortality controlled, apparently by improvements in health-related behaviour and in the outcomes of treatment. Studies of local trends in mortality due to CHD in the United States by Wing *et al.* (1987), in the Netherlands by Mackenbach *et al.* (1991), and in Norway by Aase (1989) have all demonstrated that the timing of the onset of reduction in death rates and the rate of reduction were dissimilar in areas with different socio-economic profiles. Wing *et al.* (1990) showed that, in the USA, average CHD mortality was improving at a faster rate than equity in mortality, as measured by area differences.

The same is true in the British case. Bryce *et al.* (1994) have demonstrated for England, and McLoone and Boddy (1994) for Scotland, that during the 1980s the falling trends in CHD mortality achieved in deprived areas tend to have become established later than in more affluent areas. At least for the older population, the improvements in mortality were also less marked in deprived areas, so that area inequalities had widened during the 1980s. Fig. 4.10 shows for males aged 55–74 the differential trends in CHD mortality in English District Health Authorities, grouped according to their score on the Townsend index of deprivation. Those in the fourth and fifth quintiles

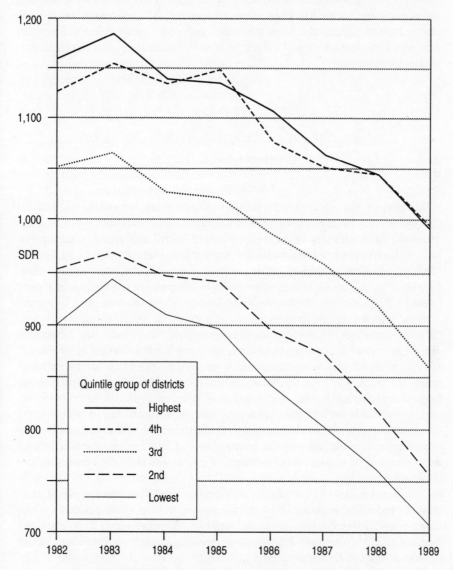

Fig. 4.10 Differential trends in Coronary Heart Disease reduction in English districts (Standardized Death Rates for males aged 55–74). (*Source:* Bryce *et al.* 1994.)

were the most deprived areas in 1981, and showed less improvement over the period than the most privileged populations in the first and second quintile of districts.

Degenerative diseases such as CHD were originally thought of as 'diseases of affluence', affecting especially those whose privileged lifestyle resulted in over-consumption of fatty foods and alcohol, smoking, and lack of exercise. However, we now find that the impact of health promotion and improved treatment has been most evident for affluent groups, who have been able to reduce their mortality due to heart disease more successfully than poorer populations. CHD is now more accurately characterized as a 'disease of poverty'.

Conclusion: changing perspectives on health inequalities

This chapter has highlighted a number of directions for recent research on health inequalities. It shows that there has been a move towards appreciation of the complexity of factors producing health difference, with a growing understanding of how factors such as economic position, sex, and ethnicity exert combined influences for different individuals and social groups. Inequality in health is an issue in all countries, but contemporary views of the nature of the problem are more sensitive to the contextual effects which can mediate the processes producing health disparities. These contextual effects operate at various geographical scales, from differences between world regions and countries to more local variation between different parts of a particular country. Changing perspectives on the epidemiological transition have made observers less confident than they once were that economic and social development in lower-income countries will bring about the same health gains as were seen in nineteenth century Europe and North America. We are beginning to understand better the global diversity of change in population health over time, and to become more cautious about trying to express these in terms of single models of health and development.

These perspectives on factors influencing health emphasize the importance of socio-economic conditions for health, as well as the health care factors which are considered in more detail in Chapters 5 and 6. The studies reviewed in this chapter are therefore strongly related to the public health perspectives which are discussed in Chapter 7. Since the factors associated with health inequality are, to a large extent, amenable to change through human action, health policies in many parts of the world are directed towards the objective of greater equity in health. For example, the European member states of the WHO have agreed a series of targets for change which would help to improve health generally and to reduce health inequalities; these strategies are discussed in more detail in Chapter 9. This chapter has shown that the target of greater equity is as applicable on the global scale, between high- and low-income countries, as it is within the individual countries of the European region, and we return to the theme of global equity in health later in this book.

|5|

Reforms in national health systems: changing strategies for equity and efficiency

Much of the discussion so far has concentrated on aspects of *health*; how it is constructed and interpreted, how it is influenced by the social and physical environment, and how it varies between human populations which are socially, chronologically, or spatially defined. This chapter introduces a more focused consideration of *health care* and, in particular, of the differences in health services which are evident at the national scale.

When thinking about health care, there is a tendency to focus on the medical professionals, who provide medical health services, and the facilities, such as clinics and hospitals, where they do much of their work. However, in order to make comparisons between national systems and to understand how they differ from each other, we also need to take account of broader structural characteristics. A national health care system depends on the social, political, economic, professional, and administrative frameworks within which health care is provided in a country. Meade *et al.* (1988: 259) suggest that basic elements of a national health care system include not only governmental, economic, and cultural aspects, but also the environment, resources, and demographic and epidemiological factors, as well as factors which can present barriers to the delivery of care.

As we shall argue in this chapter, national health care systems are difficult to define and describe because of their scale and complexity. We propose a broad working definition of a health care system as the *combination of structures which determine how health care is made available to the population of a country*. Earlier chapters have emphasized the importance of social factors as determinants of health, and it has already been noted that a large part of health care is carried out by non-professionals in the informal sector. The links between informal carers and the professional services are an important aspect of the way in which health care systems are organized. However, in this chapter, we focus on health services provided by professionals of various types.

In this chapter, we consider comparative studies of health services at the international scale, and examine the characteristics of *national* health service systems which make them distinct from each other. Because countries have a territorial as well as a social and political identity, national differences produce significant spatial variation in the organization and delivery of health care which is a fundamental aspect of the global geography of health services. This chapter also emphasizes the dynamic, developing aspects of national health service systems and their changing nature over time. We shall consider some of the important current trends bringing about change in health services, and discuss the new perspectives which we may need to adopt in order to understand international differences in health care.

The emergence of health services

Health services exist because individuals, who may not have professional health-related knowledge and skills, often require help and support to promote and protect their health and to cope with illness. This care may be provided either informally, by lay people, or else by professionals with specialized health care knowledge, equipment, and treatments at their disposal. Over time, these health care professionals have become organized into operational, institutional, and administrative systems which deliver these professional health services (or medicine) to the population. The resulting service systems are intended to ensure, as far as possible, that medicine is made available to meet approved standards of quality and quantity.

While most countries have witnessed the emergence of increasingly organized and complex health service systems, the history of this evolution will be different in each country. These historical factors frequently determine the contemporary configuration of services. Very often, the scope for developing the service in the future is also constrained by what has gone before. Thus, as we show below, the recent evolution of health services in countries like the USA, the UK, and the former USSR has followed very different patterns, reflecting the diversity of historical conditions prevailing in each country.

The historical process by which our view of health and health care has become professionalized has been described by authors such as Foucault (1973). In Chapter 3 we discussed how in many parts of the world this professionalization was associated with increasing dominance of the Western biomedical model. However, we also noted that this is not the only professional approach to health services, and the significance of complementary medicines in national health care systems is discussed below.

The objectives of health services: efficiency, equity, and effectiveness

There is no single solution to the question of how to make health services available to a population. The differences in national health systems are to a large extent the result of political and ideological decisions which are made by, or on behalf of, societies. There is much debate, internationally and within countries, over the choices which determine how health services are organized and delivered to populations. The debate is a permanent one, but in many countries, notably the UK and the USA, it has been particularly fierce in recent years because of a number of forces which have precipitated moves towards change.

Some of the most influential forces are those generating rising costs of health services – costs which are particularly difficult to meet during the period of stagnation, or even recession, which has hit the economies of many advanced industrial countries in recent years. Factors leading to increasing costs include: population change (especially the growth in the very elderly population with relatively high average health care needs); the increasing number of people with chronic illnesses, including AIDS; the inflation specific to health services which arises from increasing potential and cost of medical technology, combined with general economic inflation which increases running costs of health services; societies' increasing expectations of, and demands upon, health services.

While these factors are leading to rising costs of health services, societies have a finite amount of resources to spend. Cost reduction is an important incentive for examining the efficiency of health services. However, arguments for increased efficiency should not be interpreted solely in terms of the need for greater economy, leading to a narrow emphasis on how to spend less on health services. The *efficiency criterion* should rather be seen as one which requires that resources are used to maximize as far as possible the desired outcomes. Thus debates over cost containment in national health services should revolve around how best to achieve service objectives given the available resources.

Great emphasis is often placed on the objective of providing access to health services. The question often posed is how best to develop a system to deliver biomedical services to the population. The *equity criterion* calls for access to services which will correspond to agreed criteria of need and just desserts. It also emphasizes the objective of reducing the inequalities in health discussed in Chapter 4.

Health services delivery should not be considered as an end in itself, since it is only a means to an end (an intermediate output) of the health care system. It is also important to consider whether delivering these health services will result in the desired final outcome for the population in terms of benefits to health. The *effectiveness criterion* focuses attention on whether health services really produce the expected 'health gain' for the patient. Many medical treatments are quite controversial in this respect. There is also an important socio-geographical element to discussion of effectiveness because medical care which may be effective for a particular social group in a particular setting may work much less well in a different type of community. This issue is considered in more detail in Chapter 6, which examines questions of evaluation of health care.

There is continuing controversy over how best to meet the criteria of efficiency, equity, and effectiveness in health services. Proponents of different views often argue from different ideological standpoints, reflecting opposing socio-political or economic positions. The most fundamental of these differences concern collective versus individual responsibility; and the relative merits of state management versus market forces.

Collectivist ideologies argue that health services should be available in proportion to need (related to health status) rather than in proportion to ability or willingness to pay (related to wealth). It is argued that free markets do not ensure this type of distribution, and that therefore the state should intervene to control the distribution of health services. It is also argued that equity in health service provision is so important that all members of society have a shared social responsibility to ensure equal access for those who need it. Access to health services is not seen as part of the reward system in society, but more as a natural right for all citizens. Therefore the state should organize a national health service, and finance should be through public funds. It is argued that private charity is an unreliable and inefficient method of redistributing the costs of health care, because an individual's propensity to donate may not be in proportion to health needs, but may reflect other priorities. Similarly, it is argued that, to ensure that health services meet public goals, they should be subject to state control, rather than market forces.

Anti-collectivist philosophies stress that free markets should be allowed to govern the distribution of health services because this is the most efficient and effective way to meet the demand for health services in the population. Thus the patient is seen as a consumer who is free to exercise choice in the health services market. Since the costs of health services can be high, the usual mechanism to ensure access to health services for the patient is through health insurance. The patient plays a major part in arranging his or her insurance cover and insurance policies vary in cost and in the range of services they will cover. The ability of wealthy people to afford better cover is seen as a justifiable aspect of the reward system, so that inequalities in access are to some extent considered acceptable. Those who are unable to cover the costs of health services are not seen to be the responsibility of society as a whole, except in certain well-defined circumstances (e.g. old age or extreme poverty). Otherwise, it is argued that charity, rather than use of public funds, is a fairer way to provide health services to the less well off, because individuals can choose whether or not to contribute to charitable funds. Health service providers (professionals in surgeries, clinics, and hospitals) will compete to provide services of quality and cost that are most attractive to the patient or insurer. It is argued that this will reduce costs and is a more efficient way of providing health services than by means of state finance and control.

Following sections of this chapter illustrate how pro- and anti-collectivist views influence health service organization in different countries. Ideologies of care also depend on the influence of different views of medical professionals. These professional ideologies result in differences between countries in the extent of acceptance and use of particular biomedical therapies or of complementary medicines. The importance and acceptance of different types of medicine in different societies is therefore discussed below.

Comparing health service systems

Because health services are complex, any attempt to describe a national system involves a large degree of abstraction and simplification. Such descriptions usually obscure the detail of a health service system and should be treated as highly schematic. This is particularly true of classifications which are set up to compare and contrast health systems of different countries. In reality, the differences and similarities between national health care systems are always 'fuzzy' and approximate. Many such classifications are also static and cross-sectional, not including the dynamic character of health systems.

There is no single definitive basis for comparing national health services across countries. In fact, the literature contains a bewildering plethora of alternative schemes. This section considers some of them. The objective here is to illustrate the aspects which are thought to be important dimensions of difference between national health service systems. This section also tries to identify aspects which are common to several classification schemes, and which might be thought of as 'key' dimensions of difference. In a later section we consider examples of health services in specific countries and how they differ in terms of these 'key' aspects.

Many schemes for comparing national health systems assume a perspective from the biomedical approach to health care. The discussion in Chapter 3 demonstrated that biomedicine (also referred to as 'cosmopolitan', 'Western', or 'allopathic' medicine) is highly integrated with Western science, and features a highly professionalized and technological approach to health. Biomedicine emphasizes treatment of medically diagnosed disease and (to a lesser extent) prevention of disease. It is controlled by professionals (doctors) who practise from specialized locations for the delivery of care (hospitals and surgeries). Patients have an essentially passive role in the process, as recipients of medical intervention (Aakaster 1986). However, a comparison of health systems which concentrates only on biomedical aspects gives a narrow (and certainly ethnocentric) perspective and it is not sufficient, even in those countries where the biomedical approach is most thoroughly established. We consider several reasons for this.

Recognizing the plurality of medicines

The first reason is that the biomedical system is not equally dominant in all countries. Alternative forms of health care are often relatively influential in countries often thought of as less economically and socially advanced than Western Europe and North America. For example, Field (1989) refers to 'emergent' health services, in which the biomedical system is only partially evolved and shows only some of the features of a 'developed' health service typical of Western countries. Leslie (1980) makes the point that in many developing countries, 'cosmopolitan' medicine is not available to large numbers of people who rely entirely on traditional, complementary systems. Thus to focus on biomedicine alone would exclude a large part of the health services actually used.

However, it is not adequate to think of complementary medicine as merely a symptom of 'underdevelopment'. Some authors (e.g. Elling 1980) have viewed the introduction of biomedicine not just as an aspect of human development but as one of the instruments by which imperialist powers have sought to impose control over countries they colonized (this issue is taken up again in Chapter 9). Some countries which have gained independence from colonial rule have revived alternatives to biomedicine partly as a way of asserting their independence from Western imperialist culture. Increasingly, traditional or alternative forms of medicine are being recognized as being highly sophisticated and having some particular strengths which are not shared by the biomedical approach. These include sophisticated systems of traditional medicine, such as Chinese medicine and Ayurvedic medicine, which are discussed below.

Furthermore, as Leslie comments, 'cosmopolitan medicine' does not meet all the demand for health services even in countries like the United States, where it is particularly dominant. Here, as in most high-income countries, people use biomedical and complementary systems. Common forms of complementary medicine identified by Sharma (1992) include techniques such as homeopathy, manual healing, acupuncture, chiropractic, osteopathy, naturopathy, herbal remedies, paranormal (faith) healing, hypnotherapy, anthroposophical medicine, and reflexology. These alternative forms of medicine are becoming more widely recognized in advanced industrial countries, and in some cases integrated into the mainstream of the health services system, rather than being marginalized, as they once were.

These complementary medicines each operate according to their particular theories of disease and illness and use specific approaches to treatment. Thus, we see a re-emergence of discussion of alternative perspectives in medicine, which parallels the debate over definitions of health discussed in Chapter 3. Useful contributions to this broadening of perspectives on different medicines and their roles in health systems come from anthropological studies. Writers such as Kleineman (1978) and Janzen (1978) stress that health systems are also cultural systems. As such, they perform key tasks in most societies which include the cultural construction of illness. The processes by which illness is identified, labelled, and explained were discussed in Chapter 3, where it was argued that biomedical explanations are themselves culturally constructed and incorporate a particular, essentially subjective, set of values.

Health systems also involve the cultural construction of criteria for making choices (and allocating resources) between alternative health care practices and different types of health service practitioner. Health systems incorporate the social processes affecting health-related behaviours. In addition, health systems include the activities involved in healing those who are sick (such as therapy, medication, supportive care, and healing rituals) and the management of outcomes of these activities which may include cure, chronic illness, or death.

Thus health systems are not only important for the delivery of health services, they are also required to provide meaning to the experience of illness and to provide frameworks to guide people's reactions to health and illness. Authors such as Kleineman argue that, because biomedicine tends to focus mechanistically on the first of these roles (the treatment of sickness diagnosed in terms of specific diseases), it is often less effective than more traditional therapies in giving social meaning to illness and helping people to cope with its effects (especially when the outcome is not a cure, but continuing illness or death). This may be why most societies retain elements of more traditional medicine even when biomedicine appears to be the dominant approach. Biomedicine is also criticized because it is orientated more towards curative services than towards disease prevention and heath promotion. This aspect is given more detailed consideration in Chapter 7.

Thus, when we are comparing the health care systems of different countries, a fundamental dimension of difference needs to be recognized in terms of the relationship between different forms of medicine. Countries vary in the extent to which the Western biomedical approach to health care dominates the health system, and in the degree to which other forms of alternative medicine are established and officially recognized. Examples of varying degrees of pluralism in medicine are considered in particular countries below.

The hierarchy of health care

If we turn our attention to aspects of national difference in biomedical services specifically, we also find a large literature relating to international differences in the organization and funding of medical care. Before considering these, we need to introduce the idea of hierarchies in health services. These have developed as biomedicine has become professionalized and organized. The hierarchies are typically described as comprising levels defined as 'primary', 'secondary', and 'tertiary' care.

Primary health care is mainly provided outside hospitals to people who are living in the community. Often this type of care is delivered to people in their homes, such as when a family doctor makes a home call, or a domiciliary nurse visits a patient. Otherwise primary care is often dispensed from local health centres or clinics. Primary health care is concerned as much with keeping people well as with treating those who are ill. It covers activities like health promotion and prevention programmes such as vaccination and health surveillance and screening (e.g. child health checks, checks for breast or cervical cancer for women). Primary health care is also often the main gateway to care at the secondary and tertiary levels when people are ill. In health service systems which include general practitioners, people who have a health problem who decide to seek medical care often approach their family doctor first. Primary health care is therefore very important as the basis for a health system, and the WHO strategy for 'Health for All' encourages countries to direct particular emphasis and resources towards primary health care (see Chapter 9). Primary health care is especially relevant for promotion of public health, and figures importantly in strategies of 'the new public health movement' (see Chapter 7). Primary health care is often (though not exclusively) based on relatively simple medical technologies, and frequently requires an understanding of the social and environmental context of health, as well as professional skill in medical techniques and procedures. For example, health promotion professionals working to reduce smoking will aim to encourage individuals to change their own behaviour, and will also aim to promote change in public or private organizations, such as restrictions on smoking in public or workplaces or on tobacco advertising. This work will require communication and facilitation skills as much as medical knowledge. Since primary health care is delivered mainly in the community and needs to be used by a large proportion of the population, it must be provided in close proximity to the places where people live, and the catchment areas of health centres are typically limited in geographical extent to rather local areas.

Secondary and tertiary care has traditionally been provided in an institutional setting within hospitals. This care mainly involves treating those who are already ill, with the objective of curing them or helping to alleviate the effects of illness or injury. At the secondary level, health care is often dispensed from specialized clinics to outpatients who may live at home but travel to the clinic for treatment (e.g. radiotherapy for cancers). Hospital accident and emergency clinics also fall into this category. Patients may also visit secondary care clinics to consult doctors who are specialized in particular health problems, to obtain advice rather than treatment. Tertiary care is usually the most resource-intensive and often the most highly technological type of care, and is typically delivered to patients who are cared for as inpatients living in hospital wards for the duration of the treatment and under constant supervision by professional carers. The doctors who provide these services are often highly specialized in particular procedures, and work on a limited range of medicine.

Since secondary care and (to an even greater extent) tertiary care are only needed by a minority of the population, and require expensive resources, there are economies of scale in providing them for larger populations distributed over wider areas. Much secondary and tertiary care is delivered through district hospitals serving populations the size of a small town, but some hospital services have a regional catchment area.

The primary, secondary, and tertiary levels in the 'health service hierarchy' are not always very clearly distinguishable, and often there is overlap between them. Furthermore, the categories shift as health services develop over time. Although the traditional view was of secondary and tertiary services located in hospitals, in contrast with primary care based in the community, this distinction is actually rather blurred in many cases. Considerable progress has been made in bringing 'outpatient' services closer to the consumer. This may be achieved by changing to methods of care which make it possible to perform the treatment outside the hospital. For example, home dialysis is now a reality for many sufferers from kidney disease. There is even growing interest in the idea of providing a 'hospital at home' for some patients, especially chronically ill older people, so that care which used to be given in a hospital ward can be delivered at home instead. Another strategy is to organize specialists' work on a peripatetic basis; some specialists hold consultation surgeries in decentralized health centres, so that patients do not need to travel to hospital for this service. Similarly, the boundary between secondary and tertiary care is not fixed; simple surgical procedures which used to involve an inpatient stay in hospital are increasingly being performed in day-hospital settings.

Although this hierarchy is not a rigid one, it is important in professional and administrative terms because it indicates different types of professional work. Particularly in countries with a well-developed state health service, there is also usually a bureaucratic structure to administer and manage the health care system at each level. Furthermore, there is a strong geographical dimension to the hierarchy. Each level corresponds to different types of facility location where health care is provided, and each level is associated with a different type of catchment area; primary care is local in focus, secondary and tertiary care is often more centralized, covering a larger catchment area, and may be regional in scope.

Table 5.1 shows examples of the hierarchical structure of the health system. Britain has a system of Central and Regional Health Executives and District Health Authorities, for the local organization of the National Health Service. There is also a parallel system to administer general practitioners' services. France has a system of regional and local agencies of the national health insurance fund (*Caisse Nationale d'Assurance Maladie*). A parallel administrative system exists for regional and local planning of health services (*Directions Régionales d'Actions Sanitaires et Sociales, Directions Départementales d'Actions Sanitaires et Sociales*) and within *Départements*, hospital provision is co-ordinated in *Secteurs Sanitaires*, which correspond approximately to the catchment areas of major hospital groupings. Russia has authorities for health service planning at the level of Republics, Oblasts, and Rayons, with the polyclinic as the most local unit. In China, the urban and rural health systems are both organized in a hierarchical form, with separate urban and rural systems for primary care and a link to the mainly urban tertiary sector.

Schemes for classifying biomedical health service systems

When we look in more detail at the international diversity of health services, it becomes apparent that a number of criteria could be used to place different systems in separate categories. One example of a classificatory scheme widely used by geographers is that proposed by Joseph and Phillips (1984). They suggest that one can distinguish national systems on the basis of the following aspects: national health policy; degree of centralized control; methods of payment; relationship between physicians and patients. National health policy concerns the degree of priority accorded to health services in national policy-making, the objectives which are set for health policy, and the types of health services which are favoured by health policy. An important aspect of this dimension concerns the relative importance of public (as opposed to non-governmental) funding and ownership of health care. It is also argued that countries vary importantly in the extent to which control of health care is centralized and located within national government ministries as opposed to being decentralized to the regional or local level, or to individual health-providers. Joseph and Phillips suggest that in more centralized systems the geographical distribution of health care facilities is likely to be more even than in systems with limited central governmental control. The method of payment relates particularly to whether the cost of

Table 5.1 Hierarchies of health service organization in selected countries

Scale	Level of care	England	France		China		Russia
National	Government policy and administration	Department of Health	Ministry of Health	National Health Insurance Fund (CNAM)	Urban health administration	Rural health administration	Ministry of Health
Regional	Tertiary: specialist hospitals/clinics	Regional Authorities of National Health Service	Regional Health Authority (DRASS)	Regional insurance fund (CRAM)	Municipal	County	Oblast/City Region Health Authority
District	Secondary/primary: local hospitals and clinics, doctors' surgeries	District Health Authorities (and Family Health Service Authorities)	Département (County) Health Authority (DDASS)	Local insurance fund agency (CPAM)	District general hospital	Commune/clinic	Rayon Health Authority
Local	Primary: doctors/paramedics	Doctors' practices and health localities	Doctors' surgeries		Lane/street	Co-op./brigade	Polyclinic

health care is met mainly by out-of-pocket fees (to be paid by the patient at the point of use of health services), or through forms of insurance, or by state budget funding via tax revenues. Finally, this scheme points to international differences in the degree to which patients have a choice of doctor under their health system, and in whether patients can refer themselves directly to specialized doctors or have access to tertiary and secondary care via general physicians at the primary level.

Similar aspects of health systems are reflected in a more elaborate classification system proposed by Field (1989) which also refers to some additional dimensions. This classification scheme includes: the mix of welfare economies; the rationing system; the health care delivery hierarchy; the role of health professions relative to other partners; methods of cost containment. Under this scheme, therefore, one is led to consider the ways in which public and independent sectors of health services relate to each other and the relative importance of each. The way in which limited health service resources are allocated to patients is also relevant; methods may include rationing according to ability to pay or waiting lists. Another aspect of this scheme is the extent to which physicians dominate the health system and their power relative to, for example, government ministries, other health-related professionals, and patients. Also stressed in this classification are the different methods of cost containment being applied to health services in different countries. Field's scheme differs from some others because it tries to identify 'types' of health service systems which can be distinguished in terms of these various aspects. The five types are 'emergent' health systems (mentioned above, where comprehensive biomedical systems have not fully evolved); *pluralistic* systems with very diverse types of provision, limited control by the state, and an important role for independent health services (e.g. the USA); systems based on *insurance/social security funding* (e.g. France); the *national health service* model espoused in Britain; and health care under *socialist regimes* such as the former USSR. This classification seems to parallel earlier studies, including those by Roemer (1976) which distinguished between free market, welfare state, socialist, underdeveloped, and transitional systems and those by Terris (1980) which distinguished between public assistance, health insurance, and national health service models. The labelling of systems with relatively weak biomedical services development as 'emergent' or 'underdeveloped' implies that the move towards dominance by biomedicine is desirable, which could be criticized as ethnocentric. These models are essentially intended only to represent the organization of biomedical care, and do not reflect the significance of plurality of medicines mentioned above.

Another typology is offered in the 1993 *World Development Report* (World Bank 1993), which emphasizes finance and organization of health services in the context of different levels of development of national economies. It is argued that countries differ in their ability to afford 'essential' health services. 'Essential' clinical services are defined in terms of a minimum package of care comprising: care of mothers and children during the prenatal period and at birth; family planning; care of sick children; care of tuberculosis and sexually transmitted diseases. The authors of the World Bank report distinguish between low-income countries (such as those in the Indian subcontinent and Africa), middle-income countries (such as South Africa, Costa Rica, Turkey, Korea), formerly socialist countries of Eastern Europe (which still exhibit a rather specific economic organization although they are moving towards the free-market model), established market economies such as those of Western Europe, and the USA, which is distinctive in terms of its economic strategy and its very high national income.

The report suggests that the health service systems of countries can be distinguished in terms of these *economic groupings* and also in terms of *alternative finance channels* and *types of organization*. Four finance channels are recognized: private finance, which may be out of pocket or through private insurance; and public finance, by means of compulsory insurance schemes or from government taxation revenues. Three forms of organization are also depicted in this scheme: public organization by the state; private non-profit-making bodies; and health systems run by private for profit agencies. (The World Bank view of health and health services is considered critically in Chapter 9.)

Other classification schemes which focus on the economic structure of health services systems summarize detailed differences between health care market models. One example is Saltman and von Otter (1992), who distinguish between three types of health care market: command and control; free market; managed market. Alternatively, Ovretveit (1993) distinguishes between markets featuring: the state as purchaser and provider; purchaser competition; mixed provider competition; and patient choice.

In addition to these various dimensions, we might also add variation in the degree of *horizontal and vertical integration* which exists in the hierarchies of organization of health services. In some countries the administration of health services is vertically dominated by the central level (as in the ex-socialist regimes); in others there is more local autonomy (as in the USA). National systems also differ in the degree of horizontal co-ordination at each hierarchical level between different parts of the health service and between health services and other sectors. The degree of *consumer participation* in health policy and health service organization (as compared with the power of professionals and managers) is a further important dimension which may distinguish different health service systems.

If we consider Table 5.2, which attempts to summarize the key dimensions of health service systems stressed by various authors, we find a lengthy list of different aspects which we might seek to compare in an international study. Although this list is long, it is not comprehensive. However, it does serve to demonstrate the complexity and detail which would be necessary for a complete description and comparison of different systems. We would suggest that it is not really feasible to categorize health services so as to express fully all

Table 5.2 Dimensions of international comparison of health service systems

Health policy
Relative dominance of biomedicine
Relative importance of different forms of complementary medicine and their degree
   of incorporation in health service systems
Total resources available for health services
Ability to provide essential services (mainly primary health care)
Methods of payment
Public/private mix of sources of finance for health care
Public/private mix of health service providers
Degree of management of health care markets
Methods of health care rationing and cost containment
Degree of centralized v. decentralized control
Degrees of horizontal and vertical integration
Doctor–patient relationships
Influence of doctors v. other partners in health services (including consumers)

these dimensions of diversity. In the following sections, we therefore follow Ellencweig (1992) in adopting a 'modular' or thematic approach rather than attempting to place health service systems into distinct categories. We compare selected countries along certain broad dimensions. The following sections concentrate on the United States, England, France, Russia, India, and China as illustrations. The discussion compares countries with respect to two broad issues, which are important in contemporary health care systems and which seem to relate to a number of the attributes in Table 5.2: the relative importance of biomedicine and of complementary medicine of different sorts; and the role of the state (as opposed to independent sectors) in controlling health service resources and managing health service markets.

This discussion of particular countries is intended to reflect some of the diversity of possible perspectives on national health service systems and how they vary between countries which represent very different socio-economic, political, and historical contexts for health services.

The importance of biomedicine and complementary medicines

The influence of complementary medicines is greater in India and China than in the other countries considered here. These countries both have long established traditional medicines which receive significant official recognition and support. MacDonald (1981) has contrasted these two countries with the USA, and argues that the support of traditional medicines has been relatively strong in China and India. This support has been in part symbolic, reflected in governmental recognition of the cultural roots of these medicines and in efforts to preserve them as part of the cultural heritage of the country. Financial support has also been provided for training research and practice, and there has been legal recognition in the introduction of regulations for licensing and assuring acceptable standards of training. Such support is in contrast with the tendency, at least until recently, by governments in the USA and in Western Europe to adopt a critical or dismissive view of complementary medicine, and to offer little support in terms of finance or opportunities for training.

The result is that both China and India have rather pluralist medical systems, in which patients are likely to make use of different therapeutic strategies, often consulting both biomedical and traditional professionals. Leslie (1980) demonstrates that the traditions of Chinese medicine and of medicines indigenous to India share with biomedicine sophisticated theories of bodily function and of illness, standardized learned practices, acquired through long periods of training, and ethical codes. They also confer on recognized practitioners an enhanced social position associated with their professional status.

Chinese medicine is based on a concept of the universe as depending on the balance of elemental forces of *yin* and *yang*. In common with many traditional medicines, Chinese medicine takes a holistic approach, treating the patient as a whole rather than treating diseases as discrete entities. It also treats the patient as an active participant in regaining a healthy balance, rather than as a passive recipient of treatment (Aakster 1986). Kaptchuck (1983: 7) points out that 'the Chinese system is not less logical than the western, just less analytical'. Jewell (1983) provides a detailed description of the principles of the various forms of Chinese medicine. The conceptualization of bodily functions includes a system of twelve meridians or channels through which circulates a vital energy (*chi*). The continuous, uninterrupted flow of chi between the essential organs is necessary for good health. One application of acupuncture is to stimulate this flow. Stimulation of the appropriate meridian is also thought to help to restore the balance of *yin* and *yang*. Acupuncture may be delivered using sophisticated electronic equipment, so that Chinese medicine is not always a 'low-tech' therapy. Chinese medicine also involves the use of herbal prescriptions to restore the body's natural balance. These are prepared according to specific 'pharmaceutical' recipes (Chen 1981), which are increasingly being standardized. Chinese and allopathic techniques are often used in combination in medical facilities and Chinese health policy has promoted the integration of the two approaches (Lin and Zhu 1984). In 1983, about a quarter of medical personnel in China practiced traditional medicine and more than 70,000 hospital beds were devoted to this type of care. It seems, however, that there is some resistance to this integration on the part of allopathic doctors in China (Rosenthal 1987).

In India, it is suggested that a number of different medical systems co-exist, and that their relationship is a rather competitive one. Ayurdevic and Unani medicines are two of the more influential traditions. Ayurdevic medicine is based on Hindu concepts of health, illness, and longevity (Obeyesekere 1977). The physiological processes in the body and various diseases are explained in terms of three basic complexes of the system (motion, *vata*; energy, *pitta*; and inertia, *kapha*) and basic elements: body fluids, blood, muscular tissue, adipose tissue, bone tissue, nerve tissue, bone marrow, and generative tissue (such as sperm and ova), as well as waste products. Health is conceived of in terms of a well-balanced state of these complexes and elements, associated with gratification of the senses, mind, and soul, while illness is a state of imbalance. Thus Ayurvedic medicine takes a very holistic view of human health. Treatment of disease involves avoiding causative factors and advising medicines, diet, activity, and lifestyle to restore a healthy balance, and it requires the combined effort of doctor, nurse, and patient (Kurup 1983).

The Unani medical tradition is associated with Islam. The basic conceptual framework is the Hippocratic idea of four humours (blood, phlegm, yellow bile, and black bile). The idea of temperament is important to this type of medicine, and change in the temperament of an individual brings about a change in health. The humours are linked to different types of temperament (blood: hot and moist; phlegm: cold and moist; yellow bile: hot and dry; black bile: cold and dry). Disease is an imbalance of the humours. Drugs are assigned temperaments and are designed to restore the proper balance. At the same time, considerable emphasis is placed on the capacity of the body for self-preservation and adjustment (Said 1983). Again, this type of medicine illustrates the emphasis on viewing the patient as a whole, rather than focusing on particular diseased organs, and on incorporating the patient's natural strength and healing potential.

The practitioners of Ayurveda (vaidyas or viads) and of Unani (Hakims) are referred to as doctors in India in acknowledgement of their professional status. Government support for traditional medicines has a long history. In 1948 the Committee of Indigenous Systems of Medicine recognized that in rural areas traditional doctors were more accessible, more sympathetic, and less expensive than allopathic doctors, and recommended them as important components of the medical system. Bhardwaj (1980), in a study of the registration of different types of doctors in rural areas of India in the 1960s, has noted that other types of medicine were common, in addition to Ayurvedic and Unani therapists. The relative dominance of different types, reflected in the numbers of doctors, varied from region to region and was in part due to the religious composition of the population: Unani practitioners are more common in Kashmir, in northern India, where there is a greater Islamic influence, while in the south of the country, in regions such as Kerala and Andhra, Hindu practitioners of Ayurveda were more common. However, it seems likely that the use of traditional medicines in India depends on the context. A study in Jalgaon, cited by Bhat (1993), reported that even among the poorest people in the study, the majority (85 per cent) would use biomedicine if they were ill (mainly independently funded rather than state-provided services), while less than 4 per cent of the survey respondents reported that they would use traditional medicine.

In contrast with the well-developed and officially recognized complementary systems in India and China, we find a much more dominant role for biomedicine in Europe and the USA. Nevertheless, studies in these countries have noted a widespread informal use of complementary medicines, and there are signs that they are now being viewed more sympathetically by allopathic practitioners. In the context of the USA, Wardwell (1994) discusses the dominance of biomedicine and distinguishes between different types of 'alternative' medicines. He notes that there has been most resistance by the biomedical orthodoxy to those medical practitioners, such as chiropractors, who were seen to compete most directly with allopathy. However, he also describes how chiropractic has gained widespread acceptance today. Similarly, Fisher and Ward (1994) comment on the almost complete integration of osteopathy into Western medicine in the United States. The legal status of alternative therapies can vary regionally in the USA according to state laws. Thus, for example, Chow (1984) discusses the varying position of state legislation with respect to acupuncture, and reports that most states allowed it to be practiced by osteopathic doctors and physicians: some states allowed trained acupuncturists to practice independently, but in Utah in 1983 acupuncture was still illegal.

Several studies have demonstrated the widespread use of complementary therapies in the USA. For example, Kronenfeld and Wasner (1982) review work by sociologists which suggests that the major source of health care in the United States is 'popular medicine' (self-treatment, folk medicine, and home remedies) which goes on outside the realm of biomedicine. Many people use both 'complementary' and biomedical care. This was demonstrated by Kronenfeld and Wasner in their study of patients at a medical clinic for arthritis, which showed that 94 per cent were also using alternative practitioners or remedies which included the application of various lotions, special diets, vitamin supplements and use of jewellery of copper and other substances. Riley (1980), in a study of use of osteopaths and chiropractors in Michigan, also showed that many people used these as well as biomedical physicians. These studies show that use of complementary medicine is not common only among ethnic minorities (for whom barriers to access to biomedical services may be especially significant), nor is it solely a recourse of the less well-educated or poorer people in American society in general. Some authors (e.g. Wolinsky 1980; Kronenfeld and Wasner 1982) suggest that people are particularly likely to use complementary forms of therapy for illnesses which are considered relatively minor, or for chronic illness which biomedicine can do little to help (arthritis may be one example). Taylor (1984) argues that another reason for the growth in interest in complementary medicine is the emphasis on a close and well-established doctor–patient relationship, which is typical of more holistic medicines, and which compares favourably with the deteriorating rapport between patients and doctors in the biomedical sector.

Sharma (1992) has reviewed the growth of complementary medicine in European countries including France and Great Britain. She cites a review of surveys of use of complementary medicine from which it is estimated that at least half of the French population and between a quarter and a third of the British have used complementary medical therapies. However, it is difficult to make comparisons between studies using different research designs. It is also estimated (Aldridge 1989) that about one in six French people and one in eight in Britain use complementary medicine in the course of a year. The legal recognition of complementary medicine varies between countries. In France, for example, complementary medicine may only be practised by qualified (biomedical) doctors. In Britain there is very little legal restriction on complementary practitioners, although since 1993 it has been illegal for osteopaths to practise unless they are recognized on a statutory register of qualified practitioners. The British Medical Association is beginning to foster the idea of partnerships between biomedicine and complementary medicine, but also stresses the need for careful regulation (British Medical Association 1993). In France the state health insurance scheme will reimburse the patient for the cost of homeopathic remedies or for complementary therapies given by a qualified doctor, and 80 per cent of homeopathic medicine is dispensed on prescription (Fisher and Ward 1994). In contrast, in Britain, the state health system does not cover the cost of complementary medicine, even though biomedical doctors have been advised by the General Medical Council (representing the profession as a whole) that they may refer patients to complementary practitioners provided that they retain overall control of treatment of their patients. Fisher and Ward (1994) cite an opinion survey showing that three-quarters of the British public favour the availability of complementary medicine under the National Health Service. The availability of practitioners of complementary medicine probably varies regionally. A survey reported by Fulder and Munro (1985) of complementary practitioners in a number of different regions of England showed that they were more concentrated in some areas than in others. The ratio of complementary to general (biomedical) practitioners varied from about 11/1,000 in Cardiff and Penarth (south Wales) to about 91/1,000 in the area around Cambridge, East Anglia.

These studies therefore suggest that 'formalized' complementary medicines play a significant role in many different countries, even those with particularly well-developed biomedical systems. It would be incorrect to think of alternative medicines as indicative only of 'less developed' health service systems. However, the official recognition of complementary medicine relative to biomedicine does vary from one country to another, as reflected in legislation governing practice of complementary medicines and the allocation of state resources to make these medicines available to the population. Generally, this formal recognition is more restricted in economically advanced countries, where biomedicine is particularly dominant. In countries where biomedicine is less well developed, complementary medicines often enjoy greater recognition. This seems to be partly because these countries lack the economic resources to provide resource-intensive biomedicine to the whole population. However, the less dominant role for Western biomedicine in these countries is not simply the result of underdevelopment. It is also a reflection of the fact that complementary medicine is more sympathetic to the cultural setting in which health and illness is interpreted. Low-income countries which are seeking to assert their economic and cultural independence from Western countries often stress complementary medicines partly for this reason.

Biomedicine: the changing role of the state

Turning to a consideration of the organization of biomedicine in countries where it is well established, another important dimension of international difference relates to the role of the state in controlling resources for services and managing the ways in which services are provided to the population. Countries differ in the ways in which the role of the state relates to that of independent (non-public) services providing health care. The independent medical sector includes both commercial and non-profit-making professional health services which are not state-managed (although they may be regulated by the state).

Several models of national health service systems which have been proposed in the past have suggested that it is possible to distinguish varying degrees of state involvement in the funding and management of medical services. Thus, for example, Joseph and Phillips (1984) suggested contrasting conditions in the USA (with a small degree of public funding and public control of health services and an important independent sector), compared with countries such as the UK and the former USSR (with comprehensive national health care systems, comparatively extensive government financing and control of the health system, and less-developed independent sectors). In an intermediate position were countries, such as France and other Western European nations, where the involvement in health care of the independent and public sectors was more evenly balanced.

As we have already noted, the role of the state in health services is associated with the ideology of care espoused by the society and the government in power. Thus greater degrees of public finance and control have been seen to be associated with ideologies of care which stress collectivist approaches, with shared social responsibility for health services and socialization of the cost of medicine through national schemes of health funding ('welfarist' or 'socialist' policies, according to Jones and Moon 1987). Lesser degrees of public involvement reflect 'liberalist', more anti-collectivist ideologies, stressing individual responsibility for health services, often requiring individuals to take responsibility for insuring themselves against health risks and placing emphasis on free market forces as the means to control health resource distribution.

Associated with these ideological differences are conventional 'models' of different types of health service which can be crudely described in terms of the differences shown in Table 5.3. The following discussion focuses on the health care systems in the USA, Britain, and Russia. These three countries have been selected because they have very different health care systems, which historically represented the extremes of the range between free market versus state management of health care, and which are currently experiencing strong pressures for change. We begin by summarizing the main characteristics of the old system in each country, then consider the problems leading to change, and finally review the main aspects of reform and change.

Table 5.3 Conventional models of medical service organization

Model	Free market	Managed market	State managed
Dominant ideology	Anti-collectivist	Intermediate	Pro-collectivist
Dominant type of finance	Voluntary insurance	Compulsory insurance	General taxation
Dominant sector controlling resource distribution	Independent	Mixed	State
Sector mainly responsible for providing services	Independent (small public sector)	Mixed independent and public	Public (small independent sector)
Dominant sector managing medical service delivery	Independent	Mixed	State
Examples	USA	France	UK, former USSR

Recent policy trends in countries like these may have led to a narrowing of the differences between health care ideologies. In Britain and Russia, for example, where health care has hitherto been typified by a large degree of public intervention, the emphasis on collectivist policies has been eroded and we now find a greater element of anti-collectivist philosophies stressing the value of individual responsibility and free market forces for the organization of health services. On the other hand, in the USA, with a strongly anti-collectivist system, greater state involvement has been proposed. These changes are contributing to what some commentators have identified as a 'convergence' in the organization of national health systems. We conclude this chapter with some discussion of the convergence hypothesis.

Historic differences in the organization of medical care in three countries

The USA

The US health care system displays some of the most extreme characteristics of an anti-collectivist system. The prevailing ideology has favoured individual responsibility and patient choice, and the preferred method of funding health care has been through independent health insurance schemes, with public funding and provision only for those parts of the health system which were not adequately covered by private health insurance.

The main forms of state intervention to finance access to health care in the USA have been through Medicare and Medicaid, which were introduced in 1965. Medicare provides basic hospital and other essential health insurance for persons aged 65 and over who have a relatively high risk of illnesses requiring expensive medical care. Medicaid covers the costs of basic health care for those whose income or medical insurance is not sufficient to cover the costs of medical care they need. About 25 per cent of the population are covered by one or other of these schemes, but because of the high rate of use of health care by those who are eligible, especially the elderly, it is estimated that about half of total health care costs are state funded (World Bank 1994). This leaves protection against the risk of health care costs for the rest of the population to the independent sector, with either individuals or employers arranging private health insurance.

The state exercises very little control over the choice of doctor, and patients may refer themselves directly to specialists without consulting a primary health care doctor first. There is also very little scope for the state to control the distribution of doctors and the location of their practices.

Britain

Turning towards the other end of the spectrum of public involvement in health services, we can consider Britain as an example of a country which historically had developed a dominant role for the state in health service funding and provision. The British National Health Service (NHS) was established in 1948 with the objective of making the full range of health services accessible to all those who need them, irrespective of factors such as their income or place of residence. Funding is not based on health insurance but on a budget which is largely derived from general taxation, with a further contribution from pay-roll taxes on employees and employers. There are almost no charges at the point of use of health services, although a fixed prescription fee is charged to some patients for medicines and there are also some fees for dental and ophthalmic services. Most hospitals and clinics were managed by local District Health Authorities and a budget was distributed to these Authorities to cover the cost of the community and hospital services which they provided. This budget is cash-limited, although the separate funding for general practitioners' service and the medication they prescribe has, until recently, been more difficult to control. Expenditure on health services in Britain is lower than in similar countries (Table 5.4). (However, Baggott (1994) points out that this relatively low level of spending is not so much because of low levels of public spending but because the level of private spending is less than in other European countries.) The geographical distribution of resources is based on a formula which estimates regional health needs in a standard way, based on population numbers, demographic composition, and mortality rates (explained in Chapter 6).

A separate administrative system is operated to remunerate general practitioners who are contracted to the health service and who are paid in respect of NHS patients registered with them. The state has some control over the way that patients exercise their choice of doctor, since the patient must be registered with only one general practice to receive National Health services and access to specialists is only on the basis of referral by the GP. This means that the GP is a very important gatekeeper in access to health services in Britain (Salter 1993). It also provides the potential for doctors to

Table 5.4 Percentage of GDP spent on health services in selected countries

Country	Total health expenditure as % of GDP, 1991	Private expenditure on health as % of total health expenditure 1989–91
Canada	9.9	27.8
Denmark	7.0	18.5
Greece	4.8	23.0
Netherlands	8.7	26.9
Russia	3.0	–
Sweden	8.8	22.0
UK	6.6	16.7
USA	13.3	56.1
Average for OECD countries	9.7	39.4

(*Source:* World Bank 1994).

provide health promotion and disease prevention services because the patient has a continuing link with the doctor. Patients who for some reason find it difficult to register with a GP (for example, people who are homeless and highly mobile) are seriously disadvantaged in this system. Although doctors are independent contractors to the NHS, the contracting process does permit some government influence over the distribution and location of GPs and the ways that they practice.

Russia

Russia has inherited the health care system established under the old Soviet system which used to apply to the whole of the former USSR. The Soviet system has often been used in the past as a good example of a socialist health care system, with very strong central control over the resources allocated for health service and over health care provision (Ryan 1978; Joseph and Phillips 1984). The system aimed to provide all Soviet citizens with free health services at the point of use. The major state enterprises and public services also provided health services. Access to health care therefore depended on area of residence and on occupational status, and choice of doctor was very largely controlled by the state. Primary and some secondary health care was provided mainly through polyclinics, health centres which grouped together a number of ambulatory services for a particular district. Access to hospital care and specialists depended on referral by a polyclinic doctor. A fixed budget, determined by central government, was allocated to health services, and resources were geographically distributed strictly in proportion to population, based on fairly rigid standard norms of provision.

An important objective of the Soviet health service was to equalize provision to centrally determined standards. Several commentators (e.g. Eyles and Woods 1983; Cromley and Craumer 1990) have commented on the effectiveness with which the Soviet system increased overall health service provision and reduced regional inequalities between the Soviet Republics between the 1930s and the 1980s.

Problems leading to change

The USA

A major problem in the USA has been that the provision of health care has been very uneven, with wealthy people having access to some of the best biomedical health care in the world, while for the poor only basic services are made available through limited public schemes. For an increasing proportion of Americans it has become prohibitively expensive to pay for full health insurance. In 1992, over 37 million people in the USA were without health insurance – about 12 million more than in 1982 (Woolhandler *et al.* 1993; Summer 1994). Similarly, the geographical distribution of health care has become concentrated in richer urban areas, while poor areas and isolated rural areas have less generous service provision (Shannon and Dever 1974; Cromley 1993).

Another problem of a system based largely on independent insurance schemes is that there is little incentive for insurers or health care providers to limit the rate of use of health services, since the costs are eventually passed on to the patients or their employers through rising health insurance premiums. This is particularly true when health care is charged on a fee-for-service basis, with payment for each item of care provided. Furthermore, insurance-based systems incur heavy administrative costs because of the accounting procedures which are required. Brandon *et al.* (1991) show that 33.5 per cent of commercial health insurance premiums are spent on administration, marketing and overheads, as compared with 2.3 per cent of Medicare costs. They estimate that more efficient practices could have saved $13 billion – enough to insure 11 million people, assuming average employer costs of insurance coverage. As a result, the level of health services expenditure in the USA, as a proportion of the GDP, is considerably higher than in other developed countries (see Table 5.4). In spite of these high levels of spending, some indicators of population health are relatively poor in the USA. For example, Schieber *et al.* (1994) show that life expectancy at birth is less in the USA than the average for 23 OECD countries, and national figures for infant mortality and rates of low birthweight are among the worst for the OECD countries.

Government attempts to limit the rising cost of health care have included the imposition of payment schemes based on Diagnostic Related Groups (DRGs), whereby health providers are paid a fixed rate for treating a given type of condition rather than being paid separately for each item of care. Gay and Kronenfeld (1990) have pointed to the problems of DRGs, which include premature discharge of patients when standard resources are exhausted; retrenchment in provision for unprofitable DRGs; increased provision for profitable DRGs and reduction in use of necessary tests in order to constrain costs. Another problem is the phenomenon of 'DRG creep', as diagnosis shifts towards more lucrative conditions.

Another well-known innovation has been the Health Maintenance Organization (HMO) model. Under this system, patients enrol with a group practice of physicians and for a fixed, prepaid fee a comprehensive health care service, including treatment for illness and preventive care, is provided. It was hoped that this scheme would make the costs of health care more predictable for patients and insurers, and would encourage the HMOs to seek the best value for money in health care because they were operating within a fixed budget. However, one of the problems has been that HMOs also have an incentive to select patients who are healthier and will incur lower health care costs to the HMO. Furthermore, many HMOs have become commercial organizations little different from health insurance agencies. Navarro (1991) also criticizes HMOs for having high administration costs, being unresponsive to patients' needs, and giving rise to comparatively high levels of patient dissatisfaction.

Wimberley (1980) argues that in the USA the dominant approach to health service access has been limited, categorical, and voluntary (LCV). That is, certain types of health service are made available for certain groups of people, often depending upon voluntary enrolment in particular schemes. This is in contrast to a total, comprehensive and compulsory (TCC) approach, where the whole range of health services are made available to the whole population through compulsory subscription to nationally organized schemes. The United States has not, to date, extended state health services to produce the 'TCC' type of national health insurance for the whole population which is more common in western Europe. However, recent developments show moves towards new forms of state intervention. Many commentators have stressed that these reforms will not come before time, and that they have been advocated for years. Woolander *et al.* (1993) suggest that it is 'high noon' for US health care reform, and refer to problems of uneven coverage of the population, poor health service outcomes, and high costs of the present system, which hit the poorest most severely.

The 1993 Health Security Act also identified a number of deficiencies in the US health care system. Section 2 of the Act shows that Congress found that:

individuals risk losing their health care coverage when they move, when they lose or change jobs, when they become seriously ill, or when the coverage becomes unaffordable . . .

. . . continued escalation of health care costs threatens the economy of the United States, undermines the international competitiveness of the Nation, and strains Federal, State, and local budgets . . .

. . . an excessive burden of forms, paperwork, and bureaucratic procedures confuses consumers and overwhelms health care providers.

Britain

In Britain during the 1980s, because of the constraints placed on the state budget for health care, and because relatively few people had private health insurance, the pressure of demand on the limited supply of NHS resources became very intense. Shortages of staff and of revenue resulted in widespread hospital closures (Baggott 1994). Since the main means of rationing health care resources is by waiting lists, rather than by cost, patients have experienced long delays for treatment, especially non-urgent elective surgery. Fig. 5.1 shows regional variations in waiting lists in 1989 and 1992. The proportion of patients waiting a long time for surgery seems to have declined. However, the validity of these indicators is discussed in Chapter 6, where it is suggested that they may not be very reliable measures of the performance of the service. Public concern over delays for treatment, and clinicians' public warnings about the danger to life of cutting back on hospital care in areas such as special-care baby units, has made resource constraints on the health service a major political issue in the 1980s and 1990s. It has also been argued that the system did not allow sufficiently flexible and efficient use of resources to respond to patients' needs at the local level. Although the total budget for health services was limited, there was thought to be little incentive for individual health service providers to constrain costs of their activities, so that it was argued that there was scope for 'efficiency savings'.

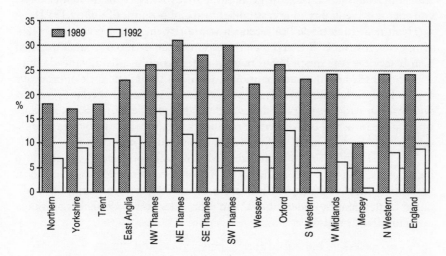

Fig. 5.1 Percentage of patients waiting more than twelve months for operations in the UK. (*Source: Regional Trends* 1991, 1993.)

Alternative systems of funding the NHS were considered at this time, as possible ways to raise additional revenue for health services: possibilities included state health insurance (with the possibility of opting out for those with private insurance), or a hypothecated tax levied specifically for health service spending. However, these were rejected. The existing system of funding of health services by means of a budget taken from general taxation was preferred, partly because it allowed for redistribution between rich and poor, and between those of working age and the very young and very old. Also, under the existing system, the collection costs were lower, and the scope for keeping an overall cap on health service spending was greater than they would be for the alternative schemes, which was very important for a government concerned to limit public spending (Baggott 1994).

Russia

In spite of the high degree of central state organization of health services, in reality the Soviet system was complex because of the different modes of provision, through district health services and through workplace services. Access was strongly dependent on occupational status, because services provided to workers in priority occupations were often of better quality than the district health service. Some services of better than average quality were funded by the state, but available only to elite groups in Soviet society. Thus there was in fact considerable inequality of access linked to social position (Telukov 1991).

Health care has tended to receive low priority in central policy and planning in the former USSR, and the proportion of the national income spent on health services was much lower than in other economically advanced countries. Because of chronic underfunding, with low status and lack of incentive for medical staff, quality and availability of health services was relatively poor, with widespread shortages of drugs and equipment (Telukov 1991). Patients often made informal out-of-pocket payments to ensure access to a reasonable quality of care, so that, in reality, access depended on ability to pay.

The rigid application of norms tended to be unresponsive to variations in local needs, and regional statistics on distribution tended to obscure local variations in provision, with rural areas continuing to be underserved compared with the cities Mezentseva and Rimachevskaya 1990; Virganskaya and Dimitriev 1992).

Reform and change in national health service systems

The USA

Proposals of reform for the US health service system, debated in 1994, were based on principles of universality, cost savings, choice, quality, simplicity, and responsibility (Zelman 1994). Section 3 of the Health Security Act set out the following objectives for reform:

- to guarantee comprehensive and secure health care coverage;
- to simplify the health care system for consumers and health care professionals;
- to control the cost of health care for employers, employees, and others who pay for health care coverage;
- to promote individual choice among health plans and health care providers;
- to ensure high-quality health care;
- to encourage all individuals to take responsibility for their health care coverage.

The proposals included the provision of secure, uninterrupted, and affordable health insurance and high-quality care which would be comprehensive, including primary and preventive care as well as specialized treatment. It also aimed to provide Americans with real choices between alternative health insurance plans while reducing escalation in health service and administrative costs. The reforms envisaged competition between providers to ensure efficiency and reasonable costs, and also that individuals would retain a responsibility for paying a share of health care costs. It is now clear that these proposals will not be carried through in their original form, because of resistance by various of the interested parties involved in the debate and the electoral defeat of the Democrats in the 1994 elections. Nevertheless, we consider them here because they illustrate a strong move towards radical change in US health service organization, and they reflect the spirit of reform which has influenced recent thinking about health care in the United States. Some commentators (e.g. Aaron 1994) have suggested that, even though the passage of legislation to enact the Clinton proposals was unlikely in the short term, the debate surrounding them might lead to more gradual processes of reform; and Rydell (1994) pointed to steps towards reform in individual states which may lead the way to more general acceptance of the need for change.

The Health Security Act envisaged that citizens under the age of 65 and not covered by Medicare would be eligible for enrolment in an insurance scheme, or 'health plan', selected and regulated by a 'Regional Health Alliance'. Health Alliances would be non-profit-making agencies, set up in each of the American states to organize enrolment and regulate the range of alternative insurance plans offered to residents. Because the Health Alliances would co-ordinate the insurance of large numbers of people, they were expected to be able to bring pressure to bear on providers to reduce the fees charged for services. Patients would themselves make a contribution to costs and (except for those who were self-employed or not in employment) employer contributions would make up the remainder (about 80 per cent) of the cost. For those who were not employed, the state would pay part of the costs of these health plans from general taxation. The amount individuals would pay would depend on the health plan which they selected: cheaper plans would provide the basic package of care, while more expensive plans would provide for extra services. A National Health Board would oversee the system and to regulate health service costs.

While this scheme fell short of a complete national health service, and still placed a good deal of responsibility on individuals and employers, it did aim to extend insurance coverage to the 14 per cent of the population who are currently uninsured and make insurance coverage more secure for many others. The amount of state funding of health services would need to be increased in order to permit this. The reforms could have also increased the degree of state control over health services by introducing public management of the health service market, while still allowing for competition between service providers. Such changes to the US health care system would thus bring it closer to the managed market type of organization shown in Table 5.3. However, critics of the proposals have raised doubts over whether they would be effective in controlling costs (e.g. King 1994). There is uncertainty over the necessity for Regional Health Alliances and what their role in the management of health care should be (e.g. Altman 1994; Patricelli 1994). There is concern that the proposed use of employer levies will jeopardize jobs for some low-paid workers, and that the reforms will have adverse short-run effects on workers in the health service industry (Fuchs 1994; Wilensky 1994). A number of consumer concerns about managed markets in the United States have also been raised (Working Group on Managed Competition 1994). These include the technical feasibility of such markets, their link to employment status, the role for insurance companies, the severing of links between insurance and income, age, or health status, and the universality of coverage. The difficulty of reaching a political and professional consensus on the appropriate strategy resulted in very complex formulations in the Clinton proposals for US health services, and illustrates the importance of the political processes involved in appeasing different ideological positions and varying interests in such a large and complex society (Rheinhardt 1994; Wilensky 1994).

Britain

The British government has sought to increase the capacity of the NHS to treat patients without significantly increasing health care costs to the state. The main means of achieving this have been seen as the introduction of an 'internal market' for health services, following ideas proposed by Enthoven (1985). It was believed that if state-managed service providers had to compete with independent providers, they would become more efficient and responsive to patients' needs. State control, it was argued, should be exercised through funding and contracting health services, not by direct management of the actual process of service delivery in hospitals, clinics, or the community. Thus the main aim of the NHS reforms, outlined in its White Paper 'Working for Patients' (Great Britain, Parliament 1989), has been to introduce an 'internal market' into the health care system. The role of purchasing health care has been separated from the provider function, and District Health Authorities now have the responsibility for commissioning the health care needed by their resident population. This is done by negotiating contracts with providers. Although there is some scope for flexibility, the intention is that most patients covered by the contract will be cared for by the providers with whom the DHA has an agreement. This purchasing role shows some parallels with the Regional Health Alliances which were proposed to manage health service markets in the USA.

Changes have also been introduced to the organization of general practice. GPs working in groups may now have responsibility for their own budgets. These fundholding practices purchase directly certain selected non-emergency services for their patients, making their own arrangements with hospitals, etc. to supply these services. Fundholding practices therefore have more freedom than non-fundholders to spend their budget as they wish, and to negotiate access to some forms of secondary care for their patients independently of the District Health Authority arrangements. Part of the inspiration for this arrangement was the original model of the Health Maintenance Organization in the USA. It has the effect of creating multiple purchasers as well as multiple providers. Critics suggest that the NHS services a patient receives now depend on whether he or she is registered with a fundholding or non-fundholding GP (Bain 1994). Other commentators have suggested that fundholding practices have been over-generously funded, compared with the money made available to District Health Authorities to assure the same services for the patients of non-fundholders (Dixon *et al.* 1994; Jones 1994; Sheldon *et al.* 1994).

The health service providers, who supply services to meet the contracts arranged with District Health Authorities or GP fundholders, must now compete with each other to provide NHS care. Independent as well as publicly managed providers are in competition, and many public hospitals and community health service units have converted to a new, semi-autonomous 'Health Care Trust' status, which frees them from managerial control by local District Health Authorities.

Thus the British health service is moving towards the managed-market model. The NHS administration is losing its old function as health service provider and developing a new role as a state contractor and regulator of health services.

A parallel strategy has been an expansion of independent health care to take up some of the demand for medical treatment. (As we noted above, the independent sector has been comparatively undeveloped in Britain com-

pared with other high-income countries.) Although most health care consumed in Britain is still funded by the state, the proportion of the population
with private health insurance is growing, and is now equivalent to 12 per
cent. The government has encouraged the growth of the independent health
care sector in various ways. For example, older people can claim tax exemp-
tions for private health insurance premiums, and development of independent hospitals was allowed to progress in a rather unrestricted fashion
during the 1980s. As a result, there has been a growth in alternative schemes
for provision of health care for patients depending on their health insurance
status. As in the USA, evidence suggests that access to private health care is
uneven, and varies both between social groups and between geographical
areas. Fig. 5.2 shows how private health insurance coverage varies between
social classes in Britain. Those most likely to have private cover are of working age and in professional and managerial occupational groups. Old people and semi-skilled or unskilled manual workers are much less likely to
have private insurance. In geographical areas with high concentrations of
people in professional or managerial classes, the level of private insurance in
the population tends therefore to be higher.

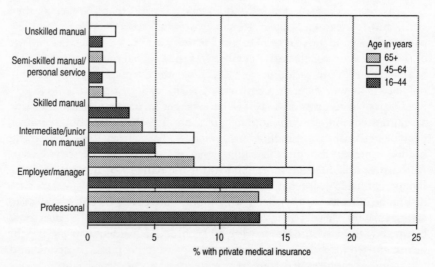

Fig. 5.2 Social-class differences in private health insurance
cover in Britain. (*Source:* OPCS, *General Household Survey*
1987.)

This is especially true of affluent residential areas where there is also a
good supply of private medical facilities and consultants. The geographical
distribution of private hospital beds shows a strongly polarized pattern,
with services concentrated around the London area and the south-east of
the country (Mohan 1988). The unevenness of distribution is much greater

than for hospital beds in the public sector. The differences in these distributions reflect the processes of resource distribution in the public and the private sector. In the public sector, health service resources are distributed to regions using a formula which estimates population health needs (see Chapter 6). The general effect has been to shift resources towards the north and west of the country, where health tends to be worse. In the private sector, market forces tend to concentrate services in the south-east, where there are most people able to afford private health services. Mohan (1988) has commented on the perverse effects of allowing the independent sector to develop in a manner which is contrary to that required by public policy. Navarro (1991) has pointed out several problems demonstrated by the American experience, which might have informed British policy and led us to expect the difficulties which observers are beginning to notice arising from the reforms. He claims that there is no evidence that HMOs result in greater patient satisfaction with doctors, and that, while market-based health services and expansion of the independent sector may be a good way to increase efficiency and raise additional revenue, this entails higher administrative costs and may reduce use of health services (particularly preventive care).

Russia

The Russians have also put in train a major programme of reform to create a health care market in place of the old centrally planned health care system (Telyukov 1991; Ryan 1993; Curtis *et al.* 1995; Sheiman 1994). The exact nature of the new system is still being worked out, but in some areas such as Moscow and St Petersburg the changes were being implemented in 1993. Finance of health care will in future be through a compulsory state health insurance scheme, paid for partly by the state and partly by employers' contributions on behalf of their workers. Commissioning of health services will be carried out by government-approved insurance agencies, who will use the health care funds to purchase health services from hospitals and clinics. At least some of these agencies will be private insurance companies, which have begun to be established in Russia in recent years. These companies are also expected to sell private health care insurance to a growing proportion of the population. There is likely to be an incentive for individuals to purchase private health cover, because the compulsory public scheme will not cover all available forms of health care.

There is also evidence of a growing private sector in health care provision, with hospitals beginning to establish beds for privately insured patients and some clinics operating independently and fixing their own fees. The users of these private services are those with the financial means to cover the costs, or else those working in companies which can afford to buy insurance for their employees.

Thus the Russian system illustrates a marked shift, from a system with strong state control and little scope for an independent health sector to one in which the independent sector plays a more prominent role in arranging finance and provision of health care. In contrast to the reformed British system, the new Russian system is insurance-based. Direct management of the insurance market is largely by private insurance companies instead of state health authorities. State involvement in the health system is likely to be reduced to one of regulator and part-financer of health care.

Conclusion: are health service systems converging?

Some authors (e.g. Field 1989) have suggested that, in spite of the historical, economic, and cultural disparities which have caused health systems to develop differently, national health systems are now tending to converge towards greater international uniformity because of rather similar socio-industrial trends affecting most developed countries. One outcome, at least in high-income countries, where the biomedical approach to health care is especially dominant and well developed, may be convergence towards managed health service markets of various types. However, other authors stress that, although there is international cross-fertilization of ideas on organization of health services, countries are normally strongly influenced by their own history and do not necessarily make changes informed by detailed evaluation of other systems (Blanpain *et al.* 1978).

The changing pattern of service organization in countries such as Britain and Russia (and also the Clinton proposals for health care reform in the USA) certainly suggests that we will need to revise the model shown in Table 5.3. There are some similarities in the motives for change in the different countries considered here: reforms are universally motivated by a concern for cost containment, combined with a requirement for basic services, at least, to be available to all. Whether the changes amount to a convergence of health systems is, however, debatable. The reasons for change were strongly influenced by the widely varying historical development of each country and the particular deficiencies of each of their respective health service systems. It is not the case that all the reforms discussed here involve introducing a very similar solution in each country. Britain is moving towards a form of managed market, but one which is still funded by general taxation. The proposals for change in the United States would also bring in a more managed market, but one based on insurance. Meanwhile, Russia seems to be moving away from funding health care by a central government budget, and allowing much more scope for the free market to operate in the allocation of funds for health services.

In each of these countries change is continuing, and we do not yet know what the reformed health service systems will eventually be like when (or if) they eventually stabilize into a new form of organization. The discussion of national health service systems in this chapter demonstrates their dynamic character, and the significant pace and scale of changes which require new perspectives on the international differences in health services. The forces for change which we have considered in this chapter, acting at the national scale, are also having profound effects at the more local level within countries, and we consider some of these effects in the following chapter.

|6|

Local perspectives on equity and effectiveness of health services

The previous chapter discussed alternative models for national health service systems. Such comparisons at the international scale tend to mask the effects of local variation in provision and access to services. Within every country, the questions of regional and local resource distribution and its correspondence with local differences in patterns of need are important for equity and effectiveness of health services for the population. This chapter begins with a consideration of the debate over equity of local variation in health services, before moving on to discuss ways of evaluating the performance and outcomes of health services.

Territorial justice or inverse care?

Underpinning debates over local health service provision is the idea that resources should be allocated in a way which will respond effectively to differences in health needs of the local population. This applies especially to the distribution of state resources for health and welfare. Davies's (1968: 39) formulation of territorial justice helped to open up the debate over local distribution of welfare resources: 'A situation in which the distribution of society's goods between areas corresponds to the varying need of these areas.'

The idealized pattern of health service provision required by this formulation of territorial justice is therefore one in which the level of provision of services varies from one place to another (i.e. it departs from perfect *equality*) but the variations in provision are in direct proportion to population need (so that there is *equity* in the distribution). Many studies concerned with territorial justice have been initiated by a concern that the pattern of territorial resource distribution, rather than coming close to the territorial justice pattern, is in fact quite opposite to it. This is the 'inverse care law thesis' (Tudor Hart 1971) which suggests that: 'The availability of good health care tends to vary inversely with the need of the population served.'

Tudor Hart, a Welsh GP, argued that the inverse care law would apply particularly in systems which relied on market forces to allocate resources for health services, although he also believed that it described the situation in Britain in the early 1970s, in spite of the fact that the National Health Service had been operating since 1948. The idea that there might be an inverse care law applying within countries has generated a large amount of research. As we saw in Chapter 5, the system in the USA results in particularly strong inequalities of health service provision, with areas occupied by less privileged and less healthy populations being the worst served, while wealthy, healthy populations are better provided for. Shannon and Dever (1974) reviewed the evidence for the inverse care law, and contrasted the inequitable situation in the USA with that in countries of Western Europe, which have a more developed national health service. Some early studies in Britain (e.g. by Noyce *et al.* 1974) also suggested that, at the regional level, NHS spending was greater in the more privileged south and east of the country than in the north and west, where the population was both poorer and in worse health.

However, detailed examination of studies of equity in health service provision have also demonstrated that the evidence for and against the inverse care law is, in fact, often difficult to describe in simple terms, and can even be rather equivocal. The picture of inequality of provision in relation to need is a complex one, reflecting the diverse dimensions of uneven social and economic development within countries.

One aspect of inequality of access to services is associated with 'underdevelopment' of rural areas. Several countries show striking evidence of lack of access to health services in less 'developed' rural areas. Phillips (1990) and Phillips and Verhasselt (1994) include a number of examples of studies in low-income countries which show levels of hospital provision to be much lower in rural areas than in urban areas, although health in rural areas is no better than in urban areas, and indeed is often worse. For example, Iyun (1994) cites data collated by UNICEF on the proportion of urban and rural populations having access to primary health care, which shows systematically that the proportion of the population with access to health services is lower in rural than in urban areas (see Fig. 6.1). Akhtar and Izhar (1994) demonstrate a similar situation in India, where the provision of hospital beds in relation to population size is on average less generous in rural than in urban areas. Their analysis suggests, however, that it is where rurality is linked with 'underdevelopment' and with rapid population growth that the level of provision is lowest. They show that the regional disparities between rural areas are more marked than between urban areas.

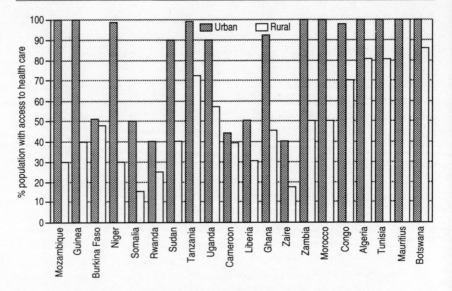

Fig. 6.1 Proportion of the urban and rural population with access to health services in selected African countries. (*Source:* UNICEF, cited by Iyun 1994.)

Lack of access to health services is also apparent as an aspect of rural deprivation in wealthy nations. In the United States, for example, there is concern over the reductions in numbers of hospital beds in rural areas (Cromley 1993; Hart 1993). There is also evidence from the USA that lower levels of health service provision may be linked with racial inequalities as well as with rural deprivation. Kindig and Yan (1993) have examined provision of physicians in rural areas categorized according to the ethnic composition of the population. They showed that provision in 1990 was relatively low in areas with large black populations and also where the proportion of Hispanics was high. This suggested inverse care in relation to need, since the health status of these ethnic groups was typically below average. However, the pattern was not consistent for all rural areas with large ethnic minority populations: in areas where proportions of native Americans were 'high' , the levels of provision were higher than in other rural areas, which suggests territorial justice. (On the other hand, the areas with the very highest concentrations of native Americans were comparatively under-provided with physicians) (see Table 6.1). This provides an illustration of the complexity of area variation of service provision in relation to population need.

Other countries also show particularly marked disparities of health service provision associated with socio-economic and racial divisions. South Africa was a country which, under apartheid regimes, developed gross inequalities in social and welfare infrastructure between predominantly

Table 6.1 Physician provision levels in rural counties according to ethnic composition of the area, USA, *c.* 1990

Ethnic composition of population for groups of rural counties	Physicians per 100,000[a], 1990
Average for all non-urban counties	73.5
High black (23.2–38.2%)	68.9
Very high black (>38.2%)	55.1
High native American (8.7–15.4%)	78.6
Very high native American (>15.4%)	66.0
High hispanic (15.9–27.4%)	70.6
Very high hispanic (>27.4%)	57.0

[a] Allopathic doctors in 1989 and osteopaths in 1987, compared with 1990 population data.
Source: Kindig and Yan (1993).

white and predominantly black residential areas. Table 6.2 illustrates the less generous provision of hospital care for blacks than for whites in regions of Transvaal, Natal, and Cape. Areas with mainly black populations (Gazankulu, Kwazulu, and Ciskei) were also much less well served than other parts of the country. The inequality of provision is even more striking when one considers that the black population shows much worse health status (for example, in 1992 the infant mortality rate in the black population was 52.8 per 1,000 live births compared with 8.6 in the white population) and also that South Africa is a relatively wealthy nation which should be able to afford a reasonable level of provision for all its people. Policies to achieve a more equitable distribution must now be among the important priorities for the new government in South Africa, which has announced a

Table 6.2 Hospital provision for children in South Africa

Area	Provision for whites		Provision for blacks	
	Children per bed	% occupancy	Children per bed	% occupancy
Transvaal (Gazankulu)	425	54	706 4,552	87
Natal (Kwazulu)	450	52	626 1,984	105
Cape	310	38	392	67
Ciskei			624	90

Source: Ransome *et al.* (1986).

new national health plan, aiming to shift resources significantly towards the primary care sector and to provide free medical services for young children, pregnant women, and nursing mothers, as well as for those who are elderly, disabled, or chronically ill. Free health treatment for children under 6 and pregnant women is already established, and a free school meals scheme has been launched in deprived areas, serving about 2 million children (Lyall 1995).

It is not only in rural areas that local populations may have relatively low levels of health service provision. Access to health services is often below average for deprived urban populations. Harpham (1994a, 1994b) points out that the rate of urbanization in low-income countries is now growing faster than in wealthier nations, and that the poor populations of urban areas represent a large and rapidly expanding category with high levels of risk to health, but without adequate access to health services. In Latin America and parts of Africa and Asia, the largest concentrations of poor people are now living in cities rather than rural settings (Harpham *et al.* 1988), and the urban slums where they live are often relatively deprived of essential infrastructure, and lacking health services which are accessible, appropriate, and affordable (Okafor 1990; Rossi-Espagnet *et al.* 1991). In richer countries, such as Britain and the United States, studies at the local scale have often demonstrated relatively low levels of health service provision in deprived inner cities and in poor suburban housing estates (Knox 1978; While 1989; Wallace *et al.* 1994).

Studies of health service provision to geographically defined populations using a limited range of indicators do therefore seem to demonstrate some crude differentials in access to health services. They show that poor populations in both rural and urban settings are relatively disadvantaged, especially in cases of extreme inequality illustrated by some of the examples considered above.

However, it is not really sufficient to analyse territorial justice in health services in such limited terms, and this is underlined by studies which have attempted more comprehensive analyses of need and provision. An early attempt at a more thorough examination, by West and Lowe (1976), investigating child health services in Britain, compared the territorial differences in a number of different indicators of need and provision. Their study illustrated positive correlations between need and provision on some indicators (territorial justice) but negative correlations (inverse care) for some other indicators. Research of this type suggests that rigorous analysis of the inverse care law is a complex task, and that early studies may have taken too simple a view. One of the most thorough examinations of the problem of researching the inverse care law is by Powell (1990). He highlights a number of difficulties in existing research on territorial justice and the inverse care law. On a purely technical basis, it is difficult to build up a cumulative impression from the studies of inverse care because the methods used in different studies are variable, and in some cases the tests used to examine association between need and provision are not very rigorous. In addition (and presenting more fundamental problems for such research), there are a number of difficulties associated with the formulation of geographical studies of need and provision. These include the following:

- lack of a consensus on what constitutes a socially just distribution;
- lack of consensus on definitions of need for health services;
- the limitations of existing indicators of population need;
- problems of measuring health service provision for the population of an area;
- the ecological fallacy;
- problems of evaluating the quality and effectiveness of health services, especially in terms of outcomes for health.

These problems deserve closer consideration, since they help to frame the current agenda for research in this area.

What constitutes a socially just distribution?

A particular view of the socially just distribution is implied by Davies's idea of territorial justice (i.e. provision proportional to relative medical need). However (as discussed in Chapter 5), this is not necessarily the consensus view, since varying ideologies of care lead to differing interpretations of the sort of distribution of health service resources which is desirable. In some societies, ability to pay, membership of a particular social category, or merit as a contributor to the general good of society might be considered acceptable criteria on which to determine the distribution of medical services. This issue is the subject of a huge literature which is outside the scope of this text to review in detail. Harvey (1972, 1992) and Smith (1994) have provided examples of explicitly geographical accounts of this debate which are particularly relevant to the issue of territorial justice. They show that there is a tendency for societies and governments to take for granted (sometimes in a quite cynical way) the political and economic structures which determine the distribution of resources in our societies. Marxist and neo-marxist critiques (e.g. Navarro 1978; Peffer 1990) of the way these structures operate in capitalist societies adopt a more questioning stance, although some aspects of their accounts have also been criticized for focusing too much on class conflict and not addressing important dimensions of social injustice such as sexism and discrimination against minority ethnic groups (Friedmann 1992; Okin 1989; Walzer 1983; Kuper 1974). A major agenda for health and social policy in the 'post-communist', 'post-modernist' era is to formulate concepts of social justice appropriate to changing conditions.

Problems of defining need

Even if there is broad agreement on the desirability of provision in proportion to health need, it is difficult, if not impossible, to find a universally agreed definition of need in the population. Again, a vast literature has accumulated on how to conceptualize the idea of need, which has been ably reviewed, for example, by Doyal and Gough (1991). Without entering into the more philosophical aspects of this debate, we note here some of the practical problems which arise when health planners and policy-makers attempt to translate poorly specified abstract ideas of need into operational terms. Definition of need, for example, depends on whether one takes a limited view of need as a requirement to meet minimum standards of health care (e.g. protection from or treatment of illnesses which can have a significant effect on fitness and ability to work) or whether one's view is more generous, allowing need for care to include health services which are not essential to survival at a basic level of fitness (e.g. cosmetic surgery). Should need include the most comfortable and acceptable form of care (hospital rooms equipped with high standards of 'hotel' amenities, for example) or only the simplest provision to meet requirements for hygiene and medical effectiveness? For some time it has been also been recognized (e.g. Bradshaw 1972) that very different views of need may be perceived by health service professionals and lay people. Some of the problems of resolving these differences in perceived priorities are discussed further in Chapter 8, which considers issues of public participation in health service planning.

The limitations of indicators for local needs assessment

Even where it is possible to propose a working definition of need, there may not be suitable indicators available to measure need in these terms. Many conceptualizations of need are essentially perspectives which stress level of illness or risk of illness in the population. Thus information on health inequalities discussed in Chapter 4 are interpreted as indicators of varying need for health services thoughout the national space. However, as demonstrated in Chapter 4, the available information on population health status is limited. We often lack information on levels of morbidity in the population, or on risk factors which might make some populations more susceptible to disease than others. Information on the amount of illness which comes to the attention of the health services is not sufficient. The idea of the 'clinical iceberg' (Last 1963) is an analogy used to express the relationship between expressed need (demand for, or use of, services) and felt needs (which are experienced in the population). These two types of need overlap, but only partially. Furthermore, patterns of health service use will be in part determined by the local supply of services, rather than by real differences in population health. An additional problem is the paucity of indicators which can be used to measure health and illness at the local scale. In practice, as surrogates for measures of morbidity and risk factors, local studies of health need often resort to using measures of mortality and material living conditions because these are the only indicators which are routinely available (see also Chapter 4).

The British literature provides one of the richest sources on the question of how to apply geographical indicators of local need to problems of resource allocation. Research on this topic in Britain has received a good deal of attention because the country has a national health service financed from the central government budget. This requires techniques for the equitable allocation of central public funds for health services to the local level, where they are used to commission health services needed by the population. Allocation of resources is intended to be on the basis of relative need, irrespective of insurance status or ability to pay for health services. The literature is interesting because it deals with the dual problem of how to define need for health service resources at the local level and how to operationalize the relevant concepts of need in order to apply them to the problem of regional resource allocation across the country. We consider here two examples of allocation formulae using local information to represent relative population need. The first is the formula for resource allocation to Health Authorities for local hospital and community service provision; the second is for payment of general practitioners.

Regional resource allocation in the British NHS: mortality as a need measure

One of the earliest attempts at matching health spending to population need was the system proposed for allocating revenue to Regional Health Authorities in England and Wales which was proposed by the Resource Allocation Working Party (RAWP) in 1976 (Department of Health and Social Security 1976). The RAWP system has been discussed in detail elsewhere (Eyles and Woods 1983; Jones and Moon 1987) . When it was introduced into the NHS in 1976, it was a radical departure from the previous system of funding. The RAWP formula determined running costs of hospital and community services (excluding the family doctor service, which is funded separately) on the basis of the size and the characteristics of the region's population. The characteristics taken to represent need for health service resources were age and sex composition (weighted according to the average patterns of use of health services nationally) and also the standardized mortality ratio, which was taken as an approximate indicator of morbidity in the regional population. Target budgets were determined for regions on the basis of formulae incorporating these indicators, and regional budgets were gradually modified over time to bring them in line with the relative need of the population. This method has attracted considerable attention internationally, and other countries have used the idea to apply to their own funding systems. In France, for example, in recent years, the potential of RAWP has been explored as a means of allocating the global budget distributed to hospitals (Meyer 1994). In Britain, however, the debate has moved on to tackle some of the shortcomings of the RAWP method, and to address the requirements of financial planning in the new internal market for health services introduced in 1990.

One part of the debate has concerned the use of data on mortality as a measure of population need for health services (Mays 1989). The original formula assumed that differences in need associated with health status would be directly proportional to the value of the standardized mortality ratio for the population of each region. However, later analysis has questioned this assumption. When the RAWP formula for England and Wales was reviewed at the end of the 1980s, an alternative formulation of relative need was proposed (Royston *et al.* 1992). This was based on an analysis of variables associated with differences in hospital use among small areas (electoral wards) in different regions of the country. In each ward, the analysis controlled for the age and sex composition of the population, and also for supply of hospital beds, since use of health services is linked to their availability. The analysis showed that area differences in use of services were statistically associated with mortality under the age of 75. The revised RAWP formula adopted by the government used the square root of premature mortality, which approximates quite closely to the statistical association demonstrated by Royston *et al*. This had the effect of targeting resources to areas of high rates of premature death. It also reduced the amount of variation in estimated need, since taking the square root of the mortality indicator has a 'dampening' effect on this part of the formula. Royston *et al* also found that differences in hospital use were independently associated with social deprivation, so that it could be argued that it would be appropriate to include social deprivation in the need element of the RAWP formula. However, this aspect of the analysis was not taken up in the new form of the RAWP formula, ostensibly because the government 'attached particular importance to keeping the formula as simple as possible' (Royston *et al.* 1992: 179).

Replacing RAWP: measures of morbidity

Not only is there debate over how to weight mortality as an indicator of relative need for health services, but a further issue is whether mortality is a valid measure of health service need at all. Its use in the RAWP formula was justified as a surrogate for morbidity, but some conditions which result in mortality (for example, accidental death) do not necessarily generate large health service demands, while there are many illnesses such as arthritis, for example, which, while they are not life-threatening, do produce considerable demand for health services. As medicine becomes more effective at reducing the mortality rates due to illnesses such as cardiovascular disease and cancer, we also see an increasing number of chronically ill people needing health services who do not figure in the mortality statistics. To assess the health service needs of such people, we need better information on morbidity, and this has generated new research designed to estimate the incidence of conditions such as stroke, for example (Malmgren *et al.* 1987). In addition, new measures of the impact of chronic illness on local populations are being developed, often based on lay assessment of functional impairment and disability rather than on medical diagnosis. For example, in 1991 the British population census (OPCS 1993) included for the first time a question on limiting long-term illness: 'Does the person have any long term illness, health problem or handicap which limits his/her daily activities or the work he/she can do? Include problems which are due to old age' (Charlton *et al.* 1994: 19). The latest proposals for a formula to determine allocation of the NHS revenue budget for hospital and community services uses information on morbidity derived from this census question, in addition to data on mortality under the age of 75 (Carr-Hill *et al.* 1994; Smith *et al.* 1994).

Measures of social deprivation as indicators of health need

Another aspect of the debate over measurement of relative need for resource allocation concerns the use of measures of material deprivation. The rationale for use of measures of socio-economic conditions comes partly from the literature on socio-economic differences in health reviewed in Chapter 5, which shows that poor people are more likely to suffer ill health than the rich. One reason for using indicators of material deprivation as health need indicators is that for small populations they may provide more statistically reliable evidence of difference than conventional health indicators such as mortality data. At the very local scale, the number of deaths occurring each year may be small, and therefore it is difficult to establish whether there are statistically significant differences between areas in the levels of mortality. In contrast, social and economic differences between populations at the very local level can be more definitely demonstrated using census data. In this sense, measures of material deprivation may provide *surrogate* indicators of health status. Several authors (e.g. Mays 1989) have also argued that for a given level of illness in the population, the difficulty of providing care will also vary in relation to conditions such as housing quality, social support, and availability of transport. Thus material deprivation indicators may be necessary *in addition* to health status measures, to provide evidence of factors likely to raise the unit cost of health service provision.

As we have seen, the 1988 revision of the RAWP formula did not include a measure of social deprivation. The latest proposals for allocation of resources for hospital and community health services in England (Carr-Hill *et al.* 1994; Smith *et al.* 1994) would include socio-economic data in addition to information on the health status of the population. Indicators of the proportion of the elderly population living alone, unemployment rates, households with a lone carer looking after a dependent, and the proportion of households headed by a manual worker have been suggested because they show an association with hospital use at the small area level which is independent of health status variation and differences in supply of services.

The national resource allocation formula is not, however, the only type of health planning application for which socio-economic data is relevant. In Britain there has been a mushrooming of research on small-area social indicators relevant to health, and these are widely used in health service decision-making. Most of these make use of the same basic data set: the decennial population census. A number of alternative socio-economic indicators have also been proposed, using different criteria of relative need. Examples of these are shown in Table 6.3. An index of 'deprivation' is calculated from the census data, using a method proposed by Jarman (1983) and colleagues. This index uses eight elements, listed in Table 6.3 with the weightings shown. The choice of elements for the index and the weightings are based on the opinions of a national survey of doctors. This index therefore summarizes the population attributes which doctors typically think will tend to increase workload per patient. The index combines information on demographic structure and social deprivation. The indicator suggested by Townsend *et al.* (1988) is based on a theory of material deprivation likely to affect health, and comprises a slightly different set of variables which are not differentially weighted. Balarajan *et al.* (1992) propose an indicator using social characteristics associated with propensity to consult a doctor reported in a national sample survey. The variables and their weightings are determined by their statistical association with GP use in the survey, so that the small-area indicators represent a form of synthetic estimation of likely consultation rates with general practitioners.

Table 6.3 Components of three alternative small-area indicators of social conditions relevant to health service needs

Component	Jarman (UPA)	Townsend	Balarajan *et al.*
Elderly alone	6.62	–	a
Children under 5	4.64	–	–
Single-parent households	3.01	–	a
Unskilled manual workers	3.74	–	a
Unemployed people	3.34	1	–
Overcrowding	2.88	1	–
Recent movers	2.68	–	–
Migrants from NCWP[b]	2.5	–	a
Lack of basic amenities	–	–	–
Lacking a car	–	1	a
Not owner-occupiers	–	1	–
Council tenant	–	–	a
Private tenant	–	–	a

[a] Component included in the indicator; weighting varies between demographic group.
[b] New Commonwealth/ Pakistan.
Sources: Jarman (1983); Townsend *et al* (1988); Balarajan *et al.* (1992).

Calculating remuneration in general practice

One application of this type of indicator is in the calculation of payments made to GPs with NHS patients, who are mainly remunerated in proportion to the number of patients registered with them. The indicator proposed by Jarman (1983) is used to calculate additional remuneration for GPs in respect of characteristics of patients likely to generate extra workload (Department of Health 1990). Because detailed information about the characteristics of each patient are not routinely available, this has to be estimated indirectly. Information on the patient's address is used to identify the residential area in which the patient lives. Residential areas are defined in terms of electoral wards, which typically have populations of 5,000–10,000 people, and for which data from the population census is available. Additional payments are made to doctors in respect of patients living in areas where the Jarman 'deprivation' score is high. The system of additional payment in respect of patients living in deprived areas is shown in Table 6.4.

This example of the application of small area census data for health planning raises a number of issues about the appropriateness of using population data to estimate relative levels of need for a doctor's services. One problem is that the method may be invalidated by the ecological fallacy (see below). Also, the use of rather crude thresholds for payment shown in Table 6.4 has been disputed by some commentators (e.g. Senior 1991) because they assume no significant extra need unless the score on the indicator is very high, which is probably not realistic. Senior (1995) also points out that

Table 6.4 Payment of additional remuneration to GPs in respect of variation in the Under-Privileged Areas Score

Type of ward of residence	UPA[a] score for ward	'Deprivation payment' in respect of each patient living in this type of ward
'High deprivation'	+50 or above	£10.20
'Medium deprivation'	+40–49.99	£7.65
'Low deprivation'	+30–39.99	£5.85
'Not deprived'	below 30	none

[a] Under-Privileged Areas Score proposed by Jarman (1983) (see Table 6.3).
Source: Senior (1995).

in 1994, the government were still using data based on the 1981 census to calculate the deprivation indicator, so that the information was very likely to be out of date.

A further difficulty in the use of the Jarman indicator is that by using the same index to estimate differences in need all over the country, it is impossible to reflect accurately the specific aspects of social deprivation which are important for health service needs in different regions. A large literature (discussed in Chapter 4) suggests that the association between health and deprivation is geographically variable. It has been suggested, for example, that the Jarman index emphasizes population attributes which are important for health service needs in deprived parts of London, but it is less relevant for some other cities or for rural areas.

More fundamentally, there are questions about the appropriateness of this measure as an indicator of relative health service need in the population. Although there have been some attempts to validate the indicator (Jarman 1984; Hutchinson 1989; Curtis 1990; Cotgrove *et al.* 1992; Jessop 1992), these are often local studies and cannot be said to represent a complete validation (Carr-Hill 1991). Thus there is uncertainty over the value of the Jarman indicator as a surrogate measure of population health difference. However, at the time when the index was developed, no suitable local data on health status was available. Mortality data are arguably not suitable because they are only approximately related to the morbidity treated in general practice. Furthermore, as we have noted, at this small scale mortality indicators are likely to be unreliable. Information on population morbidity has not hitherto been available for such small areas across the whole country.

The 1991 census question on long-term illness discussed above might in future prove useful for estimating more directly local differences in morbidity and need for services. It provides information on the prevalence of self-reported long-term illness in the population across the whole country and for small areas, so that it may help to supplement small area data on mortality as a routinely available indicator of population health status. Fig. 6.2 shows, for example, information on the relative rate of reporting of long-term illnesses for small areas used for health service planning in part of London (with population size varying from 60,000 to 90,000). As is

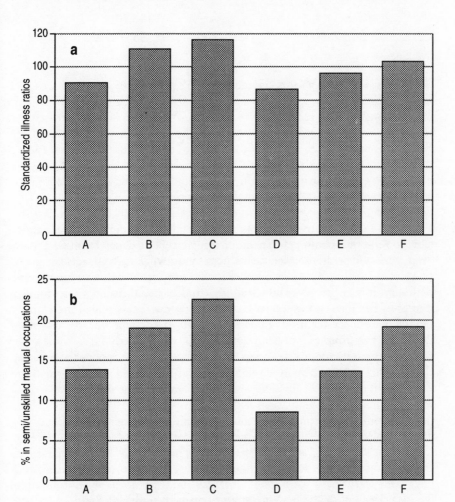

Fig 6.2 Health service planning localities, labelled A–F, in a District Health Authority in east London, showing prevalence of illness and social-class profile. (a) Relative levels of long-term illness reporting. (Ratios are standardized for age and sex and show the local rates as percentages of the average national rate. The national rate is represented here by a ratio of 100.) (b) Percentage of the economically active population in each locality with a semi-skilled or unskilled occupation. *(Data source*: OPCS, *Population Census* 1991.)

suggested by the accompanying graph of the social-class profile of the areas, longstanding illness is associated with disadvantage in the population. For the moment, however, the potential of this new indicator as a relative measure for health services is still subject to evaluation.

Measures for local needs assessment

The example of 'deprivation' payments to GPs is probably the most formalized application of data on socio-economic attributes of the population for the purposes of resource allocation in the British health service. However, these composite population indicators are also increasingly widely used in the NHS as relative measures of local population health need for other types of service. This growth in use of area social indicators is linked to the increasing demand for information at the very local scale on differences in health service need. Such data has become more necessary with the introduction of the internal market for health services, since district health authorities now have responsibility for determining the needs of their resident population, planning the allocation of their budget to meet those needs, and commissioning health services by drawing up contracts for provision of local health services. Local health purchasers often make use of small area indicators of social conditions and of health to help inform the process of commissioning services for district populations. These indicators provide one way of identifying targets for priority funding within a district. Thus the local geography of health and health service needs has become recognized as a useful tool for planners in a managed market for health services (Curtis and Taket 1989; Jones and Moon 1993).

How can territorial differences in levels of provision be described?

The measurement of the health resources made available to the population of particular areas is also difficult. One of the fundamental difficulties is that there are often flows of patients across district boundaries to consume health services provided in neighbouring districts (Clarke and Wilson 1986). Tonnellier (1990), in his study of provision of general practice in France, uses travel-to-work areas as the geographical unit of analysis, partly because these areas describe fairly well the catchment areas for local services and therefore will approximate to the zone of attraction of doctors situated in the area.

The problems are exacerbated by a lack of geocoded data on service activity. Without knowing where patients using services live, it is impossible to judge what is being consumed by the population of a particular area. For example, data collected in the British NHS has only quite recently begun systematically to include patients' postcodes as a form of geocoding. The task of linking data is made even more arduous by the fact that, in England and Wales, postcode areas do not correspond to the census areas used to compile information on the denominator population (Martin 1992) , and neither do they match perfectly with administrative divisions in the health service. In countries such as the United States, where a large proportion of service provision is in the private sector, involving many providers which are separately administered, there are even greater problems for collation of such information.

An alternative is to carry out representative population surveys which record service consumption. Several countries have national surveys of this type. In Britain, the General Household Survey (OPCS 1994) records use of GPs and hospital services in a nationally representative sample of households. In France, the decennial survey of service use (Sermet 1993) collects detailed data on health service use. In the United States, the National Health Interview Survey provides information on service use (US Department of Health and Human Services 1989). However, these are usually designed to collect representative information for the country as a whole, and do not provide comparable information for all the local areas within the country.

To further complicate the problem of assessing provision at the local level, there is the question of substitutability of public medical health services and other services provided by other agencies. In Britain and France, for example, commentators note the increasing 'medicalization' of residential homes for the elderly which are provided by public and independent operators outside the health service system (Darton 1980; Henrard 1988). A complete picture of the institutional care available to very frail or chronically ill elderly people would therefore require information on the full range of such services. In the provision of community services, the overlapping roles of health and social services can also be very significant (Ottowill and Wall 1990). With respect to acute hospital care, particularly routine elective surgery, there is also some substitution between the public and the independent sector, at least for those patients with private health insurance (Mohan 1988, 1995; Scarpacci 1990). Thus geographical variation in the availability of services may be significantly affected by the development of the independent sector, which, as we saw in Chapter 5, is often very uneven in space. Collecting information on local provision in the private sector is often more complicated than for the public sector because of the diversity of different agencies involved in such provision.

Finally, even if information on the quantity of provision is available, it may not be sufficient. As discussed below, the quality of services, their accessibility, and appropriateness for the population using them may be important dimensions of local provision from the point of view of territorial justice, but these are more difficult to measure.

Area provision and individual access

Studies of territorial justice tend to place emphasis on the geographical distribution of resources. In practice, however, it is difficult to establish how far differences in the geographical pattern of provision are influential in use of health services. Much of the early emphasis in medical geography was on the effects of distance on health service access and use. Many of these took their inspiration, for example, from 'Jarvis's Law', which was articulated in a study of psychiatric hospital use, but has been applied to other health services as well: 'The people in the vicinity of lunatic hospitals send more patients to them than those at a greater distance' (Jarvis 1850, cited in Hunter and Shannon 1985: 297).

Jarvis's Law has certainly generated some interesting discussion, and was important in helping to generate interest in health services among geographers. The idea of distance decay in use of services is still applied in many location–allocation models which are intended to summarize broad patterns of use of facilities by the population in general (Curtis 1982; Mayhew 1986; Taket 1989; Okafor 1990; Massam and Malczewski 1991). However, most models of access to health services stress social and economic barriers to access as well as distance effects (Joseph and Phillips 1984). The importance of the health beliefs of the individual or social group were also discussed in Chapter 3, which showed these to be powerful determinants of access to care and utilization. It is important to consider factors which influence individuals as well as general effects. Table 6.5, based on Puentes-Markides's (1992) model of access to health services, draws attention to factors operating at the level of whole societies (determining the general pattern of distribution of health resources), at the institutional level (effects of the local organization of services), and at the level of individuals and their families (emphasizing the differential levels of access prevailing in different family settings, even within one area).

Such models of health service access and use, as well as empirical studies of utilization, often suggest that physical accessibility of services or travel cost to reach facilities is relatively unimportant compared with factors such as cost of services, insurance status, eligibility criteria of services, information about (and perceived benefits of) services, and rapport between service providers and users. According to Hunter *et al.* (1985), even Jarvis's original data shows that a law invoking simple, universal distance decay effects is an oversimplification of the case. Thus the existence of health facilities or

Table 6.5 Factors determining access to and use of health services at different levels

Societal macro-level

- Style of development
- Macro-economic policies
- Political orientation
- Status of women
- Work/employment conditions
- Health policies
- Self-help systems
- Social support networks

Institutional level

Structure of health system	Behaviour of health professionals
Cost of services Prevalence of Western medicine Referral patterns Professional recruitment Staff training 'Race' and class structure of health system Fragmentation Ability to diagnose certain ailments Profit orientation of health system Provider consumer relationships	Feelings towards poor patients Beliefs about women's status Discrimination on basis of 'race'/ethnicity (Lack of) ability to speak client's language Willingness to treat uninsured patients Awareness of disparities in service provision Position on reproductive issues Position on expansion of gender roles

Individual/family level

Characteristics of clients	Behaviour of clients
Ability to pay for services Etiological beliefs Levels of education (Lack of) ability to speak provider's language Ability to recognize health problems Circumstances (e.g. homeless, battered, imprisoned, migrant, refugee, nomad) Type of health problem (e.g. specific diseases, mental illness, AIDS) Age group (e.g. adolescent, elderly)	Participation in decision-making Help-seeking behaviour Perception of severity of situation Perception of effectiveness of service Previous experience of services Fear

Source: adapted from Puentes-Markides (1992: 622 copyright 1995, with kind permission from Elsevier Science Ltd, The Boulevard, Langford Lane, Kidlington OX5 1GB, UK.)

medical personnel in an area do not of themselves guarantee that all the local population will have access to those resources. Perspectives on territorial justice should approach problems of distribution of health services with this in mind. It is important to assess the characteristics of the local population which may affect access, and also the quality as well as the quantity of the health services provided.

The ecological fallacy

Even if the ecological associations between need and provision in geographical areas can be reasonably accurately tested using the indicators available, there is still the possibility that the results may be misleading because of problems of interpretation referred to as the 'ecological fallacy'. There are at least three aspects of ecological studies which have given rise to concern about this 'fallacy', which are reviewed by Schwartz (1994). The first concerns the erroneous attribution of the *average* characteristics of the group to *individuals* in the group. The second concerns the difficulty of controlling confounding variables in ecological studies, so that ecological associations between two variables in the population may in fact be due to correlations with other, intervening variables. This could also occur in studies of individuals, but some epidemiologists argue that it is particularly likely in ecological studies (e.g. Lilienfeld and Lilienfeld 1980). The third weakness of ecological studies is held to be that they are only indirect ways of studying the factors which are the most immediate causes of illness and death in individuals.

The ecological fallacy is particularly important if information on aggregated populations is being used to target resources towards populations in need. In such a situation, relatively more resources will be allocated to areas where needy people form a large proportion of the total population. This is desirable because it is equitable to compensate for the additional costs of providing for large numbers of people needing to place relatively heavy demands on local services. However, needy people living among generally more privileged populations will not receive as much attention. It may indeed be the case that a large part of the needy population in fact lives outside the areas of greatest concentration of disadvantage, so that area information may be a very blunt instrument for directing resources towards relative need. Thus, for example, Berk *et al.* (1991) suggest that directing extra resources towards poor communities in the United States may not be an efficient targeting strategy; the actual number of deprived people may be more pertinent as an indicator of relative need than the average income in the area. Even in areas which are given additional funding for deprived populations, it may not be the most underprivileged people who benefit most from the additional funding. Resources targeted towards particular places of residence will, for example, only have limited benefits for highly mobile populations such as travellers or homeless people and those in temporary accommodation, although several studies have suggested that they have significant health service needs which are often unmet (Lowry 1990; Smith 1993; Pilkington 1994).

Relative needs assessment methods based on aggregated population data often try to improve the accuracy of resource targeting by using information on rather small areas, on the basis that the population within them will be more homogeneous. (Thus, for example, data on electoral wards were used in the determination of resource allocation in the British examples discussed above.) This will certainly help to identify very local 'pockets' of deprivation where resources may be usefully channelled, but it does not fully resolve the problem. It can still be argued that where needy populations are very sparsely distributed, resource targeting on an aggregate basis may be rather ineffective. This may be argued in the case of deprived rural populations, for example.

In spite of the problems posed by the ecological fallacy, we should not conclude that area information is not useful to help us to interpret the factors which contribute to health differences and need for health services. Indeed, some commentators (e.g. Susser 1994; Schwartz 1994) have stressed that it is necessary to have information about the geographical context which may influence the effects of individual characteristics (see also Chapter 4). Nevertheless, in the territorial justice debate, careful thought needs to be given to how the attributes of area populations relate to the characteristics of individual users of health services. The articulation between mechanisms for distribution of resources at the aggregate level and systems for allocating health services to individuals is often very unclear, and this can undermine health policies aimed at equity in health service access and use.

New perspectives on the quality of health services

We noted above that models of access to health services stress aspects of quality as well as quantity. Powell (1990) shows, for example, that, in the case of primary health services for London, there was inequity in terms of quality, since areas with poorer health status also seemed to have poorer quality of care on some indicators. Jarman (1981) also argued that, in parts of London where needs were greatest, the quality of provision of general practice was relatively poor. These studies suggest that variations in the quality of care may be as important as differences in quantity of provision in generating inverse care distributions. However, their conclusions are based on quite crude indicators of service quality. For example, a high proportion of doctors in single-handed practice is taken as an indication of poor quality (in contrast to group practice, which is thought to facilitate provision of good-quality care). While this may correspond to accepted professional thinking, it does not necessarily match patients' perceptions. A study in east London, for example (Curtis 1987), showed that patients whose doctor practised single-handed preferred this arrangement, and that important aspects of quality for patients included the rapport and sense of trust which they could establish with their doctor. Such attributes of services are difficult to measure using systematic indicators. While it is important to consider quality in debates over territorial justice, the issue of how to assess service quality in a meaningful way is clearly problematic.

The growing debate over how to assess quality of health services has demonstrated how complex the idea is. Donebedian (1980) and Maxwell (1984) have pointed to the multidimensional aspect of quality. A comprehensive quality assessment should, for example, include a focus on *effectiveness* of health services in improving population health and *efficiency* in using available resources to deliver services likely to produce such improvement. It should consider the point of view of the patient as well as the health professional, assessing issues of the *accessibility* and *acceptability* for patients, and the *relevance* of services to the patient's perceived needs.

In the following sections we focus on some questions of quality in health services, which have been the subject of a very wide-ranging debate in recent years, particularly in North America and Western Europe. There has been an enormous increase in emphasis on these issues, using methods variously referred to as 'quality assurance', 'audit', and 'performance review'. We discuss the potential and the limitations of these methods to change the perspectives on health service which influence decisions over resource distribution in such societies.

The 'bottom line' in discussions of quality of health service is (or should be) the issue of effectiveness, linked to appropriateness (Sheldon 1994). To what extent is provision of resources likely to have a desirable effect on population health status? Should we consider that populations 'need' services even though medicine has not been proven to make much impact on their health status, or if the cost is very great in comparison with the health gain which can be achieved? Much of the health care offered by modern medicine has been subjected to little or no evaluation in terms of its benefits for health. Even where information from evaluations is available it may be difficult to ensure that it is used to inform decisions on use of resources (Drummond 1994). This brings us to a consideration of the methods available to evaluate services in terms of health outcomes.

Conventional strategies for evaluating outcomes of specific procedures

Economists make a distinction between different types of evaluation, distinguishing, for example, *cost-effectiveness* analysis (the extent to which deployment of resources achieves a particular objective) and *cost–benefit* analysis (to determine which objectives should be pursued on the basis of the balance of benefits accruing compared with costs of provision and other opportunities forgone) (Mooney *et al.* 1980). Elements of both enter into health services evaluation work and, as we shall see, they are sometimes confused with each other in practice. In the following discussion we concentrate particularly on the problems of measuring the outcomes of health services resources, and give less attention to the equally difficult and important task of investigating costs.

Contemporary medicine, especially biomedicine, is constantly developing new techniques and technologies. These are often costly in resource terms, and as societies increasingly come to realize that resources for health are finite, there is growing concern that we examine carefully whether, in general, expenditure on different types of medicine represents good value in terms of improved health. There is also the possibility that some medical procedures may be damaging to the health of certain types of patient. Iatrogenic disease (Illich 1977) (also discussed here in Chapter 10) may be the result of activities of charlatans, but it may also result where medicines which are safe and effective for some patients are found to be dangerous for others (for example, antibiotics can be dangerous for those who are allergic to them) or where procedures at first considered safe are later found to have unexpected side-effects (for example, the effects on foetuses of the drug thalidomide, or the recent concern over damaging effects of silicone breast implants offered to women after surgery).

The typical biomedical approach to evaluation of specific medicines and procedures is through clinical trials, which have varying designs but which normally involve specially mounted research programmes. The classic case-control study requires comparisons of matched groups of patients, who are as similar as possible from the medical point of view, some of whom (cases) are treated with the medical procedure being tested while the others (controls), who are not treated, provide a group for comparison. The main aim of this type of study is to determine the typical effects of a particular treatment on disease in individual patients, while controlling for the non-treatment factors which might also have an effect on health outcomes. This is difficult to achieve because, in practice, the confounding variables are so numerous and so complex that it is impossible to control all of them. The requirement that treatment which may be beneficial is withheld from the 'control' patients, and the fact that some trials are conducted with a 'double blind' strategy (keeping patients and doctors ignorant of who is receiving the treatment), can also lead to difficult ethical issues.

The earliest clinical trials were further criticized because they focused mainly on the biomedical outcomes of intervention, such as increased life expectancy and physiological state as assessed by professionals. However, the range of measures of outcome have been expanded to include information relating to more 'subjective' aspects such as quality of life as perceived by patients. Not only are these measures more sensitive to the views of patients, but they also incorporate more information about the psychological, as opposed to physiological, effects of medical procedures. Jenkins (1992) argues that quality of life measures have value because of their potential to provide information on outcomes in terms of *feelings* (emotional impact and the effects of pain, for example), *functions* (especially activities for daily living including performance of social and home-based roles as well as work), and *futures* (the impact of knowledge of prognosis of a health condition which may affect the patient even before the physical symptoms of incipient disease are experienced).

Several quantitative indicators of quality of life have been developed (reviewed e.g. by Hunt *et al.* 1986; Bowling 1993; Walker and Rosser 1993). Davies (1994) points to the development of a number of multinational programmes to develop instruments applicable in international drug trials, such as the EUROQOL questionnaire (EuroQOL Group 1990) and 'Short Form 36' survey (Ware and Sherbourne 1992). Such attempts raise important issues of cross-cultural comparability and validity. Hadhorn (1991) reviews the potential of several of these instruments, arguing that the value of this type of measure lies in their potential to summarize the aggregated views of populations. While several of these indicators were derived from clinical trials and were originally intended for people who are ill, several of them have value for studies of health and illness in the general population. For example, the Nottingham Health Profile (Hunt *et al.* 1986) was originally developed in a clinical setting, and is still applied in this way (e.g. Bardsley *et al.* 1992), but has also been widely used as a tool for population survey work (e.g. Curtis, 1987b).

Innovative approaches to evaluation of health services

Recent developments in evaluative techniques have expanded the repertoire of methods for assessing outcomes of health services. This is partly because the conventional structures of clinical trials are not always appropriate for the type of activity being evaluated. Clinical trials were originally conceived to test drugs and medical procedures, especially those carried out in the secondary/tertiary health service sector. However, there is a growing appreciation of the need to evaluate services in the primary sector, which includes illness prevention and health promotion. This becomes more important as community-based services become more sophisticated, and as many countries are trying to shift resources for health towards the primary sector and towards long-term care in community settings.

Distinctive features of community services which have implications for methods of evaluation include the fact that some of the outcomes (especially in the case of health promotion and illness prevention) are long-term and difficult to attribute to particular interventions. Sometimes it is difficult to specify precisely the expected outcomes. In addition, the geographical distribution and residential mobility of the population served can make aggregate outcome analysis complicated. Because intersectoral collaboration is often particularly important for success of primary care (see Chapters 7 and 8), it may be necessary to evaluate the joint input of a range of service providers, and it may be difficult to identify the contribution of individual services to health outcomes.

A further challenge for evaluation in the field of community care is posed by health promotion projects which are based on community development approaches, as opposed to conventional service provision. Taket (1993) discusses the specific characteristics of this kind of project, which typically aims to improve the health of those involved, seeks active involvement of the target group, and involves working with the target group in the community to articulate needs and explore ways of meeting them. In such projects, objectives are not defined in advance by 'experts' and issues of relevance, participation, accessibility, and acceptability are very important. Such projects present challenges to conventional evaluation strategies in several respects: they lack well-defined *a priori* objectives; each project will be idiosyncratic; there are likely to be strong worker effects; there will be confounding effects associated with the environment in which the project is carried out.

Thus a traditional evaluation framework is inappropriate. For example, the identification of case-control groups from other districts is often impossible, even if affordable, because of specificity of local conditions, and furthermore it is often not acceptable on ethical grounds, or indeed feasible, to exclude local 'controls' from the project. Usually it is impossible to evaluate such projects through simple questionnaire surveys due to the complexity of issues. In addition, the target group is often not homogeneous and many of the salient variables are not amenable to measurement.

Taket suggests that an appropriate strategy is likely to involve examination of the development and refinement of the programme as it is going on (formative evaluation) as well as assessment of achievement of objectives and of final outcomes (summative evaluation). It is often desirable to examine appropriateness and acceptability of the programmes as they evolve and to modify them where necessary. It will be necessary to recognize the difficulty of separating project effects from other effects which are not due to the project. Evaluators will also have to be careful not to blame difficiencies *revealed* by the project *on* the project.

Very often an appropriate strategy will involve a triangulation approach using qualitative as well as quantitative methods (Taket 1995). Many of the criteria of validity which apply particularly to qualitative methods will be relevant. For example, it is quite likely that evaluators will be unable to take a neutral position and that they themselves will influence the development of the project and its outcomes. It may be more important to recognize these evaluator effects than to attempt to eradicate them (which is arguably impossible). Such studies will be very sensitive to the fact that evaluation is not value-free, and will be particularly concerned, for example, about issues of language and culture.

Qualitative approaches to evaluation

These considerations have led many researchers to argue for the value of approaches based on more 'qualitative' methodologies, to complement the more statistical approaches often favoured by conventional evaluation strategies. Fitzpatrick and Boulton (1994) discuss the application of methods such as in-depth interviews, focus groups, observational study, and case studies which can illuminate some of the likely reasons for the variations in health services outcomes which may be described, but not very satisfactorily explained, in more quantitative analyses.

It is often argued that the potential of qualitative techniques has not been sufficiently exploited. For example, Brannstrom *et al.* (1994) provide a framework for evaluating community-based intervention programmes, using programmes for prevention of cardiovascular disease as an illustration. Their framework (see Fig. 6.3) shows the range of approaches which can be used, including community participation, socio-epidemiological study of health-related behaviour, studies based on key informants, and social, cultural, and political description. While some of these would lend themselves to production of statistical data, most depend on other types of information.

St Leger *et al.* (1992) also discuss alternatives to case-control studies such as descriptive methods and use of testimony. Descriptive studies involve assessment by one or more evaluator, often from within the service, who portrays the service, its objectives, and the context of service delivery. The advantages of such descriptive evaluations include the fact that they may not require large additional resources (although they may have considerable real costs in terms of time taken by health personnel to carry them out), and

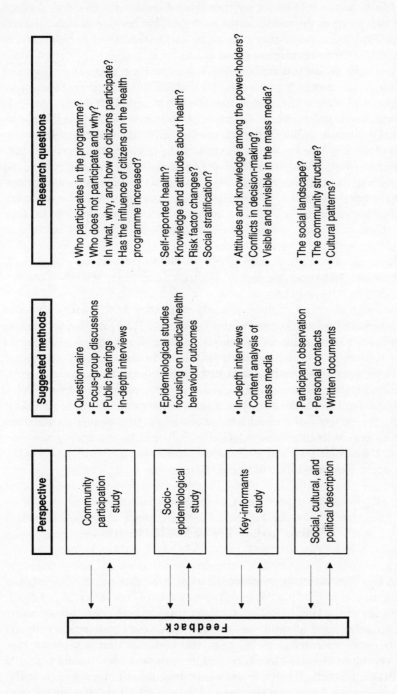

Fig. 6.3 A framework for outcome assessment of health interventions. (*Source:* Brannstrom *et al.* 1994: 129; by permission of Oxford University Press.)

that health service personnel have a sense of ownership of the evaluation and often perceive the results as the more credible because of their involvement. Description can clarify objectives, and is repeatable. The disadvantages include the requirement for observation and interpretation skills which not all health professionals have, the 'soft' or 'subjective' nature of the data which may be perceived as 'unreliable', and the fact that such studies can reveal aspects which those involved would prefer hidden.

Testimonial studies involve collation and interpretation of evidence from selected individuals. Quite often assessors are recruited from outside the service so that they will be seen as neutral by those giving testimonies. These can be rich in ideas and provide a basis for detailed work. Information can be produced rapidly. Such studies may be especially relevant where behaviours are being evaluated, or if some of the actors find it difficult to articulate their views. Special skills are required to encourage testimony. In some cases these studies are best performed in groups, but these may be difficult to organize and it is difficult to record all that is said. There are also difficulties in interpreting comments of others, and the studies may be difficult to repeat. As with other qualitative methods, testimonial studies are sometimes criticized for their lack of 'objective' information.

One example of an evaluation study employing both quantitative and qualitative methods was described by Taket *et al.* (1990b). This focused on an innovative scheme for the provision of professional advice and support on child-protection issues to health visitors and school nurses. The use of structured questionnaires demonstrated the benefits of the scheme in terms of reduced staff stress and greater confidence and satisfaction. Depth interviews of a sample of staff and their managers provided detailed case examples of the benefits of the scheme and positive effects on outcomes for clients. These interviews also enabled a clear understanding of those features of the scheme which had contributed to its success. Four years later the scheme is still in operation (Bass 1994)

Measuring service performance

The evaluation methods considered so far are typically *ad hoc* studies focused on one specific part of a health service. However, a number of methods are now available to evaluate service performance on a routine basis, using systematic and quantifiable criteria which facilitate statistical comparisons between different parts of the service system. There is evidence that more and more countries are beginning to introduce these to aid resource allocation decisions. This is particularly true of high-income countries, which have the information systems and know-how to conduct such evaluation. In low-income countries, where the biomedical service system is only partially developed, there may be more concern with trying to ensure a basic coverage of the total population by hospital services, rather than with examining the quality and effectiveness of the health services delivered. Arguably, however, it is precisely in these countries with most resource constraints where the outcome and effectiveness of hospital services should be given most careful consideration.

Part of the objective of performance review is to assess the standards of care being provided by different facilities and different professionals. The concern to ensure a consistently good standard of care throughout the health service is based on the assumption that if the process of treatment and care meets approved standards, it is also likely to produce the expected outcomes in terms of improved health for patients. This approach often involves the development of indicators which can be used routinely to assess whether services match required standards.

The recent development of systems to monitor performance in high-income countries are interesting because they show how much attempts to assess the quality of care have been dominated by the type of evaluation which emphasizes standards of health service delivery, and how much it is influenced by biomedical and bureaucratic criteria. There has been an enormous expansion of activity in monitoring or 'auditing' the quality of care, and this has taken the form of a huge range of different initiatives. Many of these are relatively small-scale local activities, based in particular units or groups of facilities. Here we concentrate on national-scale attempts to evaluate and compare performance of different parts of the health service. The following sections discuss examples of performance measurement in the United States and Britain and then review the potential and the limitations of such methods, including the scope they offer for incorporation of patients' views.

Performance review in the United States

Formal methods of quality assurance which are national in scope, focused particularly on standards of care, have been established in the United States since the 1970s. As is typical of other aspects of the US health system, the mechanisms for monitoring health service standards are diverse, operating in the private and public sector, and they are not well integrated. The most developed system is aimed at evaluation of services provided under the Medicare state programme for the population over 65 (explained in Chapter 5) (Lohr 1990; Palmer *et al.* 1991). This is conducted by 54 state-organized, independent Peer Review Organizations (PROs) contracted to the central Health Standards and Quality Bureau, part of the administration which finances Medicare. Peer review is based on the principle of evaluation by colleagues. The members of PRO committees are therefore doctors like those they are assessing, and local doctors representing all the major specialties must be members of each PRO. There are a number of possible strategies for peer review: Grol (1994) explains that it can be used at the local level to assess a range of different types of service, employing various methods including consensus development, observation, industrial-style quality circles, and small-group education. However, the function of the PROs in the USA is to carry out systematic evaluation of performance, or 'audit', and they are particularly focused on hospital services.

In addition to the PROs, there are a number of other bodies responsible for performance review in health services. The Joint Commission on the Accreditation of Healthcare Organizations is an independent organization which assesses standards in many types of provider institution, especially hospitals. Accreditation is mandatory for hospitals wishing to take Medicare or Medicaid patients. There are also organizations specializing in the assessment of ambulatory and primary care, such as the Accreditation Association for Ambulatory Health Care, American Association of Preferred Provider Organizations, and the National Committee for Quality Assurance. State agencies are also required to monitor the Medicaid services (see Chapter 5) which they contract. Thus the USA has a plethora of private and public mechanisms for quality assurance which to some extent compete with each other.

Although they are made up of local doctors in each state, the approach of the PROs is quite closely controlled from the centre by their contract. Wareham (1994a) describes how the role of the PROs has evolved as contracts have been renewed and revised. Between 1984 and 1986 they concentrated on reducing costs through cutting unnecessary utilization. Since 1986 the contracts have emphasized monitoring of standards of care to a greater degree, although they still tend to focus on 'picking out the bad apples' by identifying cases of malpractice affecting individual patients, rather than trying to improve the generality of practice. The PROs began to make use of 'generic' quality measures which could be used to check the quality of the health service process in most specialties. These included discharge planning, medical stability of patient at discharge, unexpected deaths, nosocomial infections, unscheduled readmission to hospital, and trauma in hospital. The measures were restricted to those aspects which could be readily quantified for comparison, and based on data which could be collected routinely. The National Institute of Medicine was recently asked to examine the work of the PROs. Their recommendations helped to shift the emphasis of PROs towards quality assurance to improve mainstream care, rather than picking up malpractice. A uniform clinical data set was proposed to enable assessment of the broad pattern of care provision, rather than case-by-case assessment (Jencks and Wilensky 1992) .

While the PROs focus especially on hospital institutions, performance indicators have also been produced for ambulatory care. In the independent sector, the National Committee for Quality Assurance has launched an information package, the Health Plan Employer Data and Information Set (HEDIS), to enable employers (who pay the health plan premiums on behalf of their employees) to judge the value for money offered by alternative Health Maintenance Organizations (see discussion of HMOs in Chapter 5) (Wareham 1994b). It includes indicators of a range of aspects of HMOs including 'quality', 'access', and 'patient satisfaction'. The quality measures cover preventive services (immunization of children and adult health screening), prenatal care (frequency of low birth weight and of early prenatal care), and management of illness (asthma inpatient hospital admission, retinal examination of diabetic patients, and follow-up of patients after hospitalization for severe mental disorders).

The Clinton proposals for reform (see Chapter 5) also placed a good deal of emphasis on the need for a more co-ordinated and public system of assessment of quality (Boufford 1994). One of the suggestions was for a 'quality report card' for health insurance plans, which would carry information on the services covered and also information on standards of care and outcomes. Thus we see a trend towards trying to monitor local variation in health service quality in terms of a limited range of quantifiable comparative indicators, available in a similar format for a large number of providers.

Health services audit in Britain

In Britain there has been a corresponding growth of interest in quality assurance ('audit') often inspired by methods already in use in the USA. The more integrated and comprehensive nature of the British NHS makes it easier in some ways to identify organizational units on which audit might focus, and to find information sources which are suitable for comparison. Already during the 1980s, in line with the new emphasis on managerial approaches to public services, the government was beginning to develop performance indicators applied to District Health Authorities. In 1983 'Health Service Indicators' (HSI) were developed by a team at Birmingham University. They were based on routinely collected information available for all districts, and were accompanied by careful notes on how to interpret the data which warned against simplistic assessments of differences in single indicators (Yates and Davidge 1984; Lowry 1988). They initially included indicators of hospital activity such as waiting lists, numbers of patients, length of stay and patient throughput, attendances as outpatients, and data on the socioeconomic profile of the district population. District Health Authorities were required to publish information on their performance on these indicators, and how it compared with the situation which was typical nationally. By

1990 hundreds of different sorts of information were included in a much expanded HSI data set, although by 1994 the number of indicators had been reduced and some items were rather controversially excluded, such as socio-economic details of the population (Crail 1994). These rapid changes in the composition of the HSI suggest that the history of use of performance indicators in Britain is more closely linked to variations and shifts of emphasis in government policy than to a clear rationale concerning the relevance and value of particular indicators.

However, central government objectives are not the only sources of demand for performance indicators in the British NHS. Shaw (1993) describes how at the end of the 1980s reforms of the NHS led to an even greater emphasis on consumerism and on managerial and market-based approaches (see also Chapters 5 and 8). The internal market produced a demand for audit methods among purchasers needing criteria by which to choose between providers. The clinical professions, wishing to keep control over direction of the clinical sector, have taken various initiatives to ensure that their preferred approaches are adhered to. Thus various parts of the health service bureaucracy have taken a part in developing audit, and Britain has seen a proliferation of the number of organizations involved in quality assessment in health services similar to that observed in the United States. In 1990 the Audit Commission extended its role in monitoring value for money in the public sector to include the NHS (Foster *et al.* 1994). The British Standards Institute also has criteria by which some aspects of the standards of services can be assessed. The government's NHS reforms required the establishment of medical audit committees at the national and regional levels, and locally for individual hospitals. The government accepted the view of the medical profession that audit of clinical activity should be undertaken through peer review; the professional colleges and independent associations representing medical and allied professions have issued guidelines and have overseen national audit studies on specific topics, such as stillbirths and infant deaths and perioperative deaths (Department of Health 1990; Campling *et al.* 1990). As the creation of semi-autonomous provider units has gathered pace in the health service (see Chapter 5), it becomes more difficult for District Health Authorities to control directly standards of care in hospitals they are no longer managing. A number of schemes have also begun to develop for accreditation of health service providers, especially hospitals (Shaw and Brooks 1991).

As we have seen, District Health Authorities had been called upon to publish information on the local situation in their district using the HSI data. Now the focus is at an even finer level of resolution. In 1994, for the first time, 'league tables' of individual hospitals were published, showing information on 23 indicators. Reflecting the government's preoccupation with reducing waiting times for hospital treatment, 16 of the league table indicators related to waiting times in various types of elective surgery and the remainder included the length of time patients wait to see a doctor in outpatient and accident and emergency, and the amount of day surgery activity in the hospital (Guardian 1994a). The tables included a 'star rating' system intended to facilitate comparison between hospitals. The government claims this as part of its drive towards consumerist goals for the reformed NHS, as set out, for example, in the 'Patient's Charter' (Department of Health 1992) (discussed in Chapter 8).

Potential and limitations of performance indicators

These developments in 'performance indicators' illustrate a clear trend towards treating quality assessment not just as a source of information for providers but also as a means by which information on health service facilities could be disseminated to users. The publication of 'performance indicators', intended to summarize key aspects of the quality of care in a fairly digestible comparative format, is consistent with the ideal of operation of market forces in hospital services. Consumers (individuals, health authorities, or employers purchasing health services or insurance cover) need information on the different services available in the market in order to select those which meet their needs in the most economic way.

However, in spite of these arguments in favour of such published performance indicators, they are in fact controversial, given questions over the limitations of the information which is made available and the possibility that it may be incorrectly interpreted. For example, there was a considerable outcry in the USA when the PROs began to publish information about 'outcome' indicators for particular institutions including mortality rates (Dubois *et al.* 1987; Greenfield 1988). Indicators of mortality are particularly striking measures of the outcome of care in an institution such as a hospital and, because they produce such an emotive response, they attract a good deal of attention. In fact, the problems associated with using mortality indicators to measure hospital performance can also affect most performance indicators if they are viewed in isolation, without considering other information which is relevant to their interpretation.

The problems which can arise include, first and foremost, the fact that it is difficult to adjust performance indicators of this type to allow for the different risks faced by service providers due to the nature of the patients they care for. An institution which typically receives patients in a severe state of illness, or who are for one reason or another particularly difficult to treat, may inevitably have a relatively high rate of mortality amongst its patients compared with another hospital dealing with less severe and 'difficult' cases. In such a situation, differences in risk of death may have little to do with quality of care. Similar issues of case mix and severity can also affect indicators which seem less 'dramatic' than mortality rates. For example, length of stay for hospital inpatients may depend on the severity of their condition and their home circumstances: a hospital treating large numbers of poor patients, or elderly people who live alone, may find that they need to spend longer in hospital because their home circumstances are not adequate for them to be discharged earlier. The performance indicator for primary health services mentioned above, based on frequency of low birth weight, might be associated not only with the quality of prenatal care but also with the socio-economic and ethnic profile of the practice population. It is not reasonable to make unfavourable judgements of providers if their lower performance on a particular indicator has more to do with the clients they serve than with the services they are providing. Furthermore, too much emphasis on performance indicators which are not adjusted for risk might produce the perverse incentive of encouraging providers not to take on the burden of care for the most challenging cases. Given what we have already established about socio-economic inequalities in health and access to health services, it is evident that this would particularly disadvantage already under-privileged populations.

Another difficulty with the type of performance indicators described above is that they are based on a limited range of data, so they only provide a partial view of the quality of care. Typically, the indicators which are chosen are those which can readily be quantified from routinely collected data; for example, mortality data are quite often used to assess health outcomes, but information on quality of life for surviving patients is unlikely to be available in a general and comparable form suitable for performance indicators. 'Soft' information on factors such as the understanding and rapport between carers and patients is much more difficult to incorporate into such a scheme, although such factors may be very important, especially to patients. Furthermore, routine statistics are not necessarily complete, reliable, and designed so as to give the best possible measures of the dimensions being assessed. Some of these difficulties are well exemplified in the British experience of development of systems for quality assurance.

Various authors (Baghust 1994; Frankel and West 1993; Sheldon 1994) have, for example, criticized the emphasis on waiting times in the NHS audit criteria. The data on waiting times relates to a relatively small proportion of the total work of hospitals and does not allow for severity or the benefit likely to be derived from treatment received. A large part of waiting lists may consist of people waiting for inappropriate treatments, so that waiting lists do not distinguish between inefficiency and inappropriate demand. Nor do they allow for the fact that, even if treatment is appropriate, some of those waiting need treatment more urgently than others and some will benefit more than others from the treatment they eventually receive. Furthermore, waiting lists may offer perverse incentives in encouraging providers to 'play the system'; for example, there has been criticism of waiting lists for failing to take account of time between referral by the GP and first consultation with a hospital specialist, and a suspicion that consultants have been delaying first appointments in order to avoid long waiting lists for treatment in their departments.

There seems to be increasing emphasis, in the information which is published for British and American consumers, on 'amenity' aspects of quality (Sheldon 1994), so patients may be influenced by the 'wrong' information. It is argued that performance indicators should instead provide more understandable information on effectiveness. As we have suggested here, this is much more challenging. One indicator which has been seriously considered in Britain as a measure of outcome of hospital care is the rate of unplanned readmission, which could be taken to be indicative of unsuccessful outcome in the initial treatment received. An indicator of readmission could be calculated for District Health Authorities (e.g. Chambers and Clarke 1990) using data which is collected routinely. However, it has been shown in practice that this is not likely to be reliable as a measure of quality of care. Clarke (1990), for example, showed that, when assessed by a panel of experts, less than half of readmissions were judged to be 'avoidable' and associated with the quality of treatment. This, once again, illustrates the limitations which often apply to apparently simple measures of performance.

The consumer's viewpoint on performance

Although professional ideas about the quality of care tend to dominate the formulation of performance measures, there has been some development in methods for assessing the view of health service consumers. Most of these are a form of market research: consumer satisfaction surveys adapted for health services, with emphasis on patients' views of the non-medical aspects of the care which they receive (waiting times, courtesy of staff, quality of meals, and 'hotel' services). These methods can be criticized as being still too much determined by professionals and managers and not sufficiently revealing of the patient's view of the quality of services they receive (Heyden 1993).

Wedderburn-Tate *et al.* (1995) report on a national survey of over 5,000 hospital patients treated in 36 randomly selected hospitals in Britain. The method was originally developed at Harvard Medical School and has been applied in both the USA and Canada, as well as in the UK. The method is interesting because it uses various strategies to try to get closer to patients' real experiences. For example, questions were based on issues which patients themselves had originally identified as important, including preadmission, admission, and discharge, communication with hospital staff, physical care, tests and operations, the help provided, and management of pain. Patients were asked how they were treated, not whether they were satisfied with their treatment, since patients are thought to be very reluctant in general to be critical. The surveys were carried out by trained interview staff at home after the treatment, since it was felt that this would be more likely to encourage a full and unconstrained response by patients. While it may suffer from not being specifically adapted to local conditions, this approach at least allows hospitals to compare their own institution with others. However, even well-considered surveys of this type only permit consumers to express their views within limits determined by the methodology, and the choice of instrument for evaluation is made by the service providers and administrators, not the patients. Chapter 8 devotes more detailed discussion to the problem of how to ensure real public participation.

Evaluating progress towards targets for population health gain

The recent development of techniques for assessment of health service outcomes reflects the growing concern of health policy-makers in many countries to evaluate the results of health service expenditure in terms of *health gain* in the population generally. This requires a rather different perspective from that of the clinical trial approach, focused on outcomes for individual patients or groups of patients who are treated by (usually curative) health services. It certainly cannot be assessed only in terms of the quantity of health services delivered (although rates of use of services known to be effective in improving health, including some health-promoting and illness-preventing interventions, might be relevant as surrogate indicators). We consider the wider implications of this shift of emphasis in more detail in the following chapters, which discuss the professional, social, and political reorientations which are necessary in order to achieve real improvements in population health. This focus on health gain has also had a major impact on approaches to measurement of health service outcomes, which we consider here.

We find, for example, a growing preoccupation with assessment of population health trends and comparison of changes in health with targets for health improvement set by health policies. This does not usually aim to measure the health outcomes of specific interventions, but rather to identify the combined effects of health services and other factors on population health. In some cases, these may result in levels of health which are poorer than the average, or significantly below the targeted level. Such populations are often seen as priority groups for concentration of resources and for attempts to make interventions more productive of health improvement.

This requirement to be able to assess trends in health of the population in comparison with policy objectives raises a number of technical and conceptual problems in terms of outcome measurement. Three of these issues are discussed here. As we saw in Chapter 4, there is debate over the extent to which trends in health can be attributed to health services. There is also a growing realization that existing sources of information are insufficient to assess trends in the aspects of population health status which are important and relevant to policies emphasizing health gain, so new information must be generated. In addition, the problem of scale of analysis is important to the interpretation of trends.

How far is health difference attributable to health services?

The problem of attributing measurable variation in population health to health services is well illustrated by the debate over the proposition that a proportion of illness and death is due to diseases which may be amenable to medical intervention (Rutstein *et al.* 1976, 1980), while other causes of death are largely due to other factors and cannot be reduced much by provision of good health services. Following this idea of amenable mortality Charlton *et al.* (1983) suggested that it should be possible to examine geographical variation in *avoidable mortality*. Avoidable mortality indicators are derived from data on cause-specific mortality due to conditions which, according to the authors, are amenable to health service intervention (particularly curative services). Charlton *et al.* selected 'avoidable' causes of death (see Table 6.6) from a longer list proposed by Rutstein *et al.* Charlton *et al* showed that there was variation between area health authorities in mortality due to these conditions, even when social conditions were controlled for. They argued that this variation might in part reflect different health service inputs in different areas and 'provide early warning signals of possible short comings in health care delivery', although they also warned that the performance of health services in an area should not be judged solely on the basis of differences in 'avoidable mortality'. They suggested that avoidable mortality might suggest areas towards which extra resources should be targeted: 'it could be argued that if mortality were avoidable either the heath services should be expanded in areas of higher disease incidence or social conditions giving rise to diseases should be addressed' (Charlton *et al.* 1983: 696).

A number of studies have employed the idea of 'avoidable' or 'amenable' mortality (see e.g. the study by Boys *et al.* (1991) discussed in Chapter 4). However, there has also been a lively debate over the extent to which avoidable mortality can really be used as a measure of effectiveness of health services. In their study of avoidable mortality in Belgium, Humblett *et al.* (1986) point out that, while effectiveness of curative services may influence

Table 6.6 Causes of death considered amenable to medical intervention

Hypertensive disease
Cancer of cervix
Pneumonia and bronchitis
Tuberculosis
Asthma
Chronic rheumatic heart disease
Acute respiratory infection
Bacterial infections
Hodgkin's disease
Abdominal hernia
Acute and chronic cholecystitis
Appendicitis
Maternal deaths
Deficiency anaemia
Perinatal deaths

Source: Charlton *et al.* (1983).

the lethality of illnesses, mortality also depends on the incidence and sever-ity of cases. Incidence and severity of illness in the population is more strongly associated with health behaviour, social and physical environment, and primary prevention. They question whether curative services have much independent effect on mortality when these conditions are controlled for. Carr-Hill (1994) also concludes that socio-economic conditions are the more important in determining mortality, even for conditions for which treatment is available, so that it is not reasonable to attribute mortality vari-ations directly to variations in health services. In an analysis of the factors associated with trends in population health it is very difficult, if not impos-sible, to distinguish effects which are specific to health services, as distinct from other influences. This is one reason why health policies which focus on health gain present such a major challenge to conventional health services. All organizations need to be able to identify clear goals and measure their performance with respect to their aims; but in the case of population health gain, it is very difficult for health services to establish how much they are achieving directly through their own activity.

The need for better measures of health gain and risk factors

In order to have a reasonably comprehensive impression of health outcomes in a population, we also need a much wider repertoire of information than is typically available at present. The ideal requirement is for data on a range of measurable health outputs; morbidity and mortality; positive health indicators; subjective perceptions; functional capacity and coping ability (Raeburn and Rootman 1989). A number of alternative indicators have been suggested as measures of health outcome in the general population. Routinely available mortality and morbidity indicators are commonly used. In addition, measures may be employed which demand data which can only be collected in population surveys. Some of these are designed to give a more accurate impression of health status in terms of the impact of morbidity; for example, measures of disability-adjusted life years and life expectancy without disability (discussed in Chapter 4). The quality of life measures discussed above can also offer information of this type. Measures of fitness and biometric indicators are also included regularly in some national population health surveys. Fig. 6.4 shows examples of information from the *Health Survey for England* (Breeze *et al.* 1994), illustrating the types of data which are now being collected in a regular survey of health of the British population and which are particularly intended to assess prevalence of heart disease and of risk factors associated with cardiovascular illness. This type of measure may provide more useful data on differences in morbidity and health than mortality data, especially as more people survive with conditions such as coronary heart disease. The indicators include both physical and mental health measures of characteristics such as obesity, high cholesterol, and stress levels which are risks for heart disease. Although the emphasis is on prevalence of factors which increase the risk of poor health, at least these indicators include 'positive' aspects of fitness or low risk as well as more 'negative' high-risk dimensions. In addition to biometric indicators, they provide data on health-related behaviour which is relevant to the objectives of services for health promotion as well as medical treatment of people who are ill.

The *Health Survey for England* demonstrates how the development of methods for monitoring population health outcomes is closely linked to national health policy development. The government has set out in its health policy document, *The Health of the Nation*, priorities for improvement of population health in England which comprise five aspects of health in particular: cardiovascular disease; cancer; sexual health, HIV, and AIDS; mental health; and accidents (Great Britain, Parliament 1992). These conditions were selected as priorities because they were important causes of ill health or death, and because they were conditions for which action could be taken through the Health Service and for which change was measurable. *The Health of the Nation* sets targets for the national population with respect to these five priorities, and also for change in risk factors associated with these conditions. The latter include specified reductions in consumption of tobacco, alcohol, and fat, in obesity, in average blood pressure, and in shared use of injecting equipment by intravenous drug-users.

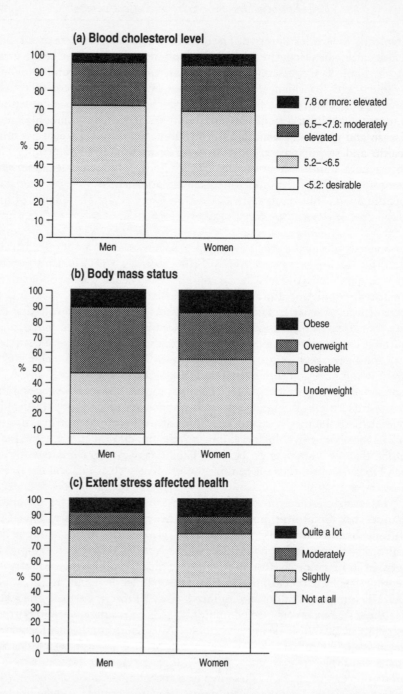

Fig. 6.4 Measures of health and fitness relevant to cardiovascular health from the *Health Survey for England,* 1991–2. (*Source:* Breeze *et al.* 1994.)

In other countries the type of information being collected in population health surveys is also geared to the requirements of policies for health gain. For example, in Quebec, the provincial government's 'policy on health and well-being' (Gouvernement du Québec 1992) identifies nineteen objectives for health improvement, relating to: social adjustment (including sexual abuse, violence, and delinquency); physical health (including cardiovascular disease and cancer); 'public health' (including infectious diseases); mental health; and social integration of the elderly and of disabled people. The government-sponsored survey organization Santé Québec carries out population surveys which provide information relating to these objectives (Santé Québec 1988, 1994).

The importance of scale for measures of health gain

Information based on representative samples of national or regional populations is designed to provide data on the trends in health of the population of the country considered as a whole. In countries such as England, targets for health gain have been set in terms of improvement to the average health of the whole national population, but are less clear about aims for achieving a reduction in inequalities in health among different groups in the population (see Chapter 9). Studies of subregional geographical variations in health trends demonstrate that an improvement in general average mortality is not necessarily accompanied by reduced local inequality.

The trends in reduction in coronary heart disease (CHD) mortality observed in several of the wealthier nations of the world during the 1970s and 1980s provide an interesting illustration. Bryce *et al.* (see Fig. 4.10) (1994) point to the tendency in England towards growing geographical inequality in CHD death rates for some age groups, which was associated with the average reduction in mortality. They suggest that health gain targets set with respect to *equity* (reduction of area differences) as well as *average* improvement across the whole national population would have been desirable. Otherwise there is a risk that national targets for health gain will be achieved mainly through gains by the most privileged and healthiest groups in the population, with relatively little improvement for the more deprived, least healthy groups. National sample surveys such as the *Health Survey for England*, described above, do not provide sufficient detail to allow the monitoring of local variations in trends in health behaviour across the whole country. In contrast, the surveys carried out by Santé Québec (admittedly for a smaller total population) are designed specifically to provide sufficient respondents in different regions of the province to make estimates of local conditions and to develop local health targets.

There is therefore an important link between the conceptualization of health gain and the types of indicator which are used to measure health change in the population. If the range of measures being used to assess health gain is too restricted, it may give an incomplete and misleading impression of the progress being made. Furthermore, target setting and assessment of outcomes cannot always be dictated centrally, but should be sensitive to local conditions and the patterns of health status and health trends in different parts of the country.

The politics of measuring health service outcomes

It is interesting to reflect on these developments in performance review, medical audit, and outcome measurement in terms of what they tell us about the political relationships in health systems of countries such as the United States and Britain. Pollitt (1993), for example, points out that the hierarchy of influence in biomedicine is clearly reflected in the methods used to evaluate services.

In both the USA and the UK, governments have accepted that medical audit should be carried out through peer review, which means that senior doctors, the most powerful actors in the health system, retain control over the process. This is justified in terms of the perceived need to preserve clinical freedom. (Sheldon (1994: 42) refers to 'the sentimental, though largely unethical, attachment to the notion of clinical autonomy'.) This means that the only part of the service which is considered legitimate ground for assessment by others in the system (nurses, managers, and patients) is that part which relates to non-medical aspects of care, so that it is impossible to carry out a real evaluation of 'total quality' and it is difficult to allow the views of the public to influence the development of medical practice.

Pollitt also points out that, while in the USA there is at least a system of external peer review, in Britain medical audit is internal, local, voluntary, non-judgemental, and confidential. Thus it is often ineffective in leading to real changes in the quality of care. It appears that medical audit in Britain may also reflect the hierarchical structure of British biomedicine quite strongly. A qualitative survey of doctors' views of audit showed that in several cases it was seen to be a way of checking the performance of junior medical staff (Black and Thompson 1993), rather than aiming to improve the work of consultants. It would be wrong to suggest that many doctors are not genuinely motivated to improve the quality of the care they provide, but it is also evident that several aspects of the way in which quality assurance has developed, at least in Britain, illustrates biomedicine seeking to maintain its dominant position in the health service and resisting attempts at change imposed from the outside.

The task of monitoring the nursing care of patients and the 'amenity' aspects of the service is left to the less dominant professions such as nurses and managers. For nurses there were some advantages in taking on this type of activity, since it stressed their independence from doctors, However, Pollitt suggests that, in adopting managerial perspectives on quality of care (e.g. amenity aspects), nurses may be undermining their position in relation to management and laying themselves open to having their activities in caring for patients determined by managerial rather than professional caring criteria.

Health service audit tends to be carried out using a perspective which is often determined by the professionals rather than by the patients. Pollitt (1993: 163) comments: 'This would seem strange to some quality experts from the private sector, where quality is not infrequently defined as "fitness for purpose" and it is explicitly understood that the purpose is that of the consumer, not the provider.' If we consider the methods of evaluation of health gain currently proposed in Britain, we find that they reflect quite strongly the priorities of central government. They tend to emphasize, for example, national targets and limited notions of individual consumerism, and are rather insensitive to local variation in needs and to issues such as inequalities in health, which have not been popular with the Conservative administration.

Equally, we find that in the United States, for example, approaches to assessing the quality of services have been framed in such a way as to exclude the issue of access, leaving aside the problems of millions of Americans for whom services are not available (see Chapter 5). Attempts by the Clinton administration to address this issue were blocked partly because of concern over possible reductions in the quality of care for those who *did* have health cover, given that resources for health are finite.

All in all, we find that various aspects of the power relations within the British and USA health services are being reproduced in the system of quality assurance. Hierarchies of health service administration, professional hierarchies, the subordination of patients' interests to those of health professionals, and the subordination of the interests of poor patients to those of the rich all seem to be difficult moulds to break!

Clearly, given such a limited interpretation of quality assurance, the scope for change is quite restricted. Several critics of the process in Britain have commented that the methods and the indicators currently used are unlikely to have much effect in improving quality, partly because the assessment methods and indicators being used are not appropriate, and also because the method of quality assessment does not enable the lessons of evaluations to be fed back into the system to inform better practice (Black 1992; Smith 1993; Buxton 1994; Sheldon 1994). Although it is true that some progress has been made in developing new perspectives on service quality and outcomes, societies in high-income countries like Britain and the United States still have a long road to travel before health services can be considered really effective in responding to the needs of the public which pays for them and makes use of them.

The agenda for research on territorial justice

This chapter opened with a discussion of territorial justice and of how the debate prompted by the idea of the inverse care law helped to focus attention on the issue of whether local resource distribution was proportional to relative needs. This debate has raised a number of issues which are very difficult to resolve, largely because of the limited progress which has been made in the area of health service evaluation and measurement of outcomes for population health. Researchers are still far from being able to answer the apparently simple questions: What constitutes relative need in different parts of a country? How do we measure the resources provided to a local population? What are the outcomes of these resources in terms of access to health services or impact on population health? All of these questions still represent a major challenge to research, and they represent part of the agenda for continuing work in this field.

On the other hand, some advances have been made in our thinking about how to assess the health and health care of populations, which may inform policies and interventions aimed at achieving health gain. These policies are the subject of the following chapters, which examine in more depth the changing perspectives necessary in order to improve public health.

|7|

Action for health gain: the agendas set by public health models

The previous chapter highlighted the debate over whether health resources are used in an efficient, effective, and equitable manner, and especially whether deployment of resources for health services produces the desired outcomes in terms of health gain in the population. We also saw in Chapter 4 that profound and persistent inequalities in health of populations are evident at all geographical scales, from the international to the local level, and that these are associated with differences in socio-economic development and living conditions. Conventional medical services certainly have a role to play in the production of health gain, but other factors operating outside the health services also seem to be very significant in creating conditions conducive to health improvement. This seems to be true in many different countries even though, as discussed in Chapter 5, they may have widely differing arrangements for organizing, funding, and delivering health services. Conventional biomedicine, with its focus on treatment of illness in individual cases, can only indirectly address the problem of promoting action to produce good health in whole populations. In this chapter, we consider the impact of public health approaches, which take as their starting-point the question of how to achieve health gain in populations, and which therefore stress action that would traditionally have seemed beyond the domain of health services.

The 'new' public health movement

We have already commented on the increasingly vocal and widespread questioning of the dominance of biomedicine over health policy and health-related intervention. This has been associated with the realization that basic problems of public health are persisting even in countries which consider themselves advanced in social and economic terms, which had undergone the epidemiological transition, and had invested in complex and expensive national health systems of the types described in Chapter 5. Chapter 4 showed how, even in high-income countries such as the United States, some groups in the population are suffering morbidity and mortality of a type and level more typical of the middle- or low-income countries. These problems stem from similar factors in all parts of the world: low income, inadequate diet, poor housing, environmental pollution, and lack of access to health care. The examples considered in earlier chapters show that populations suffering most from these problems tend to be concentrated in deprived areas within our major modern cities, as well as in poor rural areas which might be considered less economically and socially advanced, so that inequalities in health and health care can be viewed as part of the more general problem of uneven development.

This body of evidence on health inequalities and deprivation has prompted a change in perspective on health and health care over the last couple of decades which is often characterized as the 'new' public health movement. In some respects this movement is no more new than the long-standing problems which it is trying to address. It campaigns for approaches which go 'back to the future' (*Lancet* 1988), taking up again the fundamental objectives of public health and policies for sanitary reform which were being stressed in previous centuries, especially in the nineteenth century when the growth of major industrial cities began to present new challenges in terms of health and health care. Ashton (1992) reminds us of the formulation in the nineteenth century of the 'Sanitary Idea' – that collective action was necessary in the newly expanding industrial cities to tackle problems of overcrowded housing conditions which increased the risk of spread of infectious disease, the lack of sanitary facilities for provision of safe water and food or disposal of sewage, the poverty which made it impossible for large numbers of people to have adequate nutrition, clothing, or medical attention when they became ill, inadequate education which led to uninformed health-related behaviour and reduced the earning potential of the population, and the dangers of working conditions which resulted in extreme fatigue and exposure to hazards. Evidence of these problems was compiled by nineteenth century campaigners for reform in Britain (e.g. Chadwick 1842 (1965); Booth 1892) and also in America (Griscomb 1850; Shattuck 1850) and in other countries where large industrial cities were beginning to develop.

In Britain the movement for sanitary reform in the nineteenth and early twentieth centuries was led by predominantly middle-class activists: doctors, clergymen, journalists, politicians, and demographers. Campaigners such as Edwin Chadwick, Charles Booth, Octavia Hill, Beatrice Webb, Seebohm Rowntree, and William Farr collected and published information on poor health in the impoverished populations of cities of the time which helped to sway the opinion of influential groups in British society. While their efforts produced major achievements in the form of improvements in public health in major cities, these reformers did not, on the whole, seek to change the social and economic structures which underlay the health problems which concerned them. Indeed, they were more often motivated by a wish to reduce the worst degradations of urban life before these became so severe and widespread that they posed a threat to the stability of the existing social structures. The approaches which were adopted to improve the lot of poor working-class people were frequently very paternalistic, imposing the middle-class norms of social behaviour prevailing at the time. Thus, for example, Jones (1994) discusses the moral pressures to conform to middle-class notions of morality and motherhood which underlay efforts to improve conditions for working-class women in Edwardian Britain. Furthermore, capitalism was seen as the route to salvation, with many schemes for improved housing and urban infrastructure based on private investment with the expectation of a dividend for investors, described by some commentators as 'philanthropy at 5 per cent' (Mann 1953; Fishman 1982).

Recognition of the need for collective action resulted in the setting up of organizations such as the Health of Towns Association, founded in England in 1844, which provided a basis for co-ordinated action to improve environmental and sanitary facilities. Accounts by authors of the development of welfare services (Lewis 1952; Briggs 1961; Fraser 1973; Chave 1987) show that much of the resulting action was spearheaded through municipal agencies. Thus during the nineteenth century in England, and elsewhere, public health was essentially a function of local government, rather than an initiative originating from the medical sector. This pattern was repeated in other major cities in Europe and the United States (Leavitt and Numbers 1978; Knox 1983). The key elements of the Sanitary Idea are summarized by Ashton (1992: 3):

- the legitimacy of working locally
- resourcefulness and pragmatism
- humanitarianism and a strong moral tone
- recognition of the need for special skills and qualifications
- appropriate research and enquiry
- the need to focus on positive health
- the value of reports on population health
- populism
- health advocacy
- the need to work persistently and adapt to current trends
- the need to organize
- the importance of making public health the responsibility of a democratically accountable body

Thus the work of the original sanitary reformers was typified by collective, pragmatic efforts towards local action which would improve the health of the population generally. This was informed by widely disseminated information on population health and by professionals in health and health care, but it also required a wider involvement of other members of the local community (at least those from the more privileged and influential classes) who did not necessarily have medical training.

In Canada in 1974, the issues emphasized by the original public health movement were taken up again in a report by Lalonde (1974) entitled *A New Perspective on the Health of Canadians*. This advocated a community-based perspective on health, stressing health promotion and the importance of public health. It set a trend for an approach to contemporary health problems of the late twentieth century which was based on the type of model of social epidemiology proposed by Scott-Samuel (1989). This model proposes a chain of 'causation' linking social and economic processes with disease causation. For example, the demands of the contemporary economy for concentrations of low-paid labour in urban areas leads to construction of cheap, poor-quality housing to accommodate this workforce. This housing is occupied by poor populations living in overcrowded conditions with inadequate sanitation. These circumstances (also associated with poor nutrition and lack of health care) facilitate the spread of infectious disease. The intervention which is typical of conventional biomedicine occurs mainly at the last stage of this process, once the population have become ill (i.e. treatment of infectious disease once it develops). At best it might involve preventive measures such as inoculation. Scott-Samuel argues that the public health approach tries to intervene at an earlier stage, to act on the conditions which produce illness and prevent it occurring. This might involve action to incorporate health considerations into economic policy governing employment, pay, and working conditions and environmental health policy concerned with housing and sanitation regulations, enforced through appropriate public health legislation.

Similar themes have been taken up in the US context. For example, Wallace (1990) links poor public health in New York City to reductions of municipal services during the 1970s and associated degeneration of the conditions of housing and urban infrastructure in the inner city. The Institute of Medicine (1988) reported on the condition of public health provision in the country and found that, in a nation where curative medical care takes most of the health care resources available and where public funding for local health programmes is limited and determined strongly by local political processes, public health was given low priority and was unco-ordinated and ineffective. The report recommended that local population health assessments should be carried out and that public health work should take a more proactive stance, to develop health policy and ensure that essential basic health care was provided. It also commented on the sometimes damaging effects of competitive strategies in planning and funding health care, and suggested that more co-operation and collaboration would be beneficial for public health (Flynn 1992, 1993).

In low-income countries, although the context is very different and poverty is often even more severe and certainly more widespread, similar chains of association apply. Harpham (1992) points to the problems beset-ting efforts to ameliorate environmental health in poor urban areas of such countries. These include overdependence on curative health services, lack of secure housing tenure, lack of access to land to grow food, inadequate phys-ical infrastructure in the city, and the diversity of urban communities and welfare agencies, which makes it difficult to co-ordinate strategies for improvement. Paolisso and Leslie (1995) discuss the changing health needs of women in low-income countries, and underline the requirement for com-prehensive approaches which take into consideration a broad range of bio-social determinants of women's health, and are not limited only to issues of maternal and reproductive health. They also point to the need for interdis-ciplinary perspectives to provide the necessary information and the scope for innovative interventions.

Perspectives on public health from disease ecology

There are clear parallels between the holistic type of approach proposed by the 'new' public health movement and analysis in human disease ecology, which invokes a tripartite model of the factors relevant to disease (see Fig. 7.1), stressing the links between habitat, population, and behaviour (Learmonth 1988; Meade *et al.* 1988). *Habitat* includes aspects of the environment in which people live, including the built environment (hous-ing, workplaces, communication systems), the biotic and physical condi-tions (flora, fauna, climate, topography), and the services and economic and political structures of particular societies and local communities. *Population* factors include the characteristics of the population (age struc-ture, gender, genetic predisposition) which help to determine health status

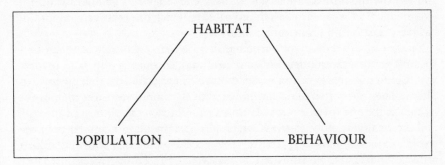

Fig 7.1 The triangle of human disease ecology. (*Source:* Meade *et al.* 1988: 32–42.)

and susceptibility to disease. *Behavioural* factors include health-related beliefs and behaviours, which, as we have seen in earlier chapters, are constrained by a range of social and economic factors in each society and local community. The triangular relationship between these factors (see Fig 7.1) is meant to suggest that there are no real boundaries between them and that population health is a result of the interaction of all three types of factor. It also indicates that action on one part of the system in isolation (e.g. introduction of vaccines to protect the population) is unlikely to be effective without complementary action on other relevant factors (e.g. information campaigns to encourage people to be vaccinated and provision of accessible and affordable services to facilitate a vaccination programme).

The disease ecology model originated in the work of the authors such as Jusatz and Rodenwaldt, who developed the idea of 'landscape epidemiology', defined by Meade *et al.* (1988: 72) as 'a geographic delimitation of the territory of a transmitted disease in order to identify cultural pathways for disease control'. Another very influential figure was May, who classified disease 'complexes' in terms of the factors associated with the disease (including the *pathogen*, which causes the disease, the *host*, affected by the disease, *vectors*, which carry the disease, and *geogens* or environmental conditions which influence the transmission process). Learmonth (1988: 8) explains that May presented a challenge to geographers studying patterns of disease: 'that to understand a disease complex it is necessary to study the ecological interactions of all the factors in the complex and therefore to be able to look at the total environment' as it affects both the agents producing the disease and the people affected.

Disease ecology is often applied to studies of communicable diseases particularly prevalent in low-income countries, which are often thought of today as 'tropical diseases'. However, while some of these diseases are genuinely restricted to tropical environments (e.g. onchocerciasis or 'river blindness' in Africa and Latin America), many of them (e.g. malaria, cholera) have in the past been common in more temperate countries which today enjoy high standards of living. Furthermore, Learmonth (1988) shows how the holistic approach of disease ecology can be relevant to, for example, cardiovascular disease and cancers which are relatively important causes of death in high-income countries.

Studies of the ecology of particular diseases can encourage a broad perspective on health and development, and may encourage efforts to involve whole communities in action which will tackle the problems that give rise to illness. They also often underline how relatively simple measures to improve living conditions for poor populations can influence the pattern of disease, and the relatively slow progress which has been made in many parts of the world towards achieving these. The recent cholera pandemic which attracted particular notice when it was detected in Latin America in 1991 demonstrated these points very clearly. Commentators at the time stressed that cholera is primarily 'a disease of slums and poverty', and that a strategy based on immunization of travellers to prevent diffusion of the disease 'has virtually no role in controlling cholera epidemics' (Breheny 1991: 1033). In Latin America, the disease posed such a major threat because of the huge numbers of people without adequate sanitation and safe water supplies (Godlee 1991). This situation reflected the uneven impact of economic and social development which, although it has brought about a reduction in the average levels of mortality due to intestinal infectious diseases in Latin America, allows these diseases to remain common causes of death in the poor population, especially among children (PAHO 1991).

It is salutary to consider that in the 1990s societies are still unable to control very preventable diseases such as cholera. There is also debate over what are the most effective measures to take, given limited resources (Feacham 1986). While curative strategies, particularly oral rehydration therapy, may effectively reduce mortality from the disease, measures to reduce morbidity require political will and economic resources to ensure basic infrastructure such as water and sanitation systems for the entire population. A clear understanding of what aspects of improvement will be most beneficial may also be important. For example, Gorter *et al.* (1991) discuss the literature which has investigated the relative importance of water *availability* and *water quality* in reducing intestinal disease morbidity. Their own study in Nicaragua examined a range of factors which might influence the spread of the disease, and concluded that having a water supply available within a reasonable distance was more important than the quality of the water or whether the household owned a latrine. In addition, mother's level of education and size of family were influential factors.

Environmental change and health strategies

Classic studies of patterns of disease based on perspectives such as landscape epidemiology or May's factor complexes tended to give a rather static view of disease ecology. However, recent concern over the processes which are producing fundamental changes in our environment are forcing disease ecologists to revise their ideas about the spatial patterning of diseases. Many of these processes are associated with various forms of environmental pollution, much of which is man-made. Meade *et al.* (1988) discuss the impact on health of what they term the 'pollution syndrome', including the problems of chemical and radioactive hazards which have directly toxic effects on human populations. The enormous literature on human impact on the environment and the links between environmental hazards and health is reviewed elsewhere (e.g. Rowland and Cooper 1983; Goudie 1993) and is outside the scope of this book to consider in detail. However, we note particularly here the recent emphasis in this literature on processes which produce global environmental change, as opposed to more local impacts around particular pollution sources. Ecological approaches to the study of health of human populations need to be revised in view of the impacts of processes such as global warming produced by the 'greenhouse effect', increasing levels of ultraviolet light associated with thinning of the ozone layer, widespread effects of land degradation and topsoil loss, effects of acid rain, reduction of biodiversity and the impact of modification of the environment associated with urbanization on a massive scale worldwide (see e.g. McMichael 1993; Bentham 1994). The problem of environmental change is not only one of technical and economic adjustment; increasingly, the political, ethical, and philosophical issues associated with the debate over environmental change are being explored (e.g. Marchant 1992; Pepper 1993; Smi 1993) and these have ramifications for the ways we think about the relationships between environment and health in contemporary societies.

Alliances for public health

In order to implement programmes of public health which can act on the full range of factors which are relevant to health requires major changes in the division of responsibilities for health in contemporary societies. Scott-Samuel (1989) was among those who set up a Public Health Alliance in Britain in the late 1980s. This informal organization was intended to demonstrate how it might be possible to bring together key people from a number of sectors, extending far beyond the traditional limits of the medical and paramedical professions. Scott-Samuel argued that such a Public Health Alliance (PHA) would:

- provide a national focus for discussion of health promotion
- help to build up collaboration for public health
- raise resources for research on and publicity of public health issues

The PHA produced a charter which is interesting to us because it underlines the breadth of areas on which one might need to act to improve public health. It is expressed in term of proposed citizens rights for the essentials for good health (Scott-Samuel 1989: 34):

- income (sufficient material means for health)
- homes (safe, warm, dry, affordable)
- food (safe, nourishing, available and cheap)
- transport (safe, accessible, environmentally friendly in terms of fuel consumption and pollution)
- public services (basic sanitary facilities and caring and welfare services)

- education and health promotion (information and confidence to act)
- comprehensive health care (free at point of use, sensitive to needs)
- equal opportunities for health (reduction in differences in ability to max-imize health potential due to age, gender, ethnicity, sexual orientation, etc.)

Such a programme would require collaboration between partners from a number of different sectors to develop community-based health services (Taket 1990). Ottewill and Wall (1990) illustrate the range of partners which might be envisaged in the British context (Fig. 7.2). Perhaps the crucial partnership in the British case is that between the health authorities (responsible for medical services) and the municipal local government authorities which have responsibility for a wide range of non-medical, social, and welfare functions. Social care of people who are too ill and frail to manage alone, or support for their family carers, protection of children whose family or living conditions put their welfare at risk, education provi-sion, social housing, and environmental health services are all examples of

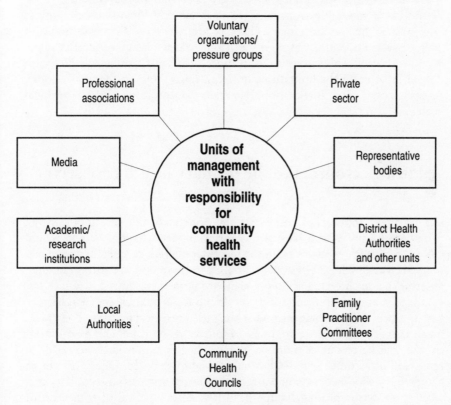

Fig. 7.2 The network of agencies engaged in the provision and development of community health services in England. (*Source:* Ottewill and Wall 1990: 402.)

the activities with relevance to health which take place in the local authority sector. Collaboration between National Health and local government
welfare services is a basic requirement in order to co-ordinate work towards
public health goals, and it is also important for policies which favour the
care of sick people in community settings (Audit Commission 1986).

Other countries have also set up new collaborative structures to aid the
advancement of the 'new' public health. Flynn (1992) reports on the
Assessment Protocol for Excellence in Public Health, a collaborative programme in the USA involving the National Association of County Health
Officials, the Association of Schools of Public Health, the Centers for
Disease Control, and the United States Conference of Local Health Officers.
This programme was set up in 1989 and produced manuals for assessment
of community health status and community participation in public health
policy development. The Protocol was applied in a number of demonstration sites around the country.

In countries concerned to develop co-ordination of public health and primary health care, stress has increasingly been put on the view that local
action is the most effective means to promote co-ordination, echoing the
emphasis of the 'Sanitary Idea'. This has contributed to the trend in health
policy toward decentralization of decision-making and management to the
very local level. This is discussed in more detail in Chapter 8, which also
considers some of the limitations of decentralization and the professional,
organizational, and jurisdictional barriers to co-ordination at the local
scale.

Shifting resources to the primary health care sector

Clearly these approaches emphasize the primary health care model being
promoted by the World Health Organization (see Chapter 9), and would
call for a shift of resources from the secondary health care sector to the primary sector. The health systems described in Chapter 5 are still ill-adapted
to achieving such change because they are largely structured around conventional, curative biomedicine. It is difficult to see that change will be very
rapid, since even in high-income countries pressures for cost constraint
mean that it is difficult to find new resources to fund public health initiatives. Focusing resources more on primary health care is likely to result in
reduction of resources for traditional secondary services. One of the most
problematic aspects of attempts to shift resources away from secondary services towards the primary sector has often been a period of transition of
funding when neither type of service is fully available, since hospital-based
provision is being closed down before new community-based provision can
be opened up. This is very damaging for the public image of primary health

care initiatives, since they become associated in the public mind with an immediate loss to the community. Any gains are likely to be seen as uncertain and long-term. Furthermore, community-based services which are only partially funded are likely to fail to deliver their full potential health gains, sowing doubts about their intrinsic potential. These difficulties are often particularly severe in inner-city areas with a legacy of large numbers of hospital institutions, no longer entirely suited to contemporary needs of the population, and rather poor provision of primary care. Maxwell (1993) comments on the problems faced in London, New York, and Paris in developing ambulatory services while rationalizing hospital provision.

Many authors also draw attention to the need to develop the primary health care sector in low-income countries which face public health problems of poor populations on a larger scale than in wealthy countries. Intersectoral primary health care development is often difficult to achieve in the face of entrenched socio-economic inequalities and attitudes which resist change, particularly when the amounts of resources available to achieve change are extremely small and are becoming more difficult to find as low-income countries attempt to restructure their economies to make them more competitive in the global capitalist system. Asthana (1994) comments that, during the 1980s, more specific primary care projects, focused on particular types of intervention such as rehydration therapy and immunization, have tended to take precedence over more general approaches directed at integrated community development for improved health. This trend is criticized by advocates of co-ordinated primary health care as a retrogressive step which makes it more difficult to tackle the root causes of disease. The development of primary health care is discussed further in Chapter 9.

The importance of the informal sector

The development of community-based health care also depends very heavily on the contribution of the informal sector; those who look after sick people or maintain the health of well people without payment and on a non-professional basis, often within the context of family, friendship, or neighbourly relationships. Support may be *emotional* (involving moral and psychological helping), *informational* (assisting with problem-solving and decision-making) or *instrumental* (help with practical tasks of everyday life) (Antonucci 1990). Reviews (e.g. by Barrera 1986; Antonucci 1990; Curtis *et al.* 1992) have examined the role of the informal sector in supporting frail elderly people in different countries. This research shows that the amount of care provided by this sector is much greater in terms of human resources than the provision made by the state or the independent professional sector. The General Household Survey in 1985 and 1990

(OPCS 1987; Green 1988; OPCS 1992) has shown, for example, in Britain that 15 per cent of adults over 16 (6.8 million people) look after or provide some regular service to a sick, handicapped, or elderly person living in their own household or another private household. It is often argued that women carry the largest burden of informal caring. The 1990 General Household Survey showed that, of those 1.4 million people in Britain devoting at least twenty hours a week to caring, 64 per cent were women. The burden of care is significant. The 1985 General Household Survey showed, for example, that 43 per cent of carers had been looking after a dependent for at least five years, 44 per cent said that no one else could replace them temporarily to care for their dependant, and 23 per cent had no help. About half the carers in this survey received no help from the statutory services. Nissel and Bonnerjea (1982) and Glendenning (1988) have also demonstrated that the economic costs of caring are considerable and can economically disadvantage carers of dependent ill or handicapped people. A study in Ireland (O'Shea and Blackwell 1993) illustrated a common finding: that the real cost of community care of elderly people would be considerably increased if the informal sector effort were costed into the total. Even routine care of healthy dependants such as children can place a considerable burden on those informal carers who are responsible. Carpenter (1980) estimated that American women with children aged 0–14 devoted a total of 275 million hours just to escorting children to use health services, and of course this does not include the other household work which goes into child health care inside the home.

Policies for community care are sometimes criticized for placing a growing burden on the informal sector (Bulmer 1987), and it has been suggested that care *in* the community may be a euphemism for care *by* the community. Policies for health gain need to be sensitive to the significant role played by informal carers and to their needs, and to find ways of involving them more directly in health policy-making.

Local frameworks for collaboration: the example of Healthy Cities

Associated with the movement to revive the 'Sanitary Idea' in industrialized countries, we find a number of new initiatives for collaborative action. For example, there has been widespread interest around the world in the 'Healthy Cities' idea. This arose from the experience of local action to achieve health gain using the 'Health for All' (HFA) framework being pursued by the World Health Organization (see Chapter 9). Ashton (1992) describes how, at a meeting in Toronto in 1984, Duhl's idea of the 'healthy city' (Duhl 1986) was taken up and used to launch what was initially expected to be a fairly limited project, based in a number European cities, to encourage action for health promotion (Tsouros 1990).

The aims of the Healthy Cities project as originally set out were:

- to produce city action plans to pursue HFA targets
- to provide models of good practice for health promotion
- to monitor and research outcomes of interventions
- to exchange ideas between cities
- to promote mutual support/learning between cities

The type of action envisaged would include:

- setting up a high-level executive group for collaboration between sectors
- setting up a technical group to monitor changes
- carrying out a 'community diagnosis'
- establishing links with educational institutions
- reviewing potential for health promotion by all agencies
- generating a debate on health
- adopting specific interventions and monitoring the results

While these aims are focused on perceived issues for promoting public health in today's cities, Ashton (1992: 10) points out that they are recalling the objectives of an earlier generation of campaigners for public health:

> . . . buried within these can be found . . . most of the activities of the local Health of Towns Association branches of 1844–5, bringing together key players in the cities, establishing a clear picture of health in different parts of the city, developing advocacy and coalition, building for change, intervention and legislation.

Although it was originally conceived as an initiative involving a limited number of European Cities, the Healthy Cities idea proved to have much wider appeal (Curtice 1993). The idea is very adaptable to local conditions. It has a strong emphasis on enabling and empowering local communities to act for themselves, calling on a range of local human, financial, and infra-structural resources. Through its potential to provide a focus for local action involving many different partners, it has provided inspiration in many cities in both the developed and the developing world. The international dimension of the Healthy Cities movement is considered in more detail in Chapter 9; here we consider some examples of specific Healthy City initiatives to illustrate the types of action which were undertaken locally in different settings.

Healthy Cities projects in Britain include one in Liverpool (Green 1992) which eventually embraced fourteen partner agencies. A Declaration on the Right to Health was produced which stressed inter-agency responsibilities for health care.

In recognising every citizen's right to good health, we accept the
responsibility carried by all agencies, throughout our society to take
account of the public health costs of all their activities . . . Such agencies include central and local government, health and education
authorities, the non-statutory sector, employers, landlords, academic
bodies . . . churches, communicators; everyone taking part in the production and consumption of our goods and services, our values and
attitudes. (Green 1992: 93)

The Declaration stressed principles including equity in health, community
participation, partnerships for health, health promotion, primary health
care, international co-operation, and research, thus reflecting the global
aims of the WHO in the local context.

In the city of Oxford, 'Health for All' goals were also taken up, and the
local authority created a health committee to help to co-ordinate efforts for
public health improvement and also to underline the fact that local government agencies, although they are separated administratively from the NHS,
have a role to play in public health and health promotion (Fryer 1987). As
part of the 'Healthy Oxford' initiative, the local authority has set up its own
'health audit' in consultation with local people, and is examining the potential of measures such as house insulation, counselling for sexually abused
women and children, and traffic-free zones where children may play safely
(Root 1995).

Healthy Cities initiatives can often be inspired initially by the efforts of
non-statutory organizations. For example, in one of London's most deprived
inner-city areas, a Health Strategy Group was created through local concern
over public health issues which were not being adequately tackled by either
health or local government services. The Strategy Group comprised representatives of the voluntary sector as well as members of the statutory services and also members of local academic institutions (including the present
authors). The group raised funding to support co-ordinators and facilitators
who liaised with local people and with statutory agencies to initiate projects
which were often aimed at making existing services more effective or appropriate for the local community. Their action led to the setting up of a 'health
bus' which enabled health personnel to make regular visits to some of the
most deprived and isolated communities in the area. A campaign to reduce
cockroach infestation of homes employed a member of the local ethnic-
minority community to explain to householders the purpose of insecticide
treatment of their homes, and to make sure that the treatment could be effectively delivered in blocks of flats. An anti-smoking campaign, directed at the
local Islamic community, co-ordinated its efforts with the Ramadan festival,
making it more appropriate and relevant to the population and involving
local religious and community leaders in health promotional activity. These
interventions demonstrate ways to involve local communities in action for
public health, and to overcome the misunderstanding which can arise

between ethnic-minority populations and the statutory authorities, which can lead to unhelpful attitudes towards ethnic-minority groups. The Strategy Group aims to make constructive attempts to change the ideas in some quarters that, for example, ethnic-minority populations make 'inappropriate' use of health care, that they are unresponsive to health education messages, and that they are unmotivated to co-operate with environmental health projects. From the viewpoint of the ethnic minorities involved, the problems are seen to be services which are inappropriate and inaccessible to them, health education campaigns which are not adapted to their particular lifestyles, and interventions such as pest control programmes which were not presented to them in terms they could understand.

The Healthy Cities strategy does not require the involvement of a whole city; an urban district or neighbourhood can organize very locally to pursue healthy city goals. One of the first examples of Healthy Cities projects in Britain was in the Bloomsbury District Health Authority, which covers a part of inner-city London. Coombes (1993) describes how an attempt was made to adopt Healthy Cities ideas in Globe Town, which has a population of about 14,000 and is among the smallest units of local government administration in London.

In the United States, also, a number of Healthy Cities projects have been set up. Flynn (1992, 1993) reviews the diversity of Healthy Cities and allied projects, and suggests that the plurality of US society and the fragmented structure of the government system is reflected in the variety of different models and strategies being adopted. Some states, such as Indiana and California, have launched programmes designed to encourage involvement of cities in setting up local programmes. Rider and Flynn (1992) describe how in Indiana six cities were selected for Healthy Cities programmes, because their populations included groups at particular risk of health problems, there was community support for local action, and community representatives were able to agree to participation in Healthy Cities action. Rider and Flynn (1992: 200) discuss the co-ordination of these projects by Healthy Cities Committees. The membership of these Committees, designed to represent all areas of public life relevant to public health, was as follows:

- Arts and culture
- Business and industry
- Dentistry
- Education
- Employment
- Environmental campaigns
- Finance
- Health care
- Energy utilities
- Indiana Public Health Association
- Local government (including the local mayor)

- Media
- Parks and recreation
- Religion
- Transport/communication

The projects involved action to enhance leadership skills required for developing, communicating, and implementing public health policies, and also the production of community assessments of local public health using input from ordinary members of the community as well as from experts. A range of methods for public communication and consultation were used to involve the population in a debate about public health issues. The interventions arising from this initiative included setting up a walking club, broadcasting health-related items on local radio stations, street-cleaning campaigns, recycling programmes, efforts to tackle housing shortages, and action to improve education for teenagers on parenthood. Some of these could be accomplished in a short time, others required a longer period. Many were designed to empower local people to become involved directly in collective action, as well as focusing public- and private-sector agencies towards public health action. The Indiana model for Healthy Cities is therefore aimed at promoting local self-reliance and responsibility for health, by placing health on the local political agenda and requiring local decision-makers to consider the health implications of their policies. It also aimed to involve the public as well as health service professionals, including 'hard-to-reach' populations who are typically underrepresented in local decision-making processes, such as children and old people, poor people, and ethnic minorities (Flynn 1993).

Rice and Rasmusson (1992) review a number of Healthy Cities projects in low-income countries. Many of these projects involve co-operative action to provide the manual labour needed to construct the basic infrastructure which is often lacking in urban slums. For example, they report one initiative in Rio de Janeiro, Brazil, which involved community action in Rochinha, one of the poor neighbourhoods in the city, to organize environmental health projects such as a rubbish collection scheme and rebuilding of the sewage disposal system. It also tackled education issues, establishing kindergartens, developing teacher training, and renovating school buildings. The projects involved co-ordination of local voluntary effort with local government officials who could help with provision of materials. There was a strong element of community involvement and participation, with local people providing voluntary labour on the projects. The Rochinha model was later adopted in other areas of the city to tackle public health issues in other neighbourhoods. In a slum area of Karachi, Pakistan, the Baldi project involved effort by local women to co-ordinate action on sanitation by constructing soak-pit latrines. In Tegucigalpa, Honduras, a group of local women petitioned for construction of standpipes in their neighbourhood and organized supervision of the water supply, collection of fees, and maintenance of the sites. In the Popular Unity Co-operative in El Puyo, Ecuador, local people set up a low-income housing construction project which later developed into a small industry producing building materials.

There remains a need for more thorough evaluation of Healthy Cities projects. Examples of evaluation to date include: Coombes (1993) and McGhee and McEwen (1993) on work in London and Glasgow; Baum (1993) on healthy cities in Australia; Flynn (1993) on the USA; Labonte (1993) on Canada; and Curtice (1993) on European projects. The lessons emerging from these evaluative studies of specific projects may help to inform future initiatives.

Healthy Cities initiatives have demonstrated the enormous, largely untapped potential for communities to help themselves through collective action and involvement of the voluntary sector (e.g. Coombes 1993). The approaches are fundamentally different from those traditionally advocated by health education, which stress mainly individual responsibility and action to promote good health. Healthy Cities initiatives are often directed towards making such isolated action more possible through collective support. The early examples of Healthy Cities initiatives have had a snowball effect, drawing many local communities into similar action, even if not officially under the banner of 'Healthy Cities'. Labonte (1993) suggests that they can provide a vehicle for integrating the challenge of sustainable development into health policy, citing the example of 'Round Tables on Environment and Economy' which provide fora with wide-ranging membership in Canada. The extent and diversity of the movement has taken it far from the initial formulation of the project by WHO, as local communities claim ownership and adapt the model to their own ends.

Experience of these projects has also demonstrated the huge barriers and constraints which limit concerted action for public health. In some large cities, the complexity of the networks required makes them very difficult to set up. The Liverpool project, for example, took a long time to get under way, partly because of the bureaucratic and professional barriers to joint working. Uncertainty over the extent of responsibility for health is a recurring theme. It arises, for example, in Britain, where the boundaries between NHS and local authority roles are contested (Fryer 1987), and in the USA, where the responsibilities of cities and counties may overlap (Flynn 1992). The trend towards decentralization of responsibility for health and health care results in a natural local focus for such action, but also often places different communities in competition with each other for the limited national or federal resources available to help support their activities. Harpham *et al.* (1988) discuss the factors which can limit the feasibility of *scaling up* primary health care projects from the local level to apply to whole cities, regions, or countries, and *scaling down* national initiatives to make them applicable to local conditions. These constraints include the relatively high cost of demonstration projects, which makes it too expensive to generalize them, and the difficulty of centrally designing interventions which will apply to a range of different situations at the local level. Farrant (1991) argues that the challenge remains of ensuring that the potential contradictions of Healthy Cities and related initiatives are recognized. For example, their colonization by consumerist notions of participation should be resisted (as discussed in the following chapter).

The new public health: changing agendas for research, policy, and practice

The perspectives discussed in this chapter show that new approaches are required in order to develop the new public health strategies, and that new agendas need to be set for theory on which research is based, the orientation of policy, and the nature of interventions to improve health.

Kelly *et al.* (1993) argue that existing theoretical frameworks offered by both the medical model of health and the social science model are inadequate to capture the potential of the New Public Health and Healthy City movements because both models approach public health issues from the point of view of system breakdown – trying to identify a malfunction in the system which can be rectified. However, they argue that Healthy Cities approaches tend to emphasize positive health and health promotion, and that 'positive health is not amenable to conventional scientific investigation or to conventional (modern) scientific discussion' (p. 161). They suggest that postmodern approaches are more appropriate because the aspects of life which are relevant to health cannot be systematized, but are chaotic and open to a range of interpretations.

Suitable theoretical frameworks may therefore need to acknowledge that there is a diversity of views on health which are unevenly represented at present in health policy and theories about how to improve public health. Berman *et al.* (1994) suggest that we may need to give more attention to the household production of health, focusing on factors which are important at the micro-level and linking anthropological, economic, and sociological ideas about how households are composed and how they function, the perception, definition, and labelling of health and its meaning for members of households, and how these factors are associated with health-related behaviour. Dunn *et al.* (1994) have shown in the case of a community in Ontario that local perceptions of the environmental hazard presented by a toxic waste contamination differed from that which outside observers would have expected, and they suggest that the psycho-social effects of contamination may depend on the community context. The perceived nature of a community as it is viewed by local agents is also important. Moon (1990) discusses the implications of the use of ideas of community in contemporary health policy. The work of Coombes (1990, 1993) in east London shows how the perspectives and perceptions of local groups were reflected in the development of action for public health (see Chapter 1). Daker-White (1995) also shows that the perception of the East End of London has influenced the way that professionals providing services for illicit drug-users have defined need in the population.

It is clear that the sorts of policy and action which are necessary to achieve real changes in public health extend well beyond the domain of bio-medicine. Political, social, and bureaucratic action is needed in order to bring about changes in the ideology of health and welfare, to facilitate changes in health beliefs and health-related behaviour, to empower groups in society which currently lack influence over factors associated with health, and to enable new styles of health-related work to be introduced. The following chapters discuss the emerging perspectives and strategies which may help to make some of this change possible, in terms of public participation and managerial reform which may be important at the local level and in terms of international initiatives aiming to change the global context in which local action takes place.

|8|

Setting agendas: health policy formulation and implementation

Introduction

This chapter is concerned with the formulation and implementation of health policy. Because of the many different ways in which the term 'health policy' can be (and has been) used, it will be helpful to begin by reminding the reader how we shall use it here. 'Health policy' is used to encompass any policy which includes actions undertaken with the aim of maintaining or improving health and/or providing for the care, treatment, or cure of ill health. Within the socio-ecological paradigm of health discussed earlier, this encompasses actions taken to address environmental, genetic, lifestyle, or health service system factors – all of which affect health. It should thus be clear that health policy is not restricted to consideration of the health sector: we shall reserve the term 'health services policy' for use where we wish to talk solely about the health service system. This wide understanding of health policy derives from the work of several authors, notably Blum (1974), and has been developed particularly through the work of the World Health Organization (WHO/UNICEF 1978; WHO/EURO 1985a, 1993a), which will be considered further in Chapter 9.

In discussing health policy, we shall sometimes need to distinguish between different components of health policy: *policy goals*, encompassing the broad aims or specific targets of policy; the *underlying framework* of values or principles; overall strategies and specific mechanisms for policy *implementation* (which may be legislative, regulative, administrative, or organizational); the different *levels* at which policy formulation and implementation take place (international, national/federal, regional, local). It is useful to think in terms of a *policy process*, beginning with policy initiation, moving through formulation and development, adoption, operationalization, implementation, and finally review, evaluation, and monitoring of policy, which may then feed back into policy formulation. While it is possible to distinguish such stages in the policy process conceptually, this should not be regarded as a simple linear, temporal progression or even as an iterative, cyclical one. In reality, the different stages may be intertwined, predetermined, or omitted, may only be provisional, may be without causal links to each other, and may be only selectively or partially linked to policy outcomes. The study of policy implementation is a necessary part of evaluating the success or failure of policy, and cannot be separated from the other parts of the process (see Milio 1987 or Palumbo and Calista 1990 for more detailed discussion). The material addressed in this chapter ranges over all the stages in the policy process.

The chapter begins by discussing rationalist and incrementalist approaches to policy, which represent some of the earliest efforts to theorize the policy process, before introducing the notion of the policy community and its actors, and then discussing a range of different approaches for analysing their interactions. There is insufficient space here to give a detailed description of each of these different approaches, but the aim is to introduce them sufficiently to allow their various strengths and weaknesses to be critically appreciated.

The chapter concludes by discussing two particular themes that represent recent trends in health policy globally: decentralization, and public participation in health policy. Although these are dealt with in two separate sections, they are often intimately linked, with one of the underlying rationales for pursuing a policy of decentralization being its argued potential for facilitating widespread participation; the empirical studies considered later examine to what extent this expectation is borne out in practice. These two policy themes are particularly interesting to us for a number of reasons. They illustrate many of the difficulties hampering the earliest efforts to theorize the policy process and the necessity of more sophisticated frames for analysis. They also exemplify the importance of the level at which analysis is carried out, reflecting one of the major themes of this book. These themes also represent two of the policy thrusts evident in many of the most recent health service reforms discussed in Chapters 5, 6, and 7 (although neither is by any means new); a further such policy theme, the reorientation of health services towards prevention and health promotion and away from curative medicine, is examined in Chapter 9, together with the associated policy shifts in the nature of health promotion.

Rationalist and incrementalist approaches to policy

Much early work in policy analysis was built around a concentration on the component decision-making processes, informed by different theories of what these processes involved (or should involve). In particular, we can distinguish two particular theories or models: the 'rational-comprehensive' model, also known as the 'purposive-rational' model or the 'rational-deductive' model; and *incrementalism.*

Drawing on Lindblom (1959), we can describe the rational-comprehensive model as involving a process which begins with the clarification of values and objectives, then proceeds to policy formulation, identifying suitable means to achieve the ends established at the first stage, supported by suitable empirical analyses. Throughout it is (implicitly) assumed that analysis can be, and is, comprehensive, in that there is a systematic attempt to take all relevant factors into account. Leaving aside questions of the desirability of such an approach, it can be questioned on the grounds of feasibility in many different ways, focusing around the implicit assumptions that it involves at various stages. For example, it may not be the case that values and objectives can be uncontroversially agreed at the outset. The desired ends cannot necessarily be treated as conceptually distinct from the means by which they might be achieved. It may not be possible for all relevant factors to be identified and taken into account, since this necessitates full knowledge of all relevant cause-and-effect relationships, and of the consequences of all alternative courses of action. Finally, the necessary resources of time and effort to carry out all relevant analyses may not be available.

In spite of these shortcomings, the rationalist model has provided the basis for the *design* of much of the health-planning legislation that emerged from the health programmes of the 1960s in the USA, for the corporate planning system developed and introduced into the British NHS in 1976, and for the 'managerial process for national health development' advocated by WHO (see Fig. 8.1). Despite this, research has demonstrated that the model does not provide a good description of most real-world policy processes in action (Lee and Mills 1982); further examples are also discussed later in this chapter.

Incrementalism differs from the rationalist approach in its limitation of the scope and depth of analysis and action involved in policy formulation and implementation. It is based on analysing policy problems one at a time, and considering options only incrementally different from the status quo ('incremental analysis' in Lindblom's (1979) terminology). It also operates on the principle of changing outcomes by means of a series of small steps ('incremental politics' in Lindblom's (1979) terminology). Agreement is reached between different interest groups by a process of partisan mutual adjustment. Incrementalism has been found to provide a realistic description of the health policy process at local level in some instances, at least for some issues. In the USA, despite laws designed on the basis of a rationalist approach, studies indicate that health policy-makers and planners may be operating in practice using an incrementalist approach (Lee and Mills 1982). In the UK, Alaszewski *et al.* (1981) argue that incrementalism in local policy processes has permitted a policy drift towards high-technology medicine by capitalizing on the dominance and prestige of consultants in

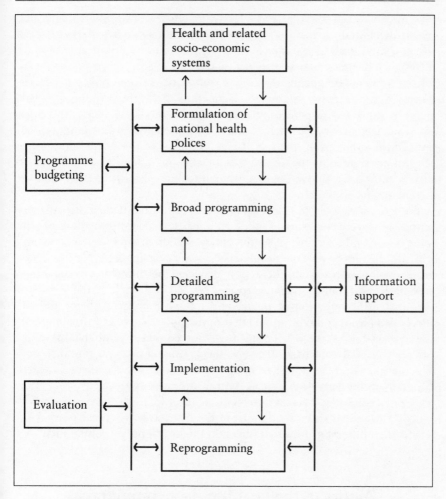

Fig. 8.1 WHO's managerial process for national health development: rationalism in theory. (*Source:* WHO 1980b: 7.)

acute specialties existing in the earliest years of the NHS. Lee and Mills (1982), studying the NHS planning system introduced in 1976, show that in practice the policy activities of local health authorities were often still characterized by disjointed incrementalism, particularly in terms of implementation. More recently in the UK, and in particular after a key national policy shift, which introduced general management in the health service, Harrison *et al.* (1990) find many instances where policy implementation at local level is still dominated by incrementalism – for example, in responses to the introduction of general management, in resource allocation, and in introducing evaluation. A number of specific factors may promote the adoption of an incrementalist approach at local level, including constraints of time and resources, vested interests in the status quo, and most particularly the existence of any or all of a number of different types of ignorance: structural, ideological, professional, or procedural (Nocon 1989). However, incrementalism is much less adequate as a description of national policy processes, where there are many examples, particularly since the 1970s, of radical changes, especially in policies relating to health service financing and organization, as identified earlier in Chapter 5 for the UK and Russia in particular. On the other hand, the recent abandonment of the far-ranging health reforms considered by the Clinton administration in the USA, also discussed earlier in Chapter 5, also points to the difficulties of moving away from an incrementalist approach.

Both these models remain unsatisfactory in terms of their explanatory power, however. For example, they do not help our understanding of why some issues find prominence on the policy agenda at a particular time, and why other issues never take prominence, or are even ignored. Neither offers much explanation of the interaction between the different groups which may be involved in the policy process, nor do they include the means to examine the distribution of power and influence between them, and the effects this has. It could be argued that both rely either on an (often implicit) assumption that there is a broad degree of consensus and shared values amongst the different parties, or else incorporate a classical pluralist view which argues that the power to influence decisions and resources is widely dispersed in modern democracies, with a multiplicity of determinants, no one predominating. As a prelude to considering some theoretical approaches which do not suffer from these defects, we move on now to review the different groups that make up the health policy community.

Actors in the health policy community

A wide range of different groups can be involved in policy formulation and implementation; these are sometimes identified as constituting the 'policy community'. The specific groups that are involved in any particular policy will obviously depend on the nature of the policy, the level at which it will operate and the stage in the policy process.

The centre of national or federal policy-making is often located within the government/state sector, strongly directed or led by the overall objectives of the government of the day (i.e. the political climate and its ideological constraints), with the legislature retaining formal powers over legislation, state expenditure, and aspects of state administration. A variety of organizational mechanisms usually exist for consultation and discussion on policy matters at national/federal level; these may involve standing advisory committees on particular topics, *ad hoc* committees, working groups, or specific and time-limited inquiries. For example, the parliamentary system in the UK delegates all major policy-making responsibility to the Cabinet, subject to parliamentary approval, while in the USA there are a variety of legislative committees with this responsibility. It can be argued that the scope of the task and unwieldy nature of parliamentary scrutiny often weakens accountability and sensitivity to local democratic pressures. In stages beyond the formulation of national or federal policy, the detailed content of policies, in terms of how they are interpreted, not to mention implemented, will be mediated through the interaction of a range of different agencies and groups (see Fig. 8.2).

Doctors are another obvious and important group within the health policy community. Their control of medical knowledge and their high prestige as a group explains and reinforces the dominance of the 'medical' or 'disease' model of illness. Professional medical organizations, such as the British Medical Association (BMA) in the UK, have been highly successful in protecting their members' interests, much more so than most trade unions, and the BMA has been extremely successful in the past in influencing health service policy at all levels (Freidson 1970; Klein 1974, 1984; Parry and Parry 1976; Jones 1981). Earlier, in Chapter 5, we discussed the key role of the general practitioner in the internal market. Besides biomedical professionals, we can also find other types of medical practitioner forming separate interest groups where some policy issues are concerned.

Health service administrators, planners, and managers at various levels within the health service organization have varying degrees of involvement in policy formulation at their level, together with implementation (and monitoring or evaluation – which may or may not be a formal requirement). Other health professional groups, such as nurses, health visitors, midwives, physiotherapists, occupational therapists, and other professions allied to medicine, are increasingly finding their voice and fighting for recognition of their professional role separate from doctors, to end the perception of them as doctors' 'handmaidens'. This has resulted in very particular policy positions being advocated by such groups. Aside from the multiplicity of professional groups, health service delivery also involves a large ancillary and support workforce, including nursing auxiliaries, care assistants, porters, and drivers. These rarely act as a cohesive group and usually play little role in policy determination at a national/federal level, but they have more recently become heavily involved in policy implementation issues at a local level, for example in resisting individual hospital closures and service 'rationalization'.

Community representation is sometimes through statutory organizations, such as the British system of Community Health Councils, intended to represent consumer interests, particularly at local level, but also acting in the national arena through the National Association of Community Health Councils. Community representation may also act through a system of

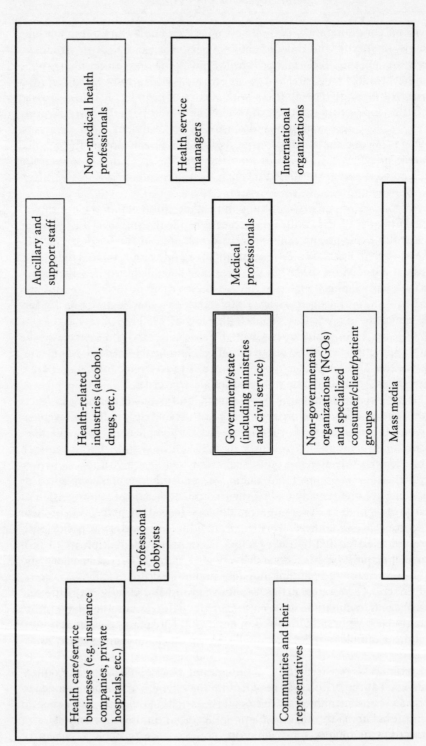

Fig. 8.2 Actors in the health policy community.

representatives on authorities and boards, for example the Health Systems Agencies (HSAs) in the United States (discussed later). There are also specific community interest groups, including groups formed around particular health problems (multiple sclerosis, cancer, etc.); such groups may be of several different types. These sometimes take the form of voluntary-sector organizations involved in providing services to a particular constituency; they are also sometimes in the form of self-help organizations concerned with particular issues (smoking, drug abuse, etc.). Finally there may be campaigning groups formed around specific issues. The focus point is either the national or federal policy arena, or the local policy scene in the case of local service-providing organizations. For example, the UK national pressure group ASH (Action on Smoking and Health), established under the auspices of the Royal College of Physicians in 1971, has been one of the players responsible for keeping smoking on the policy agenda at national level. However, ASH has not been strong enough to see all its recommendations implemented (Calnan 1984; Read 1993), particularly in the face of industrial interests promoting the tobacco industry.

Industrial and business interests form another group influencing health policy. Within some health systems this will include those private-sector components of the health service system (service providers, insurance companies, etc.). It also includes health-related industries, often omitted (or rather not admitted to), who may operate indirectly or even covertly in many cases by lobbying at parliamentary level. A classic example is that of the tobacco industry and anti-smoking policies. Milio (1981) documents a considerable variety of lobbying activity in the US health policy arena, acting to support the interests of a wide range of industrial concerns, including automobile, energy, drug, chemical, food, tobacco, and alcohol industries. In the UK context, a network of tobacco-based interests has been successful in constraining the policies adopted by successive governments, utilizing a variety of selective arguments including the revenue-generating role of tobacco and individual freedom, and being successful in limiting policy measures restricting tobacco-related advertising (Read 1993). The UK government's continued failure to institute a ban on tobacco advertising is an example of a policy failure of the sort linked by Offe (1984) to his thesis that the welfare state is committed to giving preferential treatment to the capitalist economy. More generally, recent relationships between British government and industry have been based on close understandings and a (relative) lack of public scrutiny over policies (Mills 1993; see also Baggott 1990 for a discussion of the alcohol industry's role in UK alcohol-related policies). There is also an international dimension here, since many of the industries are structured on a multinational basis; and an international dimension to health policy is provided by the environment and trans-frontier issues such as pollution. Some specific international health policy efforts in this field, and the role of international organizations like WHO, are examined in Chapter 9.

The mass media have also been included in Fig. 8.2. Although not often incorporated within the formal mechanisms for policy development, these can be influential in determining the issues that come to be defined as of concern and meriting some policy response; they thus form an important part of the policy environment. For example, a study by the Ministry of Agriculture, Fisheries and Food in the UK (MAFF 1987) illustrated the importance of media publicity in the emergence of public concern about food additives, while Grant (1989) demonstrates how the impact of various campaigning groups on food policy in the UK was aided by media access.

Theorizing interactions in the policy community

We argued earlier that both rationalist and incrementalist schools of policy analysis paid insufficient attention to the existence of different interest groups holding quite different values and assumptions in relation to policy issues. There are several distinct schools of thought about the most appropriate way to remedy these defects and gain a more adequate understanding of various aspects of the policy process. In this section we discuss selected theoretical approaches which have been used to analyse the interactions between actors in the policy community: *pluralist* approaches; *structuralist* approaches; and *poststructuralist* approaches (the micropolitics of policy).

Pluralist approaches

The classic pluralist model argues that power is widely distributed amongst the different groups that constitute the policy community, with no one group being dominant. Groups compete through mechanisms of local and central government to achieve their ends. The power to influence decisions and resources is widely dispersed, with a multitude of determinants of power. Particular interest groups may form and have influence on the basis of a diverse range of factors including occupation, ethnicity, geographical location, or social class. The classic pluralist model has been widely criticized as being naive and simplistic in its depiction of the power and influence of the different groups, leading to the development of a variety of neo-pluralist approaches, in which access to the policy process is recognized as not being equally available to all the interest groups who might wish to participate. Neo-pluralists also move away from classic pluralism in their view of the state, admitting stances other than neutrality or passivity. Instead, the state is viewed as occupying a role of referee between competing interests, able to tip the balance on particular issues. Studies based on this framework place their emphasis on the process of bargaining and negotiation (e.g. Willcocks 1967 on the creation of the NHS in the UK; Pater 1981 on policy formulation prior to the setting up of the NHS in the UK; Rivett 1986 on the London hospital system over the period 1823 to 1982; Eckstein 1960 on the operation of the BMA; Marmor 1973 on the debates inside and outside the US Congress which influenced the timing and outcome of the 1965 Medicare bill). Bomberg and Peterson (1993) argue that, in relation to health-related prevention policies in areas such as the environment and tobacco, the EC as a policy forum remains open to a vast range of interests, in a way in which national policy networks are not.

Structuralist approaches: structural interest theory and neo-marxism

The structural interest theory described by Alford (1975) arises out of his study of various different aspects of New York City's health system. Three different types of interest group are identified: dominant, challenging, and repressed. The dominant group consists of those whose interests are served by existing social, economic, and political institutions; in Alford's study, this position is occupied by the biomedical profession (which he labels the 'professional monopolists'). Some of the examples discussed earlier, including the cases of alcohol and tobacco policy, point to instances where the dominant group can be quite different. The challenging group is made up of influential groups putting forward interests which challenge those of the dominant group. The composition of the challenging group may change from time to time and from issue to issue. In Alford's study the challenging group, which he labelled the 'corporate rationalizers', included managers, administrators, planners, and even researchers. He argued that this group shared a common relationship to changes in the technology and organization of health services, resulting in a developing 'interest in breaking the professional monopoly of physicians over the production and distribution of health care' (Alford 1975: 15). Finally, the repressed group contains those whose interests are by and large submerged, and for whom there are no social institutions or political mechanisms that ensure that the repressed group's interests are served. In fact, the nature of institutions tends to ensure that their interests will not be served without extraordinary effort or circumstance. Most frequently we find the community or consumer in this position, as indeed is the case in Alford's study, which demonstrates the failure in implementation of various different policy initiatives designed to improve service availability and access to particular sections of the urban population. In identifying repressed interests, it is important to recognize that the mere existence of institutions or mechanisms charged with representing such interests is insufficient; these must also be able to be *effective* in overcoming the suppression of the concerns of this group. In the UK, for example, Community Health Councils were set up to represent community interests in the NHS, but study of the constraints, including resources, under which these operate have lead some commentators to categorize them as 'toothless watchdogs' (Levitt 1980). We consider later in this chapter the extent to which various policy moves towards consumer involvement can be regarded as merely rhetoric.

Within structuralist approaches the scale of analysis can range from the societal (national/federal policy formulation) to the very local (policy implementation within a particular health authority or even a single health service facility). There is also often an emphasis on the study of particular localities – and a recognition of the unique geography of particular places. The dominant interest group will not necessarily remain the same at different levels of analysis.

A specific kind of structural analysis is offered by the many varieties of marxism and neo-marxism. These share a concern to relate the various different interest groups to the underlying class structure, and to relate the different parts of the policy process to particular aspects of the antagonisms and struggles between classes. Thus marxist analyses of policy-making and implementation are in general less well developed below the very macro-level. They often emphasize the identity of the interests of doctors with the interests of a whole class rather than representing the interests of the medical profession alone, arguing that pluralists and structuralists analyse only the surface struggles and ignore deeper class conflicts. This type of analysis is extremely useful for offering insights into large-scale macro-issues – in analysing how the ideology of medicine (and particularly of the medical model) individualizes problems rather than laying bare the social causes of ill health, thus constraining the policy solutions produced in response to problems (e.g. Doyal 1979; Navarro 1986; Tesh 1988). They have also been particularly powerful in accounting for the lack of policy support for occupational health services (Elling 1982; Navarro 1986) in capitalist countries.

A separate, though related, strand of work, influenced by the Frankfurt School of critical theory (Held 1980), can also be identified, whose central focus is on the historicity of social action, including health policy, and which in particular emphasizes the connections between human agency and social structures existing under capitalism. Offe's (1984) analysis is of interest for his insistence on a detailed examination of the specific mechanisms and processes of the state, from which he concludes that it cannot be adequately viewed as a unitary co-ordinated institution, and that it does not possess 'the requisite foresight and analytical capacity for diagnosing the functional exigencies of capital' (p. 102). In his analysis, late capitalist societies are viewed as systems structured by three interdependent but differently organized subsystems: structures of socialization (e.g. the house-hold), guided by normative rules; the capitalist economy, i.e. commodity production and exchange relationships; and the welfare state, organized by mechanisms of political and administrative power and coercion. Thus although policy formulation will react to the 'two poles of the "needs" of labour and capital' (p. 104), the effects of policy will be influenced by the conflicts arising at the time of implementation between the different subsys-tems, and by the outcomes of these. Offe argues that the state may adopt a variety of different strategies within social policy as a response to its various contradictions, including: increasing administrative steering capacities; appeals to 'scientific' inquiry into the measurement and assessment of policy issues; and privatization. Fitzpatrick (1987) explores the relevance of Offe's ideas in the context of the British NHS, and finds each of these strategies in use in various ways, although he concludes that the schema offered by Offe is insufficient on its own to account for the particularities of the UK situa-tion, and especially for its differences from the situation in the USA.

A further example is offered by Habermas (1987), who develops his analysis of the welfare state in Western democracies out of a critique of marxian orthodoxy. He develops a general thesis of 'internal colonization', whereby the subsystems of the economy and state are argued to 'become more and more complex as a consequence of capitalist growth, and pene-trate ever deeper into the symbolic reproduction of the lifeworld' (Habermas 1987: 367). He proceeds to examine social welfare policy in terms of the interactions between the system (economy and state) and life-world (both private and public spheres), around which the roles of the 'employee', 'consumer', 'client', and 'citizen' are formed. Aspects of a Habermasian type of analysis are utilized by Jones (1995) in his study of renal policy within the regions surrounding London. Fraser, discussed fur-ther in the next section, also utilizes aspects of Habermas's analysis along-side Foucauldian ideas in her studies of social policy in the United States.

The differences between the structuralist and marxist approaches should not be exaggerated; perhaps the difference is most apparent in the level at which the analyses have been so far applied. To blur the boundaries a little further, there are also some similarities to the argumentation developed within neo-pluralist analyses, although the latter tend to focus on the fore-ground of the policy process (Harrison *et al.* 1990), rather than seeking explanation in terms of various 'deep' structural factors of demography, technology, ideology, and relationships of ownership and control. Harrison *et al.*'s study (1990) of the dynamics of British health policy during the 1970s and 1980s concludes that the best explanation for the changes observed is provided by a combination of neo-marxist and structural theo-ries (although they label the later as neo-elitism, using rather different terms from those used here). To explain the policy areas of funding, market rela-tions, and organizational structure, they find neo-marxist theory more illu-minating, although they depart from this in not necessarily inferring an imminent crisis of the welfare state. In structural terms, they find that one group of doctors, hospital consultants, were effective in obstructing policy implementation at the local level in relation to national priorities and resource allocation, in evading managerial control, and in avoiding policy debate about service evaluation. In each case their aims were achieved, not through open conflict, but through a monopoly of local legitimacy. Harrison *et al.* also conclude that it is only since 1981 that a growing managerial challenge to the dominance of consultants has existed, and further that this shift can be accounted for at a macro-level in neo-marxist terms; they thus conclude that the two approaches relate to each other in a hierarchical fashion.

The micropolitics of policy: discourses of power and needs

The final set of approaches to be considered provide a number of contrasts to the pluralist and structuralist approaches considered above. This third position argues that pluralist approaches remain naive in their views of social interaction and power relationships within society, and that structuralist approaches can be criticized as too objectivisitic, functionalist and deterministic. Here the emphasis is on trying to provide an approach capable of representing human agency, social conflict, and the construction and contestation of cultural meanings in the policy process, and which draws on both structural and interpretative approaches to the study of social life. Such studies argue that in order to understand the development and implementation of policy, we must consider the 'hidden arguments' (Tesh 1988) underlying the social and cultural construction of answers to questions about the legitimate sources of knowledge, the nature of human action and social change, and beliefs about desirable structural characteristics of society.

The view adopted within this type of framework is summarized by Hewitt (1983: 70):

> social policy constructs targets on which power is inscribed (e.g. deprived individuals and neighbourhoods); it provides a capillary through which state power is circulated throughout the social body (e.g. administrative apparatuses affecting family life); moreover by its own power social policy preserves and establishes its interests (e.g. by the self-legitimation of professional, academic and research practices).

Donzelot (1979) examines a variety of case studies of France, from the eighteenth to the twentieth centuries, illustrating the formation of social policy in areas of housing, medicine, and child-rearing, demonstrating how the family was moulded into specific functions (operating through class-differentiated mechanisms), how it came to be seen as a site for specific (state) intervention, and how the persistent failure of families to fulfil their required roles acted as a trigger for ever more sophisticated and systematic forms of discipline (state intervention). Rose (1985) traces similar processes at work in England in the formation of policy in relation to the application of the discipline of psychology to the 'problem' of maladjusted or delinquent children, resulting during the 1920s and 1930s in the creation of the specific institutional form of the child guidance clinic, together with the transformation of the role of the social worker into a therapeutic one.

Lest the foregoing sound too deterministic and fatalistic in its approach, the interpretation intended here is to recognize each node through which discipline is exercised as a potential site for contestation and resistance. One example of this is provided by Sawicki (1991), who presents a Foucauldian analysis of the discourses on the new reproductive technologies, particularly in the USA. She highlights how much of the policy critique of such technologies offered by radical feminists has focused on the dominant discourses and practices governing reproduction, arguing that it has paid insufficient attention to the resistance and struggle that is already taking place in this policy arena, thereby undermining its potential effectiveness.

Within this form of analysis, drawing on the work of Foucault, a particular view of power is often adopted, one intimately linked with knowledge, so that there is:

> [a] constant articulation . . . of power on knowledge and of knowledge on power. We should not be content to say that power has a need for such-and-such a form of knowledge, but we should add that the exercise of power itself creates and causes to emerge new objects of knowledge and accumulates new bodies of information. One can understand nothing about economic science if one does not know how power and economic power are exercised in everyday life. The exercise of power perpetually creates knowledge, and conversely, knowledge constantly induces effects of power. (Foucault 1975: 51–2)

This also involves a very specific view of the way power operates:

> in thinking of the mechanisms of power, I am thinking rather of its capillary form of existence, the point where power reaches into the very grain of individuals, touches their bodies and inserts itself into their actions and attitudes, their discourse, learning processes and everyday lives. (p. 39)

so that power is viewed as something that circulates, functioning in the form of a chain: 'never localised here or there, never in anybody's hands, never appropriated as a commodity or piece of wealth', and where 'individuals . . . are always in the position of simultaneously undergoing and exercising . . .

power. They are not only its inert or consenting target; they are always also the elements of its articulation. In other words, individuals are the vehicles of power, not its points of application' (Foucault 1976b: 98). Such analyses thus aim to 'conduct an *ascending* analysis of power, starting, that is, from its infinitesimal mechanisms' (p. 99) and study 'power at the point where its intervention . . . is completely invested in its real and effective practices' (p. 97). As a consequence, the macro-level and descending (top-down) analyses offered by the perspectives on policy formulation and implementation discussed above are found wanting.

Within such analyses the scope of enquiry is effectively widened, so that policy statements are not treated solely as the outcome of a process of policy formulation, nor are the effects of policy implementation seen as the end-points of enquiry. Rather, the processes of formulation and implementation are seen as affecting and effecting change in the realm of 'the social', defined by Fraser (1989a, 1989b) as 'a site of discourse about people's needs, specifically about those needs that have broken out of the domestic and/or official economic spheres that earlier contained them as "private matters"' (Fraser 1989a: 156); 'it is an arena of conflict among rival interpretations of needs embedded in rival chains of in-order-to relations' (Fraser 1989b: 170). Concepts of 'health needs' provide a key theme which runs throughout the policy process. The production of discourses about needs that necessarily takes place within any policy-making process (and, indeed, also in the process of policy implementation) represents an important arena in which the political claims of different groups, with different degrees of influence and legitimacy, and with different vocabularies and interpretations of need, are contested.

Tesh (1988) demonstrates how the alternative policies on the prevention of infectious disease in nineteenth-century Britain and the USA were closely linked with competing theories as to the causes of disease, which in turn were linked with particular views of the desirability and direction of social change. Examining twentieth-century policies relating to chronic disease prevention in the USA, she identifies how particular links in disease causality are conceptualized in these debates, and in particular how the term 'environment' comes to be understood. One specific example she gives is that of a group of workers exposed to a toxin like benzidine. If the environment is conceived as the passive element which contains the benzidine, policy will tend to be couched in terms of measures like the use of respirators; on the other hand, if the environment is interpreted more ubiquitously, as modified by the presence of the toxin (rather than merely being a passive container for it), then, rather than a barrier approach, prevention policy is more likely to concentrate on removing the benzidine. She identifies similar interactions at work in the debates around different views of stress in relation to the job of air traffic control, manifest in the US congressional hearings in the late 1970s and early 1980s, noting a shift from the attribution of causation to the environment itself (advanced by the air traffic controllers), to particular characteristics of the individual within an environment (advanced by the Federal Aviation Authority): 'while the controllers thought of stress as one of the inherent characteristics of guiding aircraft, their employers tended to regard it as an individual phenomenon experienced only by some workers' (Tesh 1988: 110). This 'conversion of the word *stress* from a term that refers to the overall experience of guiding aircraft to one referring to the body's biochemical reaction to that experience fundamentally changes the debate. It shifts the focus from the conditions under which air traffic controllers work to the people who do the work. It makes the workers, not the job, the locus of the problem' (p. 113). This is one example of how disease prevention policy within the USA is strongly influenced by individualism, constantly framing the questions for debate in terms that accept the social structure as given, and seeking answers in the behaviour of individuals.

The political ideology of individualism is reflected in other aspects of health policy in the USA. Fraser (1989a) examines aspects of social welfare policy in the USA, identifying the processes by which the implementation of that policy constructs welfare recipients and their needs according to specific, and contestable, interpretations, whilst simultaneously supporting the facticity of these dominant interpretations. A particular concern of Fraser's analysis, which we do not have space to pursue in detail here, is how such policies are gendered. Programmes such as Medicaid position recipients as dependent clients or 'beneficiaries of government largess' (Fraser 1989a: 152), requiring considerable and frequently demeaning effort to qualify and maintain eligibility, and subject to intrusive controls and surveillance. This is clearly expressed, for example, in the discussion of Medicaid and its categories of eligibility offered by politician and primary care physician John Kitzhaber (1993): 'the United States has developed a system that makes an artificial distinction between the "deserving poor" (those who fit into a category eligible for state support) and the "undeserving poor" (those who don't)' (p. 374). In contrast, programmes such as Medicare and Supplemental Social Security Insurance (disability insurance for those with paid work records) position recipients as rights-bearing beneficiaries, constituted as receiving what they deserve, and subject to less intrusive control and surveillance (Fraser 1989a). This is also reflected in Kitzhaber's (1993) discussions, where he recognizes the basis of Medicare as 'entitlement', in contrast to that of Medicaid.

Decentralization in health policy

We now move on to consider two particular themes that represent recent trends in health policy globally, beginning in this section with decentralization. This policy theme is interesting for a number of reasons. First, it represents a policy thrust evident in many of the most recent health service reforms, although it is by no means a new policy theme. Second, it illustrates many of the difficulties with the earliest efforts to theorize the policy process and the necessity of more sophisticated frames for analysis. Decentralization can cover a range of very different possibilities. One of the difficulties of policy analysis in this field is that these are not often clearly distinguished in policy statements, so that the interpretation of what is intended is open to contest. At the outset it is helpful to explore briefly some of the different types of decentralization.

Focusing first of all on legal, administrative and political arrangements, four types can be distinguished (Mills *et al.* 1990), although in any particular policy these may exist in combination. The first of these is *deconcentration*, characterized by the handing over of some administrative authority to locally based offices, with, however, no transfer of political authority. This represents the least extensive form of decentralization. A second type is *delegation*, where there is transfer of managerial responsibility for defined functions to local level but indirect control by central authorities is maintained. Under these first two types, the scope for local discretion in decision-making is limited, and they amount to provision of information, advice, and/or services on a local basis. Third is *devolution*, characterized by the creation or strengthening of subnational levels which are substantially independent of the national level with respect to a defined number of functions. Devolved functions often include authority to raise revenue and make expenditures. Within this type, therefore, there is scope for substantial local decisions on policy and priorities. Finally, *privatization* can be distinguished, where there is transfer of functions to voluntary organizations or to private or non-profit-making enterprises, under varying degrees of government regulation, which are often locally based .

Alongside these four types, distinctions also need to be drawn in terms of separate dimensions corresponding to financial authority and the means of representation of the local community. This last dimension is related to the wider issue of participation in health policy and is considered separately later in this chapter, as it has been the subject of numerous separate policy initiatives. Hambleton (1988) also distinguishes a 'service integration dimension' to decentralization, namely the degree to which different services are integrated at local level. This might range from no integration between different services, through co-ordination and collaboration in service delivery across the different types of service, perhaps through team working, to integrated management and generic working. There are also considerable interactions between different forms of decentralization, so that the effects of each are not necessarily clearly distinct from one another in any analysis of policy implementation.

The rationale for decentralization

Decentralization is often argued for on the basis of its assumed ability to facilitate a number of different policy goals: maximizing the opportunities for participation and community involvement; matching of services to local needs; making use of local knowledge; fostering local linkages between sectors; generation of commitment to implement decisions. At the same time, it is perceived to carry the potential for a number of undesirable effects, in particular loss of equity between geographical areas and loss of accountability for the use of any centrally provided funds. The precise nature of the decentralization envisaged or implemented in any particular setting will be influenced by the perceived likelihood of each of these desirable or undesirable features, as well as any difference in importance accorded to each of them. The resultant centre–periphery relationships thus attempt some sort of balancing act. For example, within the UK, the arrangements for the 1974 NHS reorganization (DHSS 1972) made this very explicit with the slogan of 'delegation downwards, accountability upwards', an aspiration criticized by some (Mackenzie 1979) as presenting an 'utterly unintelligible proposition'.

The adoption of decentralization as a health policy goal

The policy of decentralization has been a prominent feature of the health service reforms that have been undertaken in a wide variety of countries over the past twenty to thirty years, in particular the widespread quasi-market reforms discussed in Chapter 5, designed to replace administrative by market methods of resource allocation in the public services (Hoggett 1990). This may be linked to wider changes in the economy associated with post-Fordism, and the corresponding substantial evidence of changes in the industrial structure of many advanced industrial countries towards smaller-sized units of production. The trend towards decentralization seems to be rather general (Loveman and Sengenberger 1991), and this may indicate that technological change (particularly in the area of information technology) is reducing the transaction costs associated with market exchange.

A number of countries have a longstanding tradition of decentralization in health services. In Sweden, for example, decisions on priorities and policies are the responsibility of the county councils, working within the nationally established legal framework (Ham 1992). Many local municipalities in Sweden are proceeding with decentralization of services such as care for the elderly, with decision-making responsibilities transferred closer to the consumer (Lindgren and Prütz 1994).

In the Canadian province of Quebec, decentralization of service provision dates from the early 1970s when the province was subdivided into twelve regions, with the creation of regional councils with limited responsibilities; with effect from October 1992, this has been strengthened into a 'true political rather than only an administrative decentralisation' (Pineault *et al.* 1993). This implements a system of seventeen Regional Boards, with strengthened citizen participation, through an elected Regional Assembly and Board of Directors, whose responsibilities include identification of needs, priority determination, and budget allocation.

The policy of decentralization is prominent among the health service reforms currently being undertaken in a wide variety of countries, for example: Argentina, Brazil, Bolivia, Colombia (PAHO 1993); Bangladesh, Bhutan, DPR Korea, Indonesia, Myanmar, Nepal, Sri Lanka, Thailand (WHO/SEARO 1993); Fiji, Hong Kong, New Zealand, Papua New Guinea, Philippines, Soloman Islands (WHO/WPRO 1993); Burkina Faso (Tarimo and Creese 1990); the Czech and Slovak Republics (Raffel and Raffel 1992); and finally in Poland (Wlodarczyk and Sabbat 1993), where pilot projects are investigating a range of different structural and organizational options for decentralization, again with an explicit aim of improving cost-effective use of resources and providing for the active involvement of communities in health policy.

The effects of decentralization in practice

Writing in the early 1980s, Lee and Mills (1982) concluded that although decentralization had been a strong theme in the regionalization of health services in many high income countries, the reality did not match the rhetoric. As noted above, decentralization has however continued to feature highly in the policy agenda for health service reform. In the sections below we examine some of the most recent studies which examine the effects of various decentralization policies, focusing on a number of different aspects in turn, looking at effects on local responsiveness and equity, at the wide-ranging locality planning initiatives for the decentralization of community health services in the UK, and at the issue of centre–local relations.

LOCAL RESPONSIVENESS AND EQUITY

One possible rationale for decentralization identified above is in terms of allowing locally sensitive practices to be developed to respond to particular local circumstances and needs, and thereby stimulate the provision and delivery of cost-effective care. So that, for example, in the context of the USA, Kimberley (1990) demonstrates how decentralization to state level of the authority to regulate physician investment and referral practices would in theory act to prompt the development of cost-effective competitive practices and remove the barriers to these posed by the federal statute regulation. Studies of decentralization in practice have demonstrated that such hypothetical benefits may be hard to achieve. For example, a study by McFarlane (1992) examined the effects of the Reagan administration's decentralization of federal programmes on the federal family planning programme. Its findings showed that the more decentralized programme produced less responsiveness to individual state needs for family planning, leading to increased inequity in the access of low-income women to family planning services, which became more dependent on their state of residence. These women were the ones whom the national family-planning programme was originally intended to serve, so the implementation of decentralization in this instance was counterproductive for the original policy objectives.

Similar results were obtained from a study of the effects of decentralization of public health services in Norway (Elstad 1990). In 1984 the Norwegian Municipal Health Act allocated the responsibility for primary health care to the municipalities, unifying the previously disparate primary health care providers and strengthening local public regulation and local political control. An analysis of data from seventy municipalities showed that the number of primary health service personnel expanded considerably during 1984–8, but the distribution of services did not become more equitable. Elstad emphasizes that although the formal role of local politicians in the decision-making process increased, the health sector officers and the municipal executives still controlled the evolution of the municipal health services. Furthermore, in the municipalities studied, there were no attempts to develop direct participation by user groups, trade unions, women's organizations, or other voluntary organizations in health policy at local level. Participation was limited to the occasional invitation to comment on draft proposals. Elstad concludes that decentralization does not necessarily lead to more democracy, and that an equitable distribution of public health services can become more difficult to attain – in this instance owing particularly to the use of only local data in considering staffing needs.

Similar concerns have been expressed in the UK, namely that the effect of the recent NHS reforms, in their devolution of the responsibility for funding and provision of specific services, will result in increased geographical inequity as different priorities are adopted in different parts of the internal market. Mullen (1995) discusses the case of the funding of specialist tertiary services as one that poses particular problems: conflict between equity and local priorities; conflict between local and national priorities; and the ability to control expenditure. She evaluates seven different methods for funding such services (all of which are present to some degree in the existing NHS), and finds that none of them succeeds in providing the potential for avoidance of conflict between equity and local priorities, and that, furthermore, all of them involve incentives for either over-use or under-use.

'LOCALITY PLANNING': PANACEA OR . . . ?

Interest in locality planning is by no means a new theme in health services policy in the UK or in other countries. The World Health Organization, in its global Health for All strategy (WHO 1981) and in the strategy for the European Region (WHO/EURO 1985a), recognizes the importance of intersectoral collaboration and community participation, both of which are thought to be facilitated by decentralization and locality planning. One manifestation of this has been the Healthy Cities initiative taking place throughout Europe (Ashton *et al.* 1986). The local implementation of this has already been discussed in Chapter 7, while the international policy issues are considered further in Chapter 9. Outside the UK, locality planning based on community diagnosis has formed the basis for activities in different parts of Sweden (Secretariat for Future Studies 1984; Lagergren 1987). A wide range of similar activities are taking place in other European countries, such as Spain, Portugal, Yugoslavia, France, and Finland (WHO/EURO 1987), and outside Europe in, for example, China, India, Mexico, the Philippines, and Costa Rica (Hardie and Morris 1982; Hardie 1986).

In the UK, decentralization has been part of the policy agenda, although in a variety of forms, since the 1970s (Hambleton 1988). One particularly significant policy of the late 1980s was that of 'locality planning' – the provision of community health services provided by District Health Authorities on a more localized basis. This was supported by several British policy documents which stressed a local focus to health planning: the Green Paper on Primary Care (DHSS 1986b); the Cumberlege (DHSS 1986a) Report on community nursing; and the 1987 White Paper on Primary Health Care (Great Britain, Parliament 1987). District Health Authorities administer relatively large areas, which were originally envisaged as corresponding to the catchment areas of district general hospitals. These were considered too extensive for the organization of primary health care outside hospital with a neighbourhood focus. Thus a new, very local level of health service planning, organization and delivery was emerging, which was intended to be particularly well suited to the provision of community health services. There were considerable differences, however, between the approaches to locality planning adopted in different places, and in particular in the extent to which this represented a real devolution of service planning and delivery or merely a decentralization of provision (this is discussed further below).

General aims for locality planning in the UK were wide-ranging, amounting to devolution of service provision to a very local level, with considerable service integration. However there was no ability to tap local sources of finance: services had to be provided within budgets allocated at the regional level. The overall goal of the locality planning policy initiative can be summarized as the achievement of more effective and efficient use of scarce resources (Taket and Curtis 1989). Important components of this general aim were: to provide services more appropriate to the needs of local populations; to facilitate a team approach to providing primary health care services; to increase collaboration between all those concerned with primary health care services (health authority staff, GPs, local authority staff, voluntary workers); to avoid overlap or duplication of services; to encourage local consumer participation in the planning of services and in the definition of local needs for service; and finally, to make services more accessible to consumers. The extent to which different components were adopted varied from place to place.

Within the UK, the King's Fund Primary Health Care Group investigated the extent to which District Health Authorities were planning to organize community health services on a more local basis (King's Fund 1987). Of the 161 districts replying to the questionnaire, 140 stated that they had plans to decentralize. There was, however, considerable diversity in terms of what exactly this decentralization would mean, and the schemes reported variously emphasized management, budgeting, service delivery, planning, information-gathering, community or consumer involvement. Districts also varied across a spectrum of styles of implementation for locality planning. At one end of this spectrum are what might be called 'bottom-up' initiatives in which involvement of the consumer/community was a central aim, and where, to a large extent, the initial impetus came from the grassroots of consumer or community groups. This was the case, for example, in Riverside and West Lambeth (King's Fund 1988). At the other end of the spectrum are places where the process was initiated in a 'top-down' fashion, that is, launched at district level, by senior managers, with the aim of involving professionals, for example, Wandsworth and Islington, and in other districts where consideration of community involvement was left to a later stage (King's Fund 1987, 1988).

A study of the process of moving towards decentralization in health services in one particular east London district (Taket and Curtis 1989) is instructive in its illustration of approaches adopted to local policy-making. The initial approach followed was almost a classic example of rationalism in intent. A small interdisciplinary working group met to carry out a detailed examination of different factors which affect the choice of appropriate boundaries. The group's work resulted in the identification of a set of 'super-patch' boundaries, each of which was further subdivided into a set of patches. The 'super-patches' were intended to represent the basic management units for community services, with provision of all major services from sites within the super-patch boundary. The number of these, four, was selected to represent a realistic size of area in management terms, taking into account constraints on availability of premises and the financial resources for restructuring. Within each super-patch, an overall manager for services would be appointed. The patches were intended to identify reasonably homogeneous communities within the superpatch boundaries, and also to provide the proposed geographical focus for the working of teams of staff, such as district nurses and health visitors, associated with particular health centres or GP practices. In defining patches, the group aimed to make these respect the 'natural communities' in the health authority, determined by where people live, work, and socialize together, by the socio-demographic structure of the different parts of the district, and by the nature of the transportation networks, both public and private, that affect the ease of travel between different parts of the district. The group also examined the existing working patterns and catchment areas (where these existed) for relevant groups of health service professionals. Finally, the boundaries provided by postcode sectors and census enumeration districts were also considered, because of their importance as spatial units for which information would be available on the population. So far, this followed the rational comprehensive approach.

However, while the resulting proposals were under wider discussion in the district, local elections returned a new council. The local authority then adopted a policy of devolution for its services, including social services, and housing. In deciding the boundaries to be used, the local council made recourse *only* to the old London borough boundaries that had existed prior to the creation of the current local authority (and not to the Health Authority's work on patch planning). The local authority therefore announced its decision to decentralize on the basis of a set of seven 'neighbourhoods', which did not match well with the health service proposals. In view of the importance of collaboration between local and health authority staff in achieving the aims of locality planning, and in view of the extent of devolution of policy and planning intended by the local authority, it was decided to reconsider the question of boundaries within the health authority, and so a second exercise was undertaken to produce a new set of proposals. While the first proposals were produced by an approach which was characterized earlier as representing a rational planning model, the approach followed in this second exercise, in contrast, might be characterized as 'realism and compromise' – a form of pluralism in action.

Decentralization in the community health services in particular is increasingly associated with managerialism (Exworthy 1994), and Exworthy's study of the implementation of decentralization in one district of east London demonstrates the importance of the managerialist agenda in shaping debates about decentralization. During the implementation phase, debate was largely limited to general managers and professional managers who 'debated the policy within an institutional framework which prevented others from challenging their dominance. Their debate concerned the balance of power between themselves, not whether powers should be placed elsewhere.' (Exworthy 1994: 27). Field staff and user or community groups were effectively excluded from most of the discussions. Exworthy also stresses that this was not a simple case of professional interests versus managerial interests, i.e. a struggle between two unitary positions, instead he found a multiplicity of positions within the professional group shaped by a variety of internal and external factors. His study demonstrates the incompatibility of different professional perspectives and organizational cultures which limits the scope for the sort of local co-ordination called for in the new public health models discussed in Chapter 7. Ongoing research within east London is demonstrating other drawbacks to decentralization; in Tower Hamlets, for example, decentralization has made it more difficult for the voluntary sector to participate in service delivery. In order to bid for funds for borough-wide projects, it is necessary to negotiate with seven different neighbourhoods, each with different procedures and time-scales; for many smaller voluntary organizations this represents a considerable barrier to participation.

CENTRE–LOCAL RELATIONS

Another aspect of interest is the extent of change in centre–local relations brought about through the implementation of decentralization. A study by Lee and Mills (1982) of centre–local relations following the 1974 NHS reorganization in the UK found a picture characterized by tension and conflict between the different levels of the health service. There was greater centralization at regional and national levels, while the more local levels of Areas and Districts had been able to protect their freedom of action to a considerable extent. There was strict central control of financial aspects, remaining largely unchallenged by local levels, while other aspects of policies and priorities were more locally decided, and subject to national exhortation rather than direction. Lee and Mills conclude that it is unclear whether the stated policy aim of 'delegation downwards, accountability upwards' was achieved, especially since communications between the tiers tended to take the form of bargaining. Local health authorities were certainly not acting in the role of mere agents of central government; they had rather more local autonomy than is implied by the term 'delegation'.

Later changes in the British NHS were also phrased in the rhetoric of decentralization, and in the latest reforms, enacted from April 1991 (described in Chapter 5), a further shift in the nature of decentralization was envisaged. However, the details of the reforms included a number of mechanisms whereby central political control could be exercised over local decisions on service configuration. The examples given by Paton (1993) demonstrate this, belying the reforms' rhetoric of local control. Paton also finds similar centralist tendencies (both actual and potential) in initiatives on resource management, medical audit, and resource allocation and rationing. Even in the field of human resource management, various mechanisms set up for monitoring local performance, through the Public Accounts Committee and the Audit Commission, provide the means for a centralist influence to be exercised on local employment policies. Paton concludes that, though 'there has been significant operational decentralisation within the NHS . . . [however] on many strategic issues, centralism was increased in the late 1980s and beginning of the 1990s, not diminished' (Paton 1993: 107). Exworthy's (1994) study of debates around decentralization for community health services in one district also illustrates the workings of centralism at the local level, as district management sought to exercise control over the localities into which the district had been subdivided.

In the USA, federal regulation of health services faces cultural obstacles and scepticism among policy-makers, and devolution and competitive strategies ostensibly form the cornerstones of health policy. Nevertheless, Brown (1992) shows how centralized regulation (particularly for Medicare) grew markedly throughout the 1980s, following the failure of decentralized regulation to curb costs sufficiently. Federal policy-makers extended their influence on professional behaviour (through peer review organizations and medical practice guidelines) and budgets (through Medicare prospective payment and the resource-based relative value scale). Within the Medicaid scheme, states' discretion has increased markedly since 1981, although the process of winning federal waivers to use this discretion has limited devolution. Behavioural regulation has increased the heavy micro-management faced by providers in the United States, while budgetary regulation falls well short of the fiscal macro-management (global budgets, for example) that other Western nations use. Brown notes also the inappropriateness of the pluralist model of policy change to the case of development of regulation in the 1970s and 1980s. Federal government formulated and implemented regulatory measures, initially over the strong objections of powerful provider groups, in the hopes of controlling Medicare expenditures (Brown 1992: 35).

DECENTRALIZATION – PASSING THE BUCK?

The conclusions of the studies considered above underline the difficulties in achieving, in implementation, the ostensible aims of decentralization policies. In other words, there is a gap between the policy rhetoric and the reality of implementation. They also illustrate the problematic nature of the rhetoric supporting decentralization policies. This raises concerns that the implementation of such policies may serve ends different from those that have been explicitly set out at the stage of policy formulation. One particular concern is that decentralization policies may shift the focus of attention away from decisions that are effectively made centrally by government (for example about the total level of public finance available to the health service) to their local points of implementation within the health service.

The responsibility for difficult resource allocation decisions may also be shifted without sufficient change in organizational structures to achieve aims such as local participation. Within Canada, calls for greater lay participation in health service decision-making have resulted in the decentralization of contentious resource allocation decisions (such as service priorities and cost control) to the local level, for example in the Quebec government's mental health policy (Boudreau 1991). Within the UK, McKeown *et al.* (1994) argue that decentralization serves as a means for shifting responsibility for policy decisions, particularly in respect of priorities, onto local purchasers and away from the political arena. Paton (1993) finds that the arguments used for setting up the 1988 review of the NHS that led to the creation of the quasi-market are directly supportive of this interpretation. Allsop and May (1993) point to concerns about the lack of structures for local accountability, reflected in a variety of attempts by local managers in purchasing authorities to cultivate legitimacy with local populations, some of which are discussed further in the sections on participation that follow. Furthermore, at the time of writing, the most recent publication (NHS Executive 1994) on the role of the new health authorities (to be formed from the merger of health authorities and family health services authorities) remains ambiguous on issues of local accountability.

Lay participation in health policy

The second policy theme to be examined is that of lay participation in the policy process. This certainly represents one of the common themes in the rhetoric of health policy, couched in varying terms: 'consumer participation', 'community participation', 'lay participation', and by WHO as 'community involvement'.

There have been many different schemes which can be used for conceptualizing different types or degrees of consumer participation in health policy. McGrath and Grant (1992: 77) distinguish three models of consumer involvement: a 'traditional model geared to informing and servicing'; a 'transitional stage geared to consumer involvement in order to receive information and advice'; an 'ideal type geared to consumer involvement within a structure of accountability in order to service or empower'. Similarly Beresford (in Allen 1988) distinguishes a spectrum running from a 'market research model', where service-users are passive providers of information, through to a 'democratic model' with expanded roles and control for service users. Perhaps the most fully developed schema is that set out by Arnstein in terms of a 'ladder of participation', summarized in Table 8.1. This distinguishes eight different degrees of participation: the first three incorporate varying degrees of citizen power, the second three incorporate varying degrees of tokenism, and the last two really amount to non-participation. While this is a useful overall schema for the discussion of participation, care must be taken not to underestimate the dynamic and interactive nature of decision-making in the policy-making and implementation process, in which different interests may at times be in conflict and at times congruent, making a static classification of participation unhelpful or even misleading. Different levels of participation may also be operating simultaneously in respect of different parts of the health service.

Table 8.1 Arnstein's ladder of participation

Citizen control	Citizens handle the entire job of planning, policy making, and managing programmes, e.g. neighbourhood corporation with no intermediaries between it and source of funds.
Delegated power	Citizens hold a clear majority of seats on committees with delegated powers to make decisions. Public has power to assure accountability of the programme to them (i.e. existence of veto in particular).
Partnership	Power redistributed through negotiation between citizen and power-holders. Planning and decision-making responsibilities are shared, e.g. through joint committees.
Placation	Variety of mechanisms allow (selected) citizens to advise or plan but retain for power holders the right to judge the legitimacy or feasibility of the advice, e.g. co-option of hand-picked 'worthies' onto committees.
Consultation	Soliciting of views, etc. through methods such as surveys, meetings, public inquiries. Control of use of input remains with those in power.
Informing	Emphasis on one-way flow of information; no channel for feedback guaranteed.
Therapy	Participation is limited to being the recipients of educative/curative 'therapy' (those who don't agree must be uneducated, uninformed, or 'ill').
Manipulation	Distortion of participation into a PR vehicle, e.g. placing of citizens on 'rubber-stamp' advisory bodies.

Source: Arnstein (1969).

As well as the degree or type of participation, other aspects of the context for participation are also important. Charles and DeMaio (1993) distinguish a grid defined by the decision-making domain and the role perspective adopted by participants, in addition to the degree of participation discussed above. They distinguish three different decision-making domains: broad macro-level policy at national, state, province, or district level; shaping local service delivery, including resource allocation decisions for particular service regions; and decision-making regarding treatment or services to be provided to individuals. The first decision-making domain perhaps corresponds most closely with participation in policy development, while the second and third deal with participation in policy implementation. As Charles and DeMaio recognize, these domains are not necessarily completely distinct. They distinguish between two different role perspectives: first, that of the user – which reflects a concern with the individual's narrowly defined interests; and second, the public policy perspective – where the concern is with some notion of the broader public or community good. Again, in practice, separation between these two may be difficult.

The rationale for participation is often couched in terms of calls for participatory democracy invoking particular notions of citizenship or the restoration of a 'proper' balance of power between governments and the people. This is linked in some cases with the 'new right' and with desires to reduce the role of the state in welfare provision. In other cases it is associated with a variety of ideologies of the left, in particular with the emergence of organizations concerned with rights, self-advocacy, and self-help for service-users and people with disabilities or illnesses. Croft and Beresford (1992) label this as the distinction between 'consumerist' and 'democratic' approaches to user involvement. Other rationales for participation are improving the quality of decisions and increasing commitment to implementation. Consumerist perspectives argue from a basis of concern about cost of public services, dislike of the 'nanny' welfare state (which is perceived as creating and perpetuating dependency), and public disquiet about poor quality, paternalism and lack of response of welfare services. In contrast, democratic perspectives arise from equal-opportunities struggles (both with and without professionals involved), voluntary- and community-sector initiatives demonstrating new types of service and new styles of service delivery, and the development of philosophies of involvement by health professions and by marginalized groups (e.g. the disabled, women, minority ethnic communities).

Rhodes argues that for public services 'the twin values of caring and citizenship provide the rationale for consumerism' (1987: 68), where citizenship incorporates an active participatory concept of the individual. Efficiency (the dominant value of private-sector consumerism) is important but inadequate by itself; the concept of the citizen-consumer suggests additional values such as equity, equal opportunity, and representation and participation themselves. Potter (1988) notes that it is a small logical step from the principle of representation to that of participation – but that this would mark a giant leap in the way most public services are currently run:

> the apolitical nature of consumerism, and the fact that it is grounded in economic theory, means that it is not equipped itself to develop . . . sharing or swapping of roles between the governors and the governed, the administrators and the administrated. . . . consumerism can help authorities to advance from considering individual members of the public as passive clients or recipients of services . . . to thinking of them as customers with legitimate rights and preferences as well as responsibilities. But it will rarely be enough to turn members of the public into partners, actively involved in shaping public services. For this to take place, the arguments must be shifted to the political arena, and some attempt made to develop the ideal of citizenship. (Potter 1988: 156–7)

Thus it can be argued that the full benefits of consumerism will only accrue when it is combined with participation through a strategy for local democracy and decentralization (Hambleton 1988; Pollitt 1988), and when it is 'guided by a normative model of a (potentially) active, participating citizen-consumer, concerned with a range of values, of which efficiency . . . is only one' (Pollitt 1988: 81).

The concepts of participation and consumerism thus have strong links with the debates on citizenship, which are complex and wide-ranging, and saturated with different emphases by all parts of the political spectrum. Sections of the right stress the responsibilities, obligations, and freedom from interference of individual citizens, while the centre and left emphasize the rights and entitlements of the individual, and the obligations of collective society. The spectrum of different types of participation represented by Arnstein's ladder, discussed above, can also be related to the different theoretical perspectives advanced on the role of the citizen, from the highly participative societies envisaged in the writings of Rousseau and Mill, through the more limited variants of Benthamism, to representative democracy where the role of citizens remains more limited – to exercising their right to choice at defined points through elections. Croft and Beresford (1992) note that rights and responsibilities may not always be sharply separable. They quote the example of employment, viewed by many as a responsibility of the citizen, while for some sectors of society (e.g. disabled people, single parents) it can be seen as a right which they are denied through discrimination and barriers such as lack of access and lack of childcare. Fraser (1989a) demonstrates how, in the USA, the social welfare system, through its administrative and therapeutic practices, constructs a passive form of social citizenship, in which the interpretation of individuals' needs is treated as pre-given and unproblematic, rather than fostering active participation in processes of need interpretation.

Analysing participation

In order to analyse and contrast the rhetoric and the reality of schemes for consumer participation, it is useful to consider a number of different dimensions: the scope and nature of the particular participation scheme; the mechanisms and structures involved; access to participation in the process; information requirements; monitoring arrangements. The *scope and nature* of the participation concerns the type(s) of service(s) involved, the geographical areas covered, the level of involvement intended, the type(s) of participants to be involved, and the roles, responsibilities, and rights of each in the process. The schemas of Arnstein (1969), Beresford (in Allen 1988), Charles and DeMaio (1993), Feingold (1977), and McGrath and Grant (1992), some of which were discussed earlier, may be useful in exploring this dimension.

The question of the *mechanisms or structures* to enable participation brings into consideration any necessary enabling legislation, and the nature of any mandates, rules, and regulations. It is useful to recognize explicitly that the nature of participation may vary throughout the policy process. For example, participation in local policy implementation may be constrained by the overall means of financing and providing health services, in ways different from those affecting participation in setting national level policy goals. Thus, implementation of specific local policy goals may be hampered or even prevented by the national policy context.

The third area is that of *access to participation* in the process, entailing an examination of whether those who are intended to participate are enabled to do so. Relevant factors here include provision of necessary support in order to remove obstacles to participation. This may involve taking action on issues such as transport, childcare, relevant interpreting/signing services, provision of appropriate environments for meetings, appropriate facilitation for meetings, and making participation possible in a flexible manner to meet individuals' other commitments, i.e. enabling participation through means other than attendance at fixed-time meetings. Katayama *et al.* (1994) explore the potential of interactive computer-supported media for aiding participation, utilizing various networks, particularly for people for whom it is more difficult to participate in social situations outside the home, through lack of mobility, for example. Ensuring access to participation will also involve providing the resources necessary to initiate and sustain participation (development workers, community activists, training, information, etc.). Another relevant question is how the different perspectives, insights, and inputs of the relevant groups in the population are to be included and integrated in a way which does not perpetuate the marginalized positions of non-dominant groups, whether defined on the basis of age, sex, ethnicity, culture, ability, or class.

The fourth area, that of *information*, has already been mentioned but merits some consideration on its own. Except in the situation where the participation involved is effectively limited to the provision of information by the participants in a one-way flow, the nature of the information resources available to the different parties is usually very different. As Nocon (1989) identifies, ignorance can impede 'partnership' (or indeed other, higher, forms of participation on Arnstein's ladder). What needs to be considered, therefore, is how information is presented to, and used within, the discussion and decision-making processes. Finally, the arrangements for *monitoring* the working of the participatory process can be scrutinized. What safeguards, if any, are built in to ensure the process is working as intended? For example: Do mechanisms for complaint exist? Are participants who drop out followed up to ascertain why (by analogy with exit interviews from employment).

Participation in practice: rhetorics and realities

The rhetoric of community participation or community involvement is certainly widespread. Community involvement is seen as one of the elements of the WHO Health for All strategy (discussed further in Chapter 9), and the eighth report on the world health situation states that, in 94 out of 151 countries, mechanisms for involving people in the implementation of strategies are reported either as fully functioning or as being further developed (WHO 1993a: 52). The report goes on to recognize that there is considerable variation in what is actually encompassed by 'community involvement'. Details of the different models implemented are covered in the reports from the different Regions (PAHO 1993, WHO/SEARO 1993; WHO/WPRO 1993).

In this section we consider some studies of lay participation in various contexts. Broadly following Charles and DeMaio's categorization of different decision-making domains, we first examine participation in broad macro-level policy development (at national, state, province, or district level). We then consider participation in shaping local service delivery, including resource allocation decisions for particular service regions. Finally we discuss participation in decision-making regarding treatment or services for individuals, through examination of consumer choice and participation in service programmes. The second and third of these sections deal with policy implementation as well as formulation.

PARTICIPATION AT THE MACRO-LEVEL: POLICY FORMULATION

In the USA, Checkoway (1981: ix) dates rising public demand for participation in planning and political affairs from the 1960s. Acceptance of the principle of participation was demonstrated in 1974, when Congress enacted the National Health Planning and Resources Development Act to create a network of health planning agencies, Health Systems Agencies (HSAs), which required consumer majorities on agency governing bodies. The creation of HSAs was intended to assure effective planning and consumer participation (Checkoway 1981); however, in practice neither was comprehensively achieved for a number of reasons. Morone and Marmor (1981) identify two of the major sources of difficulty as the extremely limited nature of the HSA's regulatory authority and the assumption that the system of descriptive representation used in the Act, whereby consumer majorities were to be 'broadly representative of the social, economic, linguistic, and racial populations, geographic areas of the health service area, and major purchasers of health care' (P.L.93-641 §1512(b)(3)(c)(i)), could ensure the representation of consumer interests. They argue that such a system of descriptive representation lends itself to tokenism and, further, that barriers to participation are posed by unequal interests and disproportionate resources (including information) available to the different interest groups 'represented' on HSAs. Consequences of this observed in practice include low attendance of consumers at board meetings, and dominance of board proceedings by provider representatives. They conclude that, despite the rhetoric of public accountability that accompanied the 1974 Act, many of the HSA requirements to enhance accountability to the public are necessary but not sufficient to ensure participation, and that HSA boards in practice may be no more accountable to the public than any other federal executive agency. Where health planning agencies have been prepared to adopt innovative approaches to facilitating participation, then participation at higher levels on the Arnstein ladder can be achieved (Checkoway 1981). The approaches required included advocacy approaches that demonstrate a strong orientation to the needs of those not well served by existing services, the use of selection methods for representatives designed to facilitate accountability to different community groups, and the use of community development approaches that apply social and political change principles to health planning.

The mixed success of the HSAs in the USA has not prevented later mandates for participation by consumers on local and state planning boards, for example, of mental health 'clients' and family members on planning councils in the Anti-Drug Abuse Act of 1988 and the Alcohol, Drug Abuse and Mental Health Administration Reorganization Act of 1992. According to one study (Segal *et al.* 1993), 'mentally disabled self-helpers have an increasingly visible presence on local and state systems planning boards'. Similar findings are reported by Bognar (1994) for a programme in a number of states to involve older people with developmental disabilities in a variety of policy development and implementation activities. With suitable facilitation, commitment, and training in respect of all those involved, consumer-initiated group participation could be achieved.

In the UK there are also a variety of mechanisms which exist to provide for some sort of lay voice in policy development. At regional and district levels, authorities and NHS Trusts (provider organizations) have lay members who are often identified with public participation in the NHS; however, these individuals are appointed and not elected and thus are only arguably representative of public interest. There has also been considerable recent concern about the political nature of the appointment process (*Health Service Journal*, 15 Sept. 1994, p. 3; 27 Oct. 1994, p. 3). Studies and reports of lay members' role in policy development (Strong and Robinson 1990; Pettigrew *et al.* 1992; Limb 1994) also show little success in reaching decisions against local professional or administrative wishes. An analysis of health service policy development interpreted within the Alfordian framework of structural interests suggests that the dominant interest is represented by government, and by sections of the medical and health service administrative establishment on particular issues. The challenging interests are sections of the medical and health service administrative establishment and organized consumer groups. For some policy issues this gives rise to interesting alliances of parties that are seen in opposition to one another in other policy arenas – for example, alliances between consumer interests and the medical profession in initial opposition to the policy represented in the UK 1991 health service reforms. Finally, the repressed interests were those of large sections of the community, not organized into voluntary interest groups. At the time of writing, the most recent publication (NHS Executive 1994) on the role of the new health authorities (to be formed from the merger of health authorities and family health services authorities) does not seem to offer any changes in the nature of participation through the constitution of the new authorities. In addition, the proposed code of 'openness' has been criticized as fundamentally flawed for not including sufficient requirements for: trust and purchasing authority board meetings to be held in public; provision of information; and the restriction of 'gagging clauses', whereby staff are preventing from making public disclosures (*Health Service Journal*, 14 July 1994, p. 12; 24 Nov. 1994, p. 6).

Similar lack of widespread participation and dominance by particular interest groups has been noted in a variety of studies. Paton's (1992) study of policy formulation during the 1988–9 prime ministerial review in the UK concludes that, in the course of the review, both Cabinet and government departments were marginalized, with the Prime Minister relying on advice from selected outsiders. He further argues:

> a disturbing aspect of the Review was that the terms of its debate were set by individuals with little experience in health policy and management. This does not exclude them on grounds of intelligence, but it did raise doubts about their suitability for conducting a detailed review about a complex service. The wisdom of the Review was questioned by many close to the Prime Minister. The already massive managerial

and structural changes set in train by the development of general management, if compounded by significant changes in the financing and
provision of health services, would it seemed, render the service
ungovernable. More worrying was that the interesting potential in the
NHS for the fruitful reconciliation of resource allocation and planning
to meet need could be threatened by policies which fragmented either
financing or planned provision. (Paton 1992: 38)

A recent study (Jones 1995) of the review of renal service policy in London
and the surrounding counties, carried out following the Tomlinson report
(Tomlinson 1992) provides another example of limited participation in policy formulation. This found effective exclusion of lay participation in the
policy review; debates were dominated by medical and administrative agendas. Jones argues that this was due in part to the absence of specific mechanisms designed to ensure lay participation, beyond simple seeking of
submissions from interested groups.

Even where specifically designed mechanisms for input into policy are
provided, meaningful participation may not result. Aronson (1993) reports
on the development of long-term care policies for elderly people in the
province of Ontario, Canada, and examines in detail a process of 'public
consultation' for which the rhetoric was explicit about seeking consumers'
'unique perspective on the needs the system must meet' (Ontario Ministry of
Community and Social Services, quoted in Aronson 1993: 370), although it
was much less clear about what would be done with the input received. Her
analysis of the process leads her to conclude that the inputs were sought in
specifically structured ways and were confined to a very narrow focus – discussion of specific suggestions, proposals, and strategies which had already
been formulated. The process was *not* designed to explore current or past
experience, dreams or visions, planning, or debate of priorities; neither was
it designed to facilitate work which might have led to the formulation of
new policy suggestions. The conclusion she draws is that the 'participatory
processes fall far short of their promise to give people a say and some control over policies and practices that affect their lives' (Aronson 1993: 370).

PARTICIPATION IN SHAPING LOCAL SERVICE DELIVERY

As with participation at the macro-level, one approach to participation in
shaping local service delivery is through specific organizational structures
which provide mechanisms for participation. Many of these are relatively
recently instituted, and thus longer-term studies of their effects are not yet
available. Different approaches to participation are influenced by factors
including the recent history of political development in the country and the
degree of centralization in the system. This was well illustrated by the

comparison of the implementation of participation in the Healthy Cities projects of Barcelona and Sheffield undertaken by Smith (1991).

In the Canadian province of Quebec, with effect from October 1992, mechanisms for participation have been strengthened alongside increased decentralization (Pineault *et al*. 1993). A system of seventeen Regional Boards incorporates increased citizen participation, through an elected Regional Assembly and Board of Directors, whose responsibilities include identification of needs, priority determination, and budget allocation. Various measures in the enabling legislation ensure that the prime role is played by ordinary people (those not employed or remunerated by health or social service organizations), who now compose the majority of members on Regional Boards. Within health service-providing establishments also, Boards of Directors consisting of a majority of citizens are also required (Pineault *et al*. 1993).

The 'Oregon' experiment in the United States provides another example of an attempt to allow widespread participation in the process of deciding service priorities. This is an initiative by the state of Oregon to define an appropriate package of care for those eligible for Medicaid by means of a process of ranking different procedures. There were four main components in the process used to produce the list of 709 ranked items published in 1991: collecting data about outcomes from expert panels; producing 'quality of well-being' scores for specific outcomes on the basis of telephone interviews with the public; holding community meetings to elicit the values attached to different broad categories of services; finally, subjective reordering of the list by members of the Oregon Health Services Commission. It was then proposed that conditions falling below a cut-off point, decided by the legislature on the basis of budgetary considerations, would not be paid for under the Medicaid plan. Approval was given by the Clinton administration in March 1993 (Kitzhaber 1993) for implementation of a modified version of the Oregon proposals as a five-year demonstration project. The modification introduced eliminated one of the subjective components (the ratings of the desirability of health states using the quality of well-being scores), use of which was unacceptable to the Department of Health and Human Services (Kaplan 1994). This illustrates the difficulty of gaining acceptance of the use of information based on data which is labelled 'subjective' rather than so-called 'expert' assessments. This can severely limit the scope for incorporating the 'lay' view in policy, and acts to reinforce dominance by professionally based interests.

Although representing an ambitious attempt to provide for participation in priority-setting, the Oregon experiment has been severely criticized on a wide range of grounds. First is the question of the ethics of devising a rationing system specifically for the poor, since wealthier groups could opt out of the proposed package via private health service policies (Strosberg *et al*. 1992). Secondly, there has been a lack of clarity in how community values have been used as against the values of members of Oregon Health

Service Commission and also, more generally, a lack of clarity about details of decision-making process (Fox and Leichter 1991). There is also the problem of lack of equality of access to participation in community meetings (and to participation in telephone interviews); for example, of 1,000 people attending meetings only 5 per cent were Medicaid recipients, while 40 per cent were health service employees (quoted by Warner *et al.* 1991). Kitzhaber, president of Oregon State Senate, acknowledged 'our initial efforts to involve a representative cross section of citizens can and must be improved' (1993: 376). Another criticism is that the method conflates four different dimensions of rationing (Klein 1992). First is the allocation of resources to broad sectors or client groups; second, allocation to specific forms of treatment (within the first); third, allocation of access to individual patients; and finally, allocation of different procedures, etc. to patients who are eligible for care. There have also been criticisms that the method includes implicit discrimination against particular groups of people, such as people with medical disabilities (e.g. Hadhorn 1992). Finally on technical grounds Kaplan (1994) concludes that among the four levels of judgement, the ratings of health states (eliminated in the approved application) were supported by the most evidence of reliability and validity. Thus, although the Oregon experiment is being watched with interest in the USA and abroad (see e.g. Hansson *et al.* 1994 on the case of Norway, and Hall and Haas 1992 on the case of Australia), a number of articles stress its inapplicability for wholesale transfer to health systems in other states or countries.

Within the UK, national policy documents have increasingly encouraged health authorities to elicit public views about the content of health services (Department of Health 1991; NHS Management Executive 1992; see Table 8.2). Further documents offering guidance as to suitable mechanisms have also been provided (Sykes *et al.* 1992a, 1992b, 1992c), and there has been a limited amount of experimentation with eliciting public views in a variety of ways, reviewed in Sykes *et al.* (1992b) and Donovan and Coast (1994). However, as Donovan and Coast (1994) point out in the context of priority-setting, there are many unresolved issues: how views should be elicited; how the 'public' should be defined; who should be involved; how (and if) views should be aggregated; and how preferences should be used. Purchasing authorities have carried out a limited amount of experimentation with processes designed to involve communities in the development of five-year strategies for the development of local services. Research in progress in two London health authorities has revealed that in the first round of consultation and input, participants had *not* been representative of the diversity in the communities served (participants were mainly white, professional, and middle-class). It was suggested that steps be taken to redress this balance deliberately in the next stage; however, the final decisions are still to be made within the health authorities concerned – the implication of this being that the participation involved reduces to consultation (at best). This is supported by the conclusions of McKeown *et al.* (1994) that the shift in health

Table 8.2 The UK policy rhetoric of participation and consumer choice

Working for patients

  as one of its two objectives the aim of giving patients 'greater choice of services available' (HMG 1989a: 3)

Caring for people

  choice and independence underlies all the Government's proposals (HMG 1989b: 4)

  one of the 'key components' of community care is that services should 'allow a range of options for consumers' (HMG 1989b: 5)

Local voices

  Making health services more responsive to the needs, views and preferences of local people is central to the new role of district health authorities and family health service authorities. (NHS Management Executive 1992: 1)

  The aim should be to involve local people at appropriate stages throughout the purchasing cycle: a combination of information-giving, dialogue, consultation and participation in decision-making and feedback, rather than a one-off consultation exercise. (pp. 3–4)

  Health authorities have a dual responsibility to ensure that the voice of local people is heard. Firstly they need to encourage local people to be involved in the purchasing process. Secondly, they need to ensure that providers take account of local views in their activities. (p. 6)

The Patients' Charter

  This means a service that:
  – always puts the patient first, ... in a way responsive to people's views and needs (Department of Health 1991: 5)

  Every citizen had the following established National Health service rights: ...
  to be referred to a consultant acceptable to you when your GP thinks it necessary, and to be referred for a second opinion if you and your GP agree this is desirable;
  to be given a clear explanation of any treatment proposed, including any risks and any alternatives, before you decide whether you will agree to the treatment (pp. 8–9)

  From 1 April 1992, the Government will introduce three important new rights:
  to be given detailed information on local health standards, including quality standards and maximum waiting times. (p. 10)

policy rhetoric in the UK towards the primacy of local voices and the adoption of consumer views has not been matched by the necessary changes in terms of the mechanisms and structures involved. They also identify the dangers of the rhetoric of participation and consumer choice being used as a way of offloading responsibility for policy choices to the local level, an echo of similar findings that were reported in a number of studies of decentralization discussed earlier. This illustrates the importance of the influence of the socio-political context on the implementation of any policies involving participation. Within the restructured NHS, in terms of shaping local service delivery, it is the health authority management or GP fundholder as purchaser of health services who occupies the key position, rather than the consumer (see Paton 1992). GP fundholders may be even more resistant to participative models than health authorities. Bie Nio Ong and Humphris (1994) report a number of different local initiatives in the UK in which Rapid Appraisal (RA) methodology, a participative methodology produced from development studies (McCracken *et al.* 1988), was used. RA can be described (White and Taket 1993) as a semi-structured way of learning, quickly and in a multidisciplinary team, from local people about particular issues. The interesting finding from a number of the studies that Bie Nio Ong and Humphris report was the difference between the priorities expressed by the community, the professionals' own priorities, and the professionals' perceptions of the community's priorities. This raises important questions about the extent to which health authorities or health professionals can act as 'champions of the people' if their understanding of public opinion is imperfect.

As well as the organizational mechanisms for participation, availability of appropriate information is important for successful participation (as discussed earlier). This is illustrated by Short (1989), who reports on a study of an appeal for the purchase of a linear accelerator for an Australian hospital, which was 'successful' in its achievement of widespread community participation in fundraising, resulting in the raising of $1.5 million. Her analysis reveals however that the community was not given the information that would have enabled it to make an informed choice – the community were prompted to act in response to what was presented as an urgent need for local services. After the financial target was reached, it was revealed that it would be a minimum of five years before the equipment was operational – completely undermining the basis on which the community had participated, and leading to widespread community anger when this became known. This information had been available to the medical profession and health service administrators (who represented the dominant and challenging interest groups respectively within an Alfordian framework of analysis), but it was withheld from the community (the repressed interest group in this situation) even though, superficially, high levels of participation were achieved. The study further distinguishes the rather different interests of particular geographical sections of the medical and administrator groups – those locally based in the area, who stood to benefit from the accelerator, and those in Sydney, who stood to lose. Short concludes that, owing to the absence of appropriate community development, the community was unable to challenge the view of the situation offered by the dominant interest group. Their 'participation' in this case would be more accurately described as community manipulation rather than community participation, since the community was never in the position of being able to exercise any effective choice in shaping the policy agenda.

CONSUMER CHOICE AND SERVICE PARTICIPATION

This final section on participation considers participation in decision-making regarding treatment or services to be provided to individuals, through examination of consumer choice and participation in particular service programmes. As Le Grand and Bartlett (1993) point out, there are a number of different aspects to the concept of choice. Choice exercised by some may result in reduced choice for others; for example, the choice of home care rather than residential care by an individual may imply reduced choice for familial carers. There is also a distinction between choice of service (home versus residential care, for example) and choice of provider (between alternative sources of the same type of service). There may also be differences in the underlying rationale for choice: choice as an end in itself, or choice as a means for achieving other ends, such as efficiency, quality, or responsiveness.

There is a vast and growing literature that examines participation in terms of the interaction between individual patient and health professional which is important for the exercise of individual choice (see e.g. the review by Brearley 1990). Within specific types of service, fostering choice and participation may represent part of the prevailing philosophy of care, particularly in sections of the health service under the control of non-medical personnel. In the field of rehabilitation, for example, the non-government and non-organized sectors have developed the policy concepts and practice of 'community-based rehabilitation' (Helander 1993; Peat and Boyce 1993; Peat 1994), an essential component of which is community involvement. Many non-allopathic forms of medicine with a holistic emphasis also stress active participation of the patient with the professional in the business of managing care (see Chapter 5). The remainder of this section considers some examples of policy initiatives in this field.

Morley *et al.* (1983) examine case studies of participation in primary health care (PHC) in rural and urban communities in India, Sierra Leone, Indonesia, the Philippines, the Dominican Republic, Central and South America, and Nigeria. Their analysis demonstrates the importance of achieving participation in which people gain greater control over the social, political, economic, and environmental factors determining their health in order to produce improvements in health status. In a number of the case studies the role of women was crucially important, not just in terms of participation in specific PHC activities within the context of a particular programme, but also in the much wider role of breaking down traditional social prejudices and injustices. Their analysis also demonstrates the possibility of replicating the success of voluntary agencies' small-scale initiatives on a much larger scale with government involvement. This contradicts the often-quoted argument that local initiatives are crucially dependent on charismatic local leadership for success, although Morley *et al.* do stress the importance of appropriate selection and training of project staff. Alongside this, they note the failure of programmes where 'participation' amounts to no more than the government coercing or persuading people to mobilize their own resources to subsidize a government-planned and operated programme in which people have little or no real influence. These conclusions are reinforced by case studies of a wide range of projects working with the urban poor (Harpham *et al.* 1988), which emphasize the importance of basing participation in community development approaches, the important role of voluntary-sector organizations, and the importance of adopting an intersectoral approach. Similarly, Scarpaci (1991) reports a study of non-governmental organizations' provision of PHC services to shanty-town residents in Montevideo, Uruguay, and Santiago, Chile, which demonstrated success in achieving widespread participation. Factors important in this were the non-paternalistic nature of the services provided, their initiation by residents, and their adoption of a wide agenda set by community priorities, i.e. moving outside a narrowly based service and tackling issues such as self-help, housing, and human rights.

Major opportunities for participatory approaches exist in the case of health education or health promotion programmes. Minkler and Cox (1980) contrast two health education projects. Both adopt Freire's educational approach (1968), based on notions of participation, community development, and empowerment. One project in Honduras was highly successful; this involved working with community teacher-leaders (women) who were already socially interlinked. The other project, working with the urban elderly poor in San Francisco, was relatively unsuccessful; factors implicated in this were the premature termination of funds and the lack of well-developed pre-existing kinship and social groups. The authors argue that, for these reasons, functional rather than geographical community groupings should be considered for such projects. A similar Freirean approach is found in a successful alcohol and substance abuse prevention programme in New Mexico (Wallerstein 1993), which works with adolescents in multi-ethnic low-income communities and has achieved high levels of participation.

A particularly interesting case study is provided by the UK, where the health policy rhetoric of the Conservative government during the 1980s and early 1990s has been couched in terms of increased consumer choice and consumer responsiveness (see Table 8.2). Potter (1988) locates the mid-1980s as the time when consumerism began to enter the mainstream of public sector management theory, and in health service terms notes the importance of the Griffiths report on NHS management (Griffiths 1983). Harrison *et al.* (1992) also note the Griffiths team's identification of 'managers lack of concern for consumers' views of the Service' (p. 44) as one of the four elements of their 'diagnosis' of what was wrong in NHS management. Particularly in relation to some of the earlier policy initiatives, studies are beginning to appear which enable us to examine the match between policy rhetoric and reality with respect to this consumerism.

In the UK context, early efforts concentrated on 'hotel' aspects of health services (Potter 1988), revealing a narrow interpretation of consumerism – and the relative power of the medical profession. Relatively early in the current round of interest with consumerism, Winkler (1987) identified the importance of empowerment in achieving effective consumer involvement which moves beyond the limited 'supermarket model' based on customer relations rather than patients' rights. She identified a number of initiatives that had achieved progress in this area, and argued for a key role for CHCs in furthering these developments. She argued that, in order to build on such developments, an organizational structure is required for support, the nucleus of which already existed in the CHC, but which required development and strengthening; to date this has still not occurred. Harrison *et al.* (1992), studying the initial impact of general management in the NHS between 1983 and 1988, concluded: 'the development of greater consumer responsiveness is not an issue which emerged well from our research' (p. 56), expanding:

> notwithstanding some useful progress in respect of quality of service and consumer issues, most of these attempts have proved to be of a superficial 'window dressing' variety. No real or lasting attempt has been made to involve users directly in choices about priorities or service design – even if managers had ideas as to how to go about securing such input. (p. 110)

In comparing Scotland to England they find slower progress in Scotland, which they attribute to differences in the overall socio-political and cultural settings. Pollitt (1988) (from the same study) reports that performance measurement systems are not very accessible or responsive to the citizen-consumer, being non-participative and top-down in design and use. In terms of the policy rhetoric of the 1991 health service reforms, consumer choice was supposed to be operationalized through 'the money following the patient'; in fact, as Paton (1992) amongst others notes, it is the patients who follow the money.

Following a detailed study of the implementation of the reforms in one health authority in south-western England, as well as a general review of the ways in which the quasi-market can operate, Bartlett and Harrison (1993) conclude: 'The extent to which individual patients or users are able to make choices about the health care they receive seems more likely to be restricted by the introduction of a quasi-market than to be enhanced' (p. 91). Here they are talking about choice at the point of use; participation in planning service provision (which also provides an opportunity for the exercise of choice) was considered in the previous section. Croft and Beresford (1990) report similar conclusions for the social services, finding that most of the user-involvement initiatives amount to either therapy or manipulation in Arnstein's terms. In some instances, choice has been moved from the individual service user to a care manager under the new policy for community care, 'Caring for people' (see Table 8.2). The budgets that care managers have will be funded partially by the elimination of the residential care allowance element in social security (which used to be under the individual's control). At the same time, however, choice has been increased, in that funds can be spent on residential or community care (Le Grand and Bartlett 1993). Hoyes and Means (1993) point out limitations to consumer choice of community care, depending on individuals' power to exercise choice and on the existence of genuine alternatives, and the dangers of choice by some, leading to residualized services for the remainder. In case studies in two counties they found that most consumers did *not* have a choice of providers for a given service.

The latest consumerist policy initiatives in the UK have included the Patients' Charter, first circulated in 1991, and one of a number of different consumer charters set in motion by the Conservative government (see Table 8.2). As Paton (1993) points out, this tends to undercut the decentralist moves of the 1991 White Paper reforms by the use of centrally determined norms, for example for waiting times. These may have effects that people regard as undesirable, such as lack of adequate attention to urgency in order to ensure admission before the charter limit for less urgent cases, or a shift of cases from elective to emergency workload, something that has been observed in 1994 and is causing great concern (*Health Service Journal*, 29 Sept. 1994, p. 3). In addition, other 'rights', such as the right 'to be referred to a consultant acceptable to you, when your GP thinks it necessary, and to be referred for a second opinion if you and your GP agree this is desirable', are becoming increasingly difficult to reconcile in practice with the use of contracts in the health service internal market, without unrestricted access to extra-contractual referrals.

Le Grand and Bartlett conclude (1993: 219):

> the reforms that involve the district health authority as a purchaser of health care and the social service department as a purchaser of social care do not seem to hold out much prospect of gains in terms of efficiency, choice and responsiveness, but may not have much adverse impact on equity either; whereas the housing, education GP fundholding and care-management reforms seem to hold out the prospects of real improvements in efficiency, responsiveness and choice, but, unless the incentives for cream-skimming are reduced, may have a detrimental effect on equity.

Rodgers (1994) restates the arguments that, without empowerment, consumer involvement will remain merely rhetoric, and concludes that progress so far in the health service is only limited, with more widespread success (although by no means complete) in the social and community care field, particularly focused around use of advocacy as a key part of the process.

Given these arguments about the importance of empowerment and advocacy in achieving meaningful participation (also found in the studies from outside the UK we considered earlier in this section), we conclude by look-ing at some of the studies in the UK that have incorporated these ideas. In terms of participation in local level policy content and implementation, a number of voluntary-sector-based projects have demonstrated the success of advocacy; these include advocacy for people with disabilities (Butler and Forrest 1990), the use of bilingual advocates (MORI 1994), the possibilities of achieving self-advocacy (Oliver *et al.* 1994) by people with disabilities, and also the mental health system survivors movement (Chamberlain 1988). In the field of public health, Coombes (1993), in her study of the work of three different public health groups working within an area of east London, found that it was the voluntary sector group that was most successful, in terms of the work carried out, and in the achievement of participation in the running of projects and the determination of future strategies (i.e. in terms of both policy implementation and policy formulation). In the field of com-munity care, Biehal (1993) discusses the possibilities for participation of elderly people in determining the social care they receive, and stresses the importance of a clear framework of rights for service-users as well as prac-tice strategies to ensure those rights in day-to-day contacts. She also, how-ever, identifies exclusionary practices by agencies and front line staff which work against this (see also Rodgers 1994). Similarly, Dowson (1990) describes how service-providers managed to keep 'self-advocacy' safe, so that it became a means of controlling people with learning difficulties rather than enabling them to take control. These provide a good illustration of the micro-politics of policy implementation at work, with resistance to, and contestation of, the views of traditionally dominant groups of professionals or managers not always being successful.

There are obvious tensions between the adoption of community develop-ment approaches (with positive results and participation moving towards the upper rungs of Arnstein's ladder) and the interests of hitherto dominant interest groups, particularly health service providers and the medical pro-fession. This is played out particularly in the encounter between individuals that occurs every time health services are utilized. Given a desire for empow-erment of the 'consumer' in this process, what mechanisms are available to assist this? Adopting a Foucauldian view of power/knowledge, Fox (1993) argues that if empowerment is to be achieved, and the interaction between patient and professional is not to reinforce dominance and dependency, then the basis for this relationship needs to be rethought. He uses Deleuze and Guattari's (1984, 1988) perspective on desire as a positive investment as productive (this is in contrast to the equation of desire and lack found in traditional Freudian or Lacanian psychoanalysis). Fox also draws on Cixous's notion of 'feminine desire' (which, he argues, presents a similar positive reading of desire orientated around characteristics such as generos-ity – leading to the gift of care, confidence, commitment, involvement, delight, esteem, accord, admiration, curiosity), and the Nietzchean idea of the eternal return (put crudely, 'whatever you will, will it in such a way that you also will its eternal return'). These ideas are used to suggest a way of enacting (a politics and ethics of) the relationship between 'carer' and 'cared-for' in the health service setting that is based on positive empowering qualities, rather than bound in the repetition or re-inscribing of dependency, possession, and control, which, he argues, characterizes much of existing 'professional' practice. This reformulation, he argues, makes it possible to live a commitment to the Other, to difference, and offers the possibility of co-participation in health decision-making at this most local level.

Setting and implementing policy agendas: challenges for the future

One theme running throughout this chapter has been the identification of the gaps between policy rhetoric and the reality of policy implementation, underlining the necessity for research to examine *all* parts of the policy process. We have also illustrated the inevitably problematic and contested nature of the rhetoric underlying different policies, particularly in the case of our examination of decentralization and participation. While it is common to find a shared rhetorical language among the policy statements produced by different political agendas, in implementation these become taken up in rather different ways. Ideology is a key explanatory variable in disease prevention policy (Tesh 1988; Mills 1993), and electoral short-termism can be an important determinant of policy in a liberal democracy (Hann 1993).

We have also identified how individualism underpins a number of different health policies. It lies behind much of the current health service reforms within European countries (see Chapter 5), and behind some of the policy moves towards consumerism discussed in this chapter. It also underpins disease prevention policy in the USA (Tesh 1993) and the UK (Mills 1993), graphically seen in the case of tobacco, where the focus is on measures to stop individuals smoking rather than on measures to tackle the pressures and inconsistencies involved in continuing to allow advertising (either direct or indirect in the form of brand-naming and sponsorship). There is a complex combination of political, ideological, and economic reasons underpinning contradictory, and thus not wholly effective, policies in the tobacco field.

There is also a need for detailed analysis of how policies acquire and communicate meaning. Such analysis needs to be interpretative if it is to satisfactorily explain the different features in policy implementation. A graphic example of this is provided by Yanow (1993) in a study of the Israel Corporation of Community Centers, an agency created to implement national social policies, carried out over the period 1969–81. This finds that the agency might have been seen as having failed to implement its policy mandate, according to measures accepted as objective and factual (such as gaps between two major population subdivisions). However, it was acclaimed a success by most of its stakeholders, because it acquired meaning by 'validating the "under-dog's" claim to government attention while at the same time validating the values associated with the "establishment"' (Yanow 1993: 53).

The contested nature of the policy process links back to our earlier discussions in Chapter 3 on contested views of health, as we can see many of those debates refracted and reflected in different policy disputes. Some of these are examined further in the next chapter, where we explore health policy at the international level, and in particular focus on the work of the World Health Organization and the World Bank.

|9|

The widening international perspective on health

Introduction

Recently we have seen a growing influence of international organizations concerned with health in the fields of health policy and health services system restructuring. The realization is growing that health is not an issue that can be discussed only in national terms. This is important in at least two senses, first through the increasing internationalization of health policy discourses. Here the roles of organizations such as the World Health Organization are particularly important, and we consider later in the chapter particular examples of the potential and limitation of actions such as WHO's strategy for Health for All by the year 2000.

A second type of internationalization operates through the linking of health and development, and the emphasis within international aid on health and health-related development infrastructure; here international organizations such as the World Bank and the international aid agencies come onto the scene. There are numerous different organizations involved here. Multilateral organizations, which include UN specialized agencies such as WHO and financial multilateral agencies such as the World Bank and various regional development banks, have universal areas of concern and play a key role in influencing development through the leverage of financial resources. There are also a variety of bilateral agencies of donor countries which can also influence health policy by the magnitude of their financial resources. Lastly, numerous private organizations are involved in the business of aid, with corresponding effects on policy; these include universities, foundations, professional associations, non-governmental organizations (NGOs), and industries. A good overview of the main agencies, their sources of funding, aims, key programmes, and regions of operation can be found in Umhau *et al.* (1991). The World Bank's 1993 development report provides an estimate of $4,794 million as the global total for external assistance to 'developing countries' for health in 1990 (World Bank 1993: 166); of this total some 40 per cent was administered through bilateral agencies, 33 per cent through UN agencies (including WHO), 17 per cent through

NGOs, 8 per cent through development banks, and the remaining 2 per cent through foundations. This expenditure on health represented some 6 per cent of the total official development assistance funds, a decline from the figure of 7 per cent obtained in the period 1981–5. While a detailed examination of this second aspect of internationalization is outside the scope of this book, a section in this chapter looks at some of the policy themes involved.

A further aspect of this second type of internationalization of health is the impact on health of social and economic interactions within and between countries, through impacts on the environment and consequent effects on health (see Chapter 7), through trans-frontier pollution (the effect of the explosion of a nuclear plant at Chernobyl is an example), and through the globalization of trade and industrial activity. This chapter begins with the last of these aspects of internationalization by offering a brief historical perspective on the origins of international organizations with specific roles in relation to health, before moving on to focus on the structure, organization, and functions of WHO. We then examine some of the more recent orientations in WHO's work, in particular the Health for All strategy and related programmes of policy development, and look at the debates around the role and effectiveness of such approaches. Another recent entry into international discourses on health and health policy is provided by the World Bank's 1993 *World development report*, which takes as a central focus the notion of 'investing in health'; the implications of this report are considered and contrasted with the Health for All strategy. The final section outlines some of the important challenges for the future in terms of the role of international organizations in health and health policy.

Health and internationalism: a brief historical overview

International activity in relation to health and health services has a long history; particular strands include its origins in, and links to, the history of colonialism and imperialism, and a concern with the field of quarantine and the fight against infectious diseases. The influence of colonialism and imperialism has been manifested overtly through a concern with the health treatment of the subjects of the imperial powers while they were posted abroad and, to a much lesser extent, with improving the health of colonized peoples, and covertly in a variety of measures undertaken with the effect of establishing or maintaining colonialism and reinforcing the political rule of colonial powers over subjected indigenous populations. The history of Western medicine overseas, particularly during the period from 1815 to the Second World War, illustrates the role of medicine as a 'tool of empire'

(MacLeod 1988: 2), enabling the expansion and consolidation of political rule.

To take a specific example, imperialism can be linked to the development of allopathic health services in Africa in three different ways: first, through the activities of Christian missionaries in establishing medical missions (Good 1991; Ityavyar 1992); second, through the establishment of colonial medical services, including the establishment of hospitals, initially to provide for the health of Europeans living abroad (Sanders and Carver 1985; MacLeod 1988; Itavyar 1992); and third, by the provision of a more limited service for the indigenous African populations employed, this latter being 'important only so far as it maintained a healthy labour force which enabled higher productivity and hence higher surplus value and profit for the capitalists' (Ityavyar 1992: 72; see also Sanders and Carver 1985).

Medical arguments were also used to support particular features of colonial rule. For example, Marks and Andersson (1988) and Bell (1993) document different ways in which arguments supporting segregation in the South African urban environment were supplied by medical personnel and various literatures in the late nineteenth and early twentieth centuries, utilizing the warrant of medical 'science', but in fact based on unmarked and unfounded racist assumptions linking 'native' populations with disease and degeneration. In a study of sleeping sickness in northern Zaire during 1900–40, Lyons (1992) examines how the administrative features of the campaign against sleeping sickness and public health measures implemented by the Belgian colonial administration formed an integral part of the harsh conquest and reorganization of society by the colonial power. Sleeping-sickness policy was yet another feature of Belgian domination, involving enforced labour movements and other ruthless alteration of people's lives. The design of the measures adopted was based on flimsy evidence and neglect of social and economic factors, resulting in increased hardships and suffering for the indigenous populations, who adopted a variety of strategies of resistance in order partially to circumvent the humiliation and hardship inflicted by the policy measures. Lyons concludes: 'in answer to the question whether medicine was a form of imperialism – an effective mechanism used by the Belgians to impose their authority over subject people, the answer must be yes' (p. 225). Another study, this time of Tanganyika (Little 1991), demonstrates how British administrative policies in Africa were a key factor in precipitating changes in the diet of the Sukuma ethnic group in north-west Tanganyika, resulting in declines in standards of nutrition. The analysis offered by Little further illustrates how racist attitudes on the part of the colonial administration enabled the evidence of poor nutritional status to be assumed to be a continuance of the pre-colonial situation rather than a colonial creation, using the myth of a pre-existing problem, a deficiency in 'the normal diet of the African', which was directly contradicted in fact by the data available on the nutritional value of the pre-colonial diet of the group studied. Lado (1992), in a study of selected societies in Kenya, Ghana,

Nigeria, the Gambia, and Sudan, further illustrates how adequate food security was common before the imposition of European colonialism, but was disrupted by the land use changes imposed through colonialism.

A key role in the internationalization of health has been played by bodies such as the Rockefeller Foundation, which during the period 1913–19 had public health as its major preoccupation (Fisher 1978). One of the mechanisms by which influence was exerted was through the creation and support of educational establishments. Fisher summarizes this role of the Foundation, and in particular examines its support for the creation of the London School of Hygiene and Tropical Medicine, which the Foundation regarded as particularly important in transmitting the Foundation's influence throughout the Empire that Britain then controlled, and in helping to prepare the ground for increasing investment of American capital abroad. The Foundation also exercised its influence more directly through involvement in specific programmes in different countries. Solórzano (1992) examines the Rockefeller Foundation's yellow fever campaign in Mexico over the period 1920–41. His analysis illustrates how the campaign, initiated at a time when there were strong anti-American sentiments within Mexico, served to advance the economic and political interests of the United States, as well as those of a particular section of the Mexican medical profession and the incumbent President of Mexico. Solórzano argues that the need for the campaign was highly questionable, since yellow fever was not a salient disease affecting the general population, having been effectively controlled during the period 1910–19 by a Mexican programme involving vector control, environmental sanitation, and the provision of drinkable water in homes. The epidemic of 1920, which prompted the Foundation's proposal for a campaign, primarily affected the Mexican army and was hampering the President's efforts to pacify opposition throughout the country. The Foundation's interests were twofold: avoiding the spread of the fever to the United States, and guaranteeing its own international reputation. Its campaign passed through several stages: early involvement included offers of vaccination to troops (thereby indirectly supporting and strengthening the position of the President against domestic opposition); later stages in the campaign, including daily interaction of Rockefeller Foundation doctors with the population, resulted in a transformation of anti-American sentiments, alongside the eradication of the fever. Longer-term effects were the shaping of the modern Mexican health service system along the US model, with an emphasis on biomedicine and a neglect of the social and economic origins of disease, and, Solórzano concludes, the establishment of diplomatic relationships between Mexico and the USA.

These studies discussed above and those reported in Arnold (1988), MacLeod and Lewis (1988), Falola and Ityavyar (1992), and Turshen (1984) illustrate the wealth of historical and contemporary material now becoming more widely available which analyses the colonial past and post-colonial legacies. These, in addition to Doyal (1979), Navarro (1982),

Banerji (1984), Sanders and Carver (1985), and Kanji *et al.* (1991), provide useful overall analyses of how current health concerns in low-income countries have their origins in imperialist and capitalist relations at the global scale. Andersson and Marks (1989) warn, however, against too simplistic a reading of the effects of colonialism, illustrating, with reference to southern Africa, that while most of the countries of the region are in some sense part of the 'periphery', and a product of colonialism, these labels are insufficient to explain the differences between them in terms of disease patterns and health service systems. An adequate explanation must also consider the specificities of internal social dynamics, including local class, ethnic and gender struggles, and political conflicts. Andersson and Marks conclude that, while the position of the state in the international and regional economy, its specific form, and the nature of its class relations are predictors in some sense of health and health services, a variety of micro-level political and social decisions and their interactions must also be taken into account.

Imperialism also had a particular relationship to the health situation in the home countries of the colonial powers. Cantor (1993) offers a fascinating illustration of this 'domestic dimension' in his study of twentieth-century interest in the rheumatic diseases as an example of the imperialist ideology of British medicine from the 1920s through to the 1950s. His work documents clearly how the institutional structures of the rheumatism campaigns were used to convey and reinforce imperialist sentiments in the British population, as well as how such sentiments were also embedded in the conceptions of rheumatic diseases and appropriate therapies of the time. He concludes:

> the imperialist shaping of British medicine was more extensive and more long-lived than we had thought. Even medical research and charity dealing with rheumatism, that classic ailment of cold, damp Britain was seen in an imperial context by its supporters before World War II. In the post-World War II era, Britain tried to regain its imperial past by exploiting the political and economic dependence of its African colonies . . . Once again, research into arthritis and rheumatism embodied this imperial dimension, reviving ideas and images that stretched back seventy years or more. In a world where the tide had begun to turn against British colonial rule, imperialism still retained remarkable cultural force for medicine. (Cantor 1993: 493)

While actions by individual colonial administrations were affecting the international picture of health and health services, international organizations with specific health functions were also being created to support some of the imperial concerns. In 1851 the first of a series of international conferences was called to deal with quarantine regulations and the first International Sanitary Convention was adopted in 1903 (Bennett 1991). The Pan-American Health Organization was created in 1902, initially as a Sanitary Bureau with a mandate of 'attempting to simplify the basic problem of the numerous quarantine regulations that were hindering trade among the nations of the hemisphere' (quoted by Guerra de Macedo 1991: 163–4), and in 1909 the International Office for Public Hygiene was set up in Paris, dealing only with quarantine matters (Bennett 1991).

In 1923 the Health Organization of the League of Nations was set up. This was part of the economic and social activities which, although originally subsidiary to the League's concerns with the maintenance of peace, grew in importance throughout the life of the League. The Health Organization carried out programmes of epidemic control, research, publication of health information, standardization of vaccines, sponsorship of conferences, etc., as well as providing some direct assistance to governments. It was highly regarded as one of the most successful of the League's social and economic divisions (Bennett 1991), and became the main centre for international activities in promoting improved world health standards.

The Health Organization of the League was superseded by the creation of the World Health Organization in 1948 as a specialized agency of the United Nations. Its constitution, adopted in 1946 by the International Health Conference held in New York City and signed by representatives of 61 states, contains as the objective of WHO: 'the attainment by all peoples of the highest possible level of health' (WHO 1988: 2). According to its constitution, WHO is an organization of member states co-operating among themselves and with others to protect and promote the health of all peoples. WHO is the major organization involved with directing and co-ordinating international health work; it has become one of the largest specialized agencies of the UN in terms of budget and one of the largest in membership (Bennett 1991).

WHO: structure, organization, and functions

Globally, WHO has over 180 member states; as of 31 December 1993 the exact figure was 187 members and 2 associate members (WHO 1994). The headquarters of the organization is in Geneva, where the governing body, the World Health Assembly (WHA), meets annually. The World Health Assembly is made up of a delegation from each member state, and is the body responsible for initiating, adopting, implementing, monitoring, and evaluating WHO activities; in particular it must approve WHO's programme of work and budget. The forty-sixth World Health Assembly approved an effective working budget of $822.1 million for 1994–5, a decrease of 3.5 per cent in real terms from 1992–3 (WHO 1994). Where necessary the assembly will vote on matters, with each member state having one vote; however, it tends to operate more on a consensus system. Delegations to the WHA usually comprise ministers and senior civil servants. An Executive Board consisting of thirty-one technically qualified individuals drawn from member states is responsible for overseeing the implementation of the policies of the WHA. Implementation is the responsibility of the Secretariat, which comprises technical and administrative staff and the Director-General of the organization. The Director-General is elected every five years, and has a constitutional role as chief technical and administrative officer. The post is currently occupied by Hiroshi Nakajima, who is the first Japanese to head any of the UN's specialized agencies; he was elected to the post in 1988.

Many of the activities of WHO are decentralized through the regional structure of the organization, with six regional offices: African (Brazzaville); American (Washington DC); Eastern Mediterranean (Alexandria); European (Copenhagen); South-East Asian (New Delhi); Western Pacific (Manila). The Pan-American Health Organization (PAHO) has a formal agreement with WHO to serve as the American Regional Office for WHO, while maintaining its own identity. PAHO is thus a part of the UN specialized agency structure, but also manages health programmes for the inter-American systems of the Organization of American States (Umhau *et al.* 1991). Each Regional Office has a Regional Director as its chief executive. The policy-making bodies in each region are the Regional Committees, which meet annually, either at the regional office or somewhere else in the region. Like the WHA these comprise delegations from each member state in the region, each with one vote. Regions are not organized along strictly geographical lines, as member states can apply to join any region they wish. So that for example, the European region contains Turkey, Israel, and the countries of the former USSR, covering a total of some 850 million people. Finally, in selected low-income countries, there is a Country Office for WHO.

The constitution of WHO gives it a number of distinct powers. Under Article 19, WHO has the authority to adopt *conventions* or *agreements*, which come into force for each member state when accepted in accordance with its own constitutional processes. WHO has not so far used this constitutional possibility. Under Article 21 the WHA may adopt *regulations* concerning a limited number of subjects (sanitary and quarantine regulations, other procedures to prevent the international spread of disease, nomenclature, standards with respect to diagnostic procedures, safety, purity and potency of biological, pharmaceutical and similar products, and the advertising and labelling of these products). Under Article 22, member states may give notice of rejection or reservations about regulations within a certain period. To date, only two regulations have been adopted: the International Nomenclature Regulations and the International Health Regulations (these last are sanitary in nature). Finally, under Article 23, WHO can adopt *recommendations* which do not have legal force. This is the most common mechanism in use. WHO thus does not represent pure supranationalism, having no directive powers over member states, unlike the European Community, for example, whose directives are binding on member states. Its role is mainly one of international diplomacy designed to encourage the member states themselves to set up legislation and policies beneficial to health.

The work of WHO falls into several distinct areas, introduced briefly below; later in the chapter we examine how the balance between these different areas has changed over time. The first area is direct involvement in specific programmes of work and technical co-operation, including work on nomenclature and standards. The classic successful example of such a programme is the global eradication of smallpox (Fenner *et al.* 1988). This was first adopted as a goal in 1958, made the subject of an intensified programme in 1966, and declared achieved in 1979 with the last recorded 'natural' case occurring in 1977. The only exception to eradication is the maintenance of two known stores for 'research purposes', with the associated risk of 'breakout' cases. The campaign for malaria eradication has been less successful (Learmonth 1988; Gish 1992), complicated by the intensification and spread of resistance of parasites to antimalarial drugs, by resistance of mosquitoes to insecticides, and by the impact of irrigation projects in environments where previously most mosquito vectors were unable to survive; this has now been replaced by a strategy of control (WHO 1993b). There are also many specific projects on a more local scale, for example on immunization and vaccination, health information systems, and health service system development.

Secondly, WHO has a role in facilitating exchange of expertise, technology and information, and in fostering and supporting inter-country collaboration, along with particular support and co-ordination of training and research. WHO also has a role in policy development through both global and regional activities; this is examined in detail later in this chapter. Finally, WHO is also involved in lobbying at international level for funds (from international banks and from multilateral and bilateral agencies) to support specific activities or programmes of work, and in acting as the technical agency supporting such programmes. WHO administers a substantial proportion of the budget of the UN Children's Fund (UNICEF), and also administers programmes and funds as a part of the UN Development Programme. WHO also has particularly close collaborative relationships with UNICEF, the Food and Agriculture Organization (FAO), the International Labour Organization (ILO), and the UN Educational, Scientific and Cultural Organization (UNESCO).

The balance of work carried out by WHO varies from region to region; historically there have always been much larger technical assistance programmes in non-European regions, although the scale of activity being undertaken now in relation to the Eastern European countries, where there are a large number of countries in need of direct assistance (of one or more different types: emergency relief, reconstruction, development), means that it has become increasingly apparent that it is not realistic to view the European region as uniformly highly developed.

In terms of the balance of work between the different functions, the lowest emphasis has traditionally been given to policy development. Perhaps one of the key changes that has been occurring in parts of WHO recently is that during the late 1970s and 1980s the task of policy and strategy development has been taken up in a much more proactive way. Walt (1993) characterizes WHO's approach during the period 1948–73 as being one of caution and stability. Later in this chapter we examine some of the changes in WHO's approach that have occurred from the late 1970s onwards, in particular focusing on policy and strategy development and examining some of the tensions that run throughout this work. One of these that recurs repeatedly can be characterized as a tension between radically different conceptions of 'health', exemplified in the two WHO definitions of health (discussed in Chapter 2): 'Health may be expressed as a degree of conformity to accepted standards of given criteria in terms of basic conditions of age, sex, community and region, within normal limits of variation. It is a relative concept' (WHO 1957: 14); 'Health is a state of complete physical, mental and social well-being and not merely the absence of disease or infirmity' (WHO constitution, quoted in WHO 1988: 1). The definition offered by the measurement study group is a normative definition, closely linked to the biomedical model, and could well be rephrased as: 'Health is an absence of disease.' Its emphasis is on 'objective' measurement, and it is highly problematic in terms of its use of 'normal limits of variation'. On the other hand, the WHO constitution definition emphasizes health as a positive quality, conceptualizing health as an ideal goal, not as a state but a task or process. The first definition relates to the initial domination of WHO by the medical profession (and is still strongly linked to the allopathic view), while the second definition is more associated with the rather different perspectives on health offered by other health professionals, including nurses and non-allopathic medical practitioners. The implications of this are explored later when WHO's Health for All strategy is discussed.

Initial focuses of WHO's work

As Doyal (1979) and Navarro (1984), amongst others, point out, the early work of WHO was strongly influenced by legacies of colonialism and imperialism, characterized in particular by its technical focus on response to specific disease problems (smallpox, malaria, etc.), narrow conceptions of the scope of disease eradication campaigns, and the dominance of biomedicine. A related issue is the emphasis on the mechanism of population control in addressing 'problems' of so-called third-world countries (involving other international agencies as well). There was no corresponding recognition of the social and economic forces that would destine such policies to failure, nor of the effects of such policies in terms of distortion of patterns of social life in places – see the discussion by Doyal (1979) in particular for an expansion of these arguments. WHO's early attitude was also characterized by an unquestioning acceptance of the superiority of Western allopathic medicine, and of the importance of hospital provision and the continued development of high-technology medicine. This might be seen as a direct legacy of the concerns, discussed earlier, of colonial powers to provide within their colonies the same configuration of services as was available to their home populations (and for the use of these populations while away from home).

In terms of the work of WHO, the success against smallpox but failure as far as malaria is concerned can be seen as a direct result of the concentration on technical solutions (use of insecticides in the case of malaria), rather than on general environmental health improvements. A contrast here is provided by China's success in malaria control through low-technology mass action versus the failure represented by the development of insecticide-resistant mosquitoes and drug-resistant forms of malaria elsewhere. Rising concern about the effectiveness of such programmes gradually resulted in a shift in the focus of WHO programmes away from the technically based and towards those related much more generally to wider health problems and their social and economic determinants. This changing focus implies of necessity an increasing politicization, as the links between health and social, environmental, and economic factors are explicitly acknowledged. This has resulted in increasing tension within WHO and a resistance (although not often overt) from many countries. There exists a sharp contrast between many national representatives who argue that WHO is becoming 'too political' and critics who argue that it is not political enough. This is particularly evident in reactions to the Health for All strategy (considered in the next section), which represents one of the main focuses of WHO's activities more recently. Walt (1993, 1994) summarizes other instances of the resistance to WHO's stance on certain issues – for example, the USA's withholding of financial contributions at times owing to disapproval of particular policies (e.g. the essential drugs programme), and the US vote against the code on breast milk substitutes, passed by the WHA in 1981, on the grounds that this represented interference by WHO in global trade.

Modern orientations in the work of WHO: health for all?

The basis for WHO's Health for All strategy is the objective of WHO enshrined in its constitution quoted earlier: 'the attainment by all peoples of the highest possible level of health'. This provided the policy basis for the passing in 1977 of a key resolution by the WHA, stating that the main social target of governments and WHO should be the attainment, 'by all the people of the world by the year 2000, of a level of health that will permit them to lead a socially and economically productive life'. The resolution become popularly known as 'Health for All by the year 2000', (after the phrase 'Health for All' originated by the then Director-General of WHO, Halfdan Mahler), and later abbreviated as HFA2000 or HFA.

The origins of the HFA programme have been traced by Bryant (1980), who identified a number of key influences including: changing ideas on the nature of development and in particular an increasing rejection of the 'trickle-down' concept; the increasing importance of debates over social justice and equity; and the recognition of enormous health problems still remaining, particularly in low-income countries, as illustrated in Chapter 4. An increasing focus on the importance of comprehensive primary health care (PHC) services was strongly connected to the ideas of HFA. Although the concept of PHC had existed for some time, it was only around about the late 1960s and early 1970s that health policy began to stress its particular importance. This can be seen as arising out of dissatisfaction with the technological orientation of much of Western medicine and with the rising cost of health services, and out of concerns about the lack of effectiveness of existing hospital-orientated health service systems in resolving priority health problems, together with the realization that the services particularly appropriate for low-income countries were not specialized, hospital-based care but much more basic and less technically complex forms of care, with an emphasis on accessibility. Analysis of the failure of vertical disease control programmes (i.e. programmes focused on a single discrete disease and characterized by hierarchical organization), such as WHO's global malaria campaign, contributed to the formulation of the primary health care concept as the basic international strategy for health improvement (Gish 1992).

It was also increasingly argued that the achievement of improvements in the health status of populations and the achievement of the ambitious goal of HFA2000 would only be possible through a reorientation of services towards the promotion and protection of health rather than an emphasis on the curing of ill health. Thus a strengthening of PHC services, the point at which protection and promotion effectively takes place, began to be seen as essential (see Chapter 7). Also influential was the success of a number of small-scale PHC projects (and the Chinese experience on a larger scale), demonstrating the effectiveness of PHC practice (WHO 1983). This increased emphasis on PHC applied equally to all countries (low- or high-income), and so the predominant concern became the reorganization of *all* health services with the aim of prioritizing PHC. This represented a radical change from the earlier attitude, a legacy from the colonial period, which could in essence be stated as: 'how can we continue to develop high-technology medicine without spiralling costs and how can this be made available to poorer countries?'

A key event in this recognition of the importance of PHC was the 1978 conference held in Alma-Ata, in the former USSR, sponsored by WHO and

UNICEF and attended by delegations from 134 governments and 67 UN organizations, specialized agencies, and NGOs in official relations with WHO – including countries from all different stages of development. This conference reaffirmed health care as a fundamental right, and reiterated that the inequalities that existed, both between and within countries, were unacceptable. Alma-Ata was called at the time a 'historic collective expression of political will in the spirit of social equity' (WHO/UNICEF 1978).

The report of the conference summarizes a damning critique of most conventional health service systems as increasingly complex, increasingly costly, of doubtful social relevance, distorted by dictates of medical technology, and of low effectiveness (largely because of the dominance of high-technology curative medicine). The report concluded that it was not relevant for low-income countries to continue trying to emulate such systems. It argued that a new approach was needed through the development of PHC, with an emphasis on comprehensive basic services for all rather than sophisticated medical care for a few. PHC was thus seen as the key to attaining the goal of HFA2000, and it was argued that all other sectors in health service systems should be reorientated to support the PHC sector rather than the other way round. Key features of the new PHC-based health services that were sought are summarized in Table 9.1. One important aspect of the work carried out by WHO in support of the development of appropriate PHC was the essential drugs programme. This was based around the identification of model lists of essential drugs (the first report was published as WHO 1977) and, commencing in 1981, a special action programme which attempted to tackle a number of key issues, including: selection of drugs; use of generic (rather than brand) names; efficient procurement and distribution; appropriate training in prescription. The essential drugs programme was strongly disapproved of by the pharmaceutical industry, who exerted considerable (unsuccessful) lobbying efforts to try and restrict the programme (Walt 1994).

The new conceptualization of PHC was noticeable for its wide concern with factors supporting health, not limiting itself purely to the health sector. Thus there is explicit mention of water supplies, basic sanitation, education, and the food supply. There is also the need for a multisectoral approach, involving co-ordinated action of health, education, agricultural, housing, sanitation, and industry sectors, in order to respond to the multifactoral causation of ill health, and in particular to the importance of social and environmental factors. One particular aspect, indicating a shift in the position of WHO and an increasing distance from its colonial origins, is the call for the integration of different types of medical practitioner in providing health services, based on a recognition of the contribution that non-allopathic medical systems have to make to health (WHO 1978; Vuori 1982); for example, a WHA resolution adopted in 1977 made so-called 'traditional' medicine a focus of research and study, and urged governments 'to give adequate importance to the utilisation of their traditional systems of medicine' (WHA30.49, 1977).

Table 9.1 Key features of the 'new' PHC-based health services

Service provision in relation to needs of population, available and accessible to all in the community

Should cover, promotive, preventive, curative, and rehabilitative services (with prominence given to promotion and prevention)

Community participation, individually and collectively, in planning and implementation of health care

Multisectoral approach: multifactoral causation of ill health, and the importance of social and environmental factors should be recognized; co-ordinated action in health, education, agricultural, housing, sanitation, industry sectors is necessary

Appropriate technology (low-cost, high-quality essential drugs, self-reliance, and affordability, in keeping with local culture, i.e. 'acceptable')

Integration of different types of medical practitioners (non-allopathic and allopathic)

Use of paramedics and community-based health workers

According to the Alma-Ata declaration, eight essential elements should be found in the PHC sector:

Education about prevailing health problems and methods of prevention and control

Promotion of food supply and proper nutrition

Adequate supply of safe water and basic sanitation

Maternal and child health services, including family planning

Immunization against major infectious diseases (diphtheria, tetanus, whooping cough, measles, polio, TB)

Prevention and control of locally endemic diseases

Appropriate treatment of common diseases and injuries

Provision of essential drugs

As Navarro (1984) and others have pointed out, the major recommendations put forward by the Alma-Ata conference were by no means new; it was, however, the first time that they had been endorsed in such a comprehensive way by so many countries. Critiques of the Alma-Ata report, including Navarro's, argued that the systems proposed were still to be medically dominated, that the importance of class divisions and struggle within societies was ignored, that insufficient attention was paid to how existing power relations within societies were to be changed, that development (and particularly capitalist development) was still implicitly accepted as bringing about improvement in health, and, finally, that insufficient attention was given to interrelationships between health and other sectors. Navarro (1984), addressing the Alma-Ata declaration and the global HFA goal, welcomed the major recommendation on shifting priorities towards primary health care and away from secondary and tertiary care, but found the analysis and proposed solutions set out in the Alma-Ata declaration permeated by capitalist political ideology masquerading as scientific neutrality. He concludes: 'Its recommendations reproduce, for the most part, the view of development establishments. These views are part of the problems and not

Table 9.2 Health for All: a brief chronology

1977 (May)	World Health Assembly of the World Health Organization first adopted the 'Health for All' policy goal HFA or HFA2000: 'the main social target of governments and WHO in the following decades should be for all citizens of the world to attain by the year 2000 a level of health which will permit them to lead a socially and economically productive life'
1978	Alma-Ata declaration on primary health care
1979	Resolution in support of formulation of global, regional, and national HFA strategies
1980	Adoption of Regional strategies for HFA
1981	Adoption of global HFA strategy
1984	First set of European HFA targets agreed
1984–5 1987–8 1990–1 1993–4	Monitoring and evaluations of progress towards HFA
1986	Start of Healthy Cities project Use of local-level targets
1990–1	Revisions of European targets
1991	Adoption of second set of European targets: the common European health policy

of the solution; they represent the perspective of the dominant classes in today's world' (p. 473).

In 1979, the year following the Alma-Ata conference, the WHA endorsed the conclusions of the conference and the content of the Alma-Ata declaration, and adopted a resolution calling for the derivation of national, regional, and global strategies for the achievement of HFA, based on the reorientation of health service systems along the lines suggested by Alma-Ata. Strategies were seen as necessary to expand on the meaning of the goal expressed as HFA2000 in rather more concrete and measurable terms, and to establish appropriate mechanisms for moving towards this goal. This development of regional and global strategies was very much seen as the health sector contribution to the establishment of the 'New International Economic Order' – the declaration and programme of action initiated by the 1974 General Assembly of the UN and based on ideals of equity, sovereign equality, interdependence, and co-operation (Bryant 1980).

Thus HFA is a global policy, but taken up in different ways in the different regions through regional strategies. A brief chronology is summarized in Table 9.2. Each of the regional strategies were to be based on the national strategies, where such existed, and would be particular to the special problems of the region concerned. The global strategy would very much represent whatever synthesis was possible between the strategies of the six individual regions. Once regional and global strategies had been agreed,

Table 9.3 Global HFA indicators

Endorsement of Health for All as policy at highest possible level

Mechanisms for involving people in the implementation of strategies have been formed or strengthened, and are actually functioning

At least 5 per cent of GNP is spent on health

A reasonable percentage of the national health expenditure is devoted to local health care

Resources are equitably distributed

Existence of well-defined strategies for Health for All, accompanied by explicit resource allocations, with, where appropriate, sustained support from more affluent countries for needs for external resources

PHC available to the whole population with at least the following:
- safe water in the home or within fifteen minutes' walking distance, and adequate sanitary facilities in the home or immediate vicinity
- immunization against diphtheria, tetanus, whooping cough, measles, poliomyelitis, and tuberculosis
- local health care, including availability of at least twenty essential drugs, within one hour's walk or travel
- trained personnel for attending pregnancy and childbirth, and caring for children up to at least 1 year of age

Adequate nutritional status of children:
- at least 90 per cent of infants with birth weight of at least 2,500 grams
- at least 90 per cent of children having weight for age that corresponds to appropriate reference values

Infant mortality rate for all identifiable subgroups below 50 per 1,000 live births

Life expectancy over 60 years

Adult literacy rate for both men and women exceeds 70 per cent

GNP per head exceeds US$500

---

progress in achieving these was to be monitored and evaluated. A set of global indicators was agreed for use in monitoring and evaluation, summarized in Table 9.3. As an example of one of the ways that the HFA strategy has been developed, the next section considers the approach adopted in the European region of WHO, including some of the key health policy themes involved. Following that, we examine some of the evidence on the policy impact of HFA in specific countries, drawing on the results of the evaluation and monitoring exercises and other published research.

The European health policy: targets for Health for All

Within the European region of WHO, a regional strategy for HFA was formulated and approved in outline by the regional committee meeting in 1980. The strategy was formulated in the absence of national HFA strategies (i.e. the process deviated from that envisaged initially, which suggested that regional strategies should be built up from national strategies). There were five key features in the strategy: first, a reorientation of services away from curative to preventive and promotive (i.e. endorsement of the Alma-Ata principles), coupled with, secondly, a move towards a holistic view of health, away from the medical model of health to a more socio-ecological model. Third was the recognition of the importance of social and environmental factors in the causation of ill health, and the importance of actions to remedy this (e.g. controls on tobacco advertising, changing of social and environmental conditions conducive to poor health, and choice of unhealthy behaviour, based on the recognition of the influence of environment and particularly deprivation on choice of behaviour, e.g. through inability to afford an adequate diet, or stress caused by unemployment). Fourth was the importance of health education aimed at affecting people's lifestyles and influencing choices of healthy behaviours. Finally, there was recognition of inequity, and the aim of reducing and eventually eliminating this. The strategy was very much based on a *synthesis* of the cultural/behavioural and materialist/structuralist explanations for ill health, and aimed for a balance of emphasis on individual and collective responsibility for health.

Despite the seemingly radical reorientation of the approach embodied in the strategy, some critiques (e.g. Strong 1986) found it still to be technocratic in its emphasis, ahistorical in not analysing the historical reasons for existing inequalities, and apolitical in not recognizing political constraints on action. Given the enormity of the task, namely the agreement of a common strategy amongst thirty-two member states with political and social systems as diverse as the UK, the former USSR, Spain, and Turkey, these features perhaps should not be seen as too surprising. The achievement represented by the agreement of any such common approach should not be underestimated. In fact, the European region of WHO and the member states concerned could be seen as explicitly acknowledging the vagueness and blandness of some of the very general statements in the strategy by stating the need to move beyond the strategy to formulate rather more precise targets. They argued that, in order to translate into operational terms the broad objectives expressed in the regional strategy, the way forward was for the region to agree a specific set of health targets dealing with each of the major components of the strategy. The development of regional targets, together with a set of detailed quantitative and qualitative indicators to measure progress towards the targets, can be seen as answering some of the criticisms levelled at the original strategy document.

The development of European regional targets for HFA

The period 1981–4 saw the initial formulation of targets, which, although set against the background of an epidemiological review of the health situation in Europe (eventually published as Brzezinski 1985), was essentially a political process of discussion and negotiation to obtain agreement to a common set of objectives by all member states in the region. By 1982 an initial list of 150 targets had been formulated; by 1983 this had been reduced to 82 targets, further reduced to a set of 38 targets endorsed by all 33 (by then) member states in Europe at the meeting of the Regional Committee in 1984.

A revised set of targets was agreed in 1991. Changes were made first to reflect the considerable changes and developments in the region. Secondly, some of the original targets contained implementation dates of 1990 and 1995, which obviously required attention. Finally, the mid-term point between 1980 and 2000 offered an opportunity to clarify the detailed formulation of targets in some areas, where perhaps they had not been well understood before, and where further policy development work had been carried out in the interim. The process that led up to the revision of the targets commenced in 1989 with a questionnaire to member states, followed by a review of the epidemiological situation (Van Oyen 1990), examination of the results of monitoring and evaluation in 1985, 1988, and 1991, a series of special meetings, and work by technical units in WHO and consultants, as well as sessions held during regular WHO meetings.

The changes that occurred as the targets went through successive drafts into the list agreed in 1984 and then into the revised 1991 version illustrate nicely some of the constraints operating in the work of international organizations such as WHO in terms of contestation over elements of health policy. In particular, they show clearly the influence of political and economic factors in the target-setting process. Table 9.4 summarizes these shifts in just three different areas: war; tobacco and alcohol; equity.

The first example concerns the initial proposal for a specific target on war: 'By 1990, Europe will be free from the fear of war.' At the time, this was argued for in the discussion document on the basis that '*At the present time*, it is not war itself that presents health problems, but the fear of war. It has become evident that living under the constant threat of a nuclear holocaust induces serious mental strain in many people in the European Region' (WHO/EURO 1983: 23; emphasis added). Reactions from member states were very mixed; while none wished to argue with the aim of the target, its inclusion was strongly resisted by national defence lobbies, and by those with concerns about sovereignty. It was argued that the scope of the target was outside the health sector, too political, and infeasible. This last line of argument illustrates one of the tensions that has run throughout the whole target-setting approach, that engendered by trying to produce 'a sound mixture of future dreams and today's realities' (WHO/EURO 1983: i). The outcome of this particular debate was the dropping of any specific targets in the area, but the inclusion of discussion related to the importance of peace for health within the resulting policy document in a chapter on 'prerequisites for health' (WHO/EURO 1985a).

Health and Societies

Table 9.4 Policy contests and compromises in target formulation in selected areas, 1983–91

War	Tobacco and alcohol	Equity
Draft targets, for discussion, 1983:	Draft targets, for discussion, 1983:	First set of targets, 1984:
'By the year 1990, Europe will be free from the fear of war.' (WHO/EURO 1983: 23)	'By 1990, tobacco consumption in all countries will have been reduced by at least 50 per cent and at least 80 per cent of the population will be non smokers.' (WHO/EURO 1983: 37)	'By the year 2000, the actual differences in health status between countries and between groups within countries should be reduced by at least 25 per cent, by improving the level of health of disadvantaged nations and groups.' (WHO/EURO 1985a: 24)
	'By 1990, no country or group within a country will have a consumption of pure alcohol per person aged 15 and over of more than 15 litres per year.' (p. 38)	
Reactions from some member states:	Reactions from some member states:	First evaluation, response from some member states (1985):
National defence lobby	Industrial lobby from producers	No information given
Issues of sovereignty	Concern over tax revenues	
Outside health sector concerns, too political		
Impossible		

First set of targets (1984):

No target adopted, material included in policy document in chapter on 'prerequisites of health'

First set of targets (1984):

'significant increases in non-smoking'
'significant decreases in overuse of alcohol'

Suggestions for revision (Apr. 1991) by WHO/EURO:

'By the year 2000, the production and consumption of health-damaging and dependence-producing substances such as alcohol, tobacco and psychoactive drugs should have been significantly reduced in all Member States.'

Accepted by regional committee (Sept. 1991):

'By the year 2000, the consumption of health-damaging and dependence-producing substances such as alcohol, tobacco and psychoactive drugs should have been significantly reduced in all Member States.'

First evaluation report (1986):

Information from published research used by WHO/EURO

Suggestions for target revisions (1989) from *one* member state:

Very complex area
Difficult to measure
Drop the target

from other countries:

Strong support for keeping it unchanged, even though unlikely to be achieved, as it represents the essence of the HFA policy

Accepted by regional committee (Sept. 1991):

Target retained unchanged: 'to maintain the strong political commitment to equity expressed in the original target'.

The second example is concerned with tobacco and alcohol. Here, the initial proposals were for specific numerical targets for reduction of consumption (see Table 9.4), with, in the case of alcohol, further, more detailed targets to apply for countries already below the overall target level. These very specific targets were resisted. Again, the interests of specific lobbies (in this case tobacco and alcohol producers) were threatened, and governments were concerned over tax revenues. For analysis of these lobbies at national and international level see Calnan (1984), Baggott (1986, 1990), Salvador (1990), and Read (1993). The final target versions agreed in 1984 were weaker (see Table 9.4), and the emphasis was shifted away from individual countries towards the region as a whole. This last might be interpreted as exerting less direct pressure on countries to modify relevant policies, as well as shifting attention away from progress at country level. During the revision of targets carried out in 1989–91, initial proposals contained suggestions to strengthen these targets (a lack of progress in a number of countries on each of these targets had been noted – WHO/EURO 1986; Taket *et al.* 1990), and to introduce a focus on production as well as consumption within the targets (see Table 9.4). Alongside this introduction of production came specific discussion of the range of policy measures relevant to controlling production. Again this was successfully resisted, demonstrating the preference for focusing attention on so-called individual choice of health-related behaviours, and away from economic and industrial interests, and the targets agreed in 1991 remain phrased in terms of consumption only.

The final example to be considered at this point is the case of equity in health. In agreeing the first set of targets in 1984, a number of very specific references to elimination of differences in health status between socio-economic and social groups were deleted from many targets, to be replaced by a single general target on equity, the first target. This contained a call for reductions in inequalities in health status both between and within countries, to be obtained through a process of improvement of the level of health of disadvantaged nations and groups (see Table 9.4). In reporting on progress towards HFA in 1985, a number of countries gave no information on progress in this area; information from published research studies was used in the evaluation and monitoring reports published by the regional office (WHO/EURO 1985b, 1986, 1989a, 1989b; Curtis *et al.* 1989), and this was one of the targets where very little progress was observed. During the process of target revision (1989–91), suggestions from countries for target revision were sought. One member state, in relation to this target, commented that it represented a very complex area, difficult to measure, and the specific suggestion was made to drop the target. In contrast, from other countries, there was strong support for keeping the target unchanged (even though it was unlikely to be achieved), as it represented the essence of the HFA policy. The document finally accepted by the regional committee in September 1991 agreed to keep the target unchanged, 'to maintain the strong political commitment to equity expressed in the original target'.

In general, the process of agreeing targets resulted in a final choice of targets that were less specific, less ambitious, and phrased for achievement by the European region as a whole, rather than by each country. There was a move away from the earliest drafts, which represented social engineering on a gigantic scale (European in scope and covering many aspects of life), to something less all-encompassing. None the less, the achievement of getting agreement on a single set of health-related targets for the region, which covers such a diverse range of countries, should not be underestimated. The targets are quite variable in content, some very precise and quantified, for example targets on reduction of mortality, others much less precise and detailed.

The process of target revision carried out between 1984 and 1991, and the resulting changes in targets, illustrate some interesting tensions at work in this international policy arena. First there was the question of providing an 'epidemiological' basis or justification for targets. Connected to this, a second area causing debate was around the issue of quantification of targets and whether or not this was desirable and/or feasible. There were also contrasting views about the role of targets: whether they should represent a policy vision, in the sense of providing a coherent framework for mobilizing action, or whether they should be a set of justified, quantified, feasible targets. These tensions can be particularly clearly seen in the reformulations of targets 7 and 8 (see Table 9.5).

In each of these two cases, the reformulated target moves away from a narrow focus on a single mortality indicator to deal with a specific population group. Within the justification offered for each of these targets (WHO/EURO 1993a) there is explicit mention of human rights issues, connected to the particular disadvantaged social positions occupied by children, young people, and women; this represents a considerable shift from the earlier target formulations, which had a much narrower focus. This explicit concern with human rights issues represents a marked change from the conclusion reached by Alston (1979), in his study of response of the UN's specialized agencies to the 1976 International Covenant on Economic, Social, and Cultural Rights: that WHO (along with the FAO) had neglected considerable opportunities for standard-setting in the area of human rights.

Policy responses to HFA within countries

As mentioned earlier, the first evaluation of progress towards HFA took place in 1985, followed by monitoring in 1988, a second evaluation in 1991, and further monitoring in 1994. These produced a wealth of information on trends in health status and health systems, some of which was discussed earlier in this book. This section examines the policy response

Table 9.5 Changes in European HFA targets, 1984–91

1984	1991
Target 7, Infant mortality	*Target 7, Health of children and young people*
By the year 2000, infant mortality in the region should be less than 20 per 1000 live births.	By the year 2000, the health of all children and young people should be improved, giving them the opportunity to grow and develop to their full physical, mental, and social potential.
	This target aims at achieving:
	• comprehensive support of children and their families, according to their health needs and socio-economic circumstances
	• a reduction of infant mortality rates in countries with rates currently between 10 and 20 per 1,000 live births to below 10, and in countries with rates currently above 20 per 1,000 live births to below 15
	• a reduction of 25 per cent in the differences in infant mortality rates between geographical areas and socio-economic groups
	• a reduction of 25 per cent in mortality and serious injury in children and young people, notably due to accidents
Target 8, Maternal mortality	*Target 8, Health of women*
By the year 2000, maternal mortality in the region should be less than 15 per 100 000 live births.	By the year 2000, there should be a sustained and continuing improvement in the health of all women.
	This target aims at achieving:
	• a reduction in maternal mortality to less than 15 per 100,000 live births
	• a substantial reduction in health problems that are unique to women
	• a substantial reduction in the health problems of women related to their socio-economic status and the burden of their multiple roles
	• a substantial reduction in the incidence and adverse health consequences of sexual harassment, domestic violence, and rape
	• sustained support for women providing informal health care
	• a reduction of at least 25 per cent in the differences between maternal mortality rates between geographical areas and socio-economic groups

within countries, both in terms of general HFA policy and with reference to some key policy themes.

Countries vary considerably in the extent to which they are introducing the goals of HFA and parts of the regional strategies into their own health policies and health systems. Nordic countries, especially Finland and

Sweden, incorporated the ideas in the European regional strategy and the targets fairly comprehensively into national strategies for HFA at an early stage (WHO/EURO 1986), and considerable changes in health services organization and orientation have been seen. Other countries completed national strategies for HFA at a later stage, for example Ireland, Malta, and the Netherlands (WHO/EURO 1989a, 1989b). In the USA, a detailed set of disease prevention and health promotion objectives were set up in 1980 which were updated ten years later (US Department of Health and Human Services 1990); these are based on three broad health goals: increasing the span of healthy life, reducing health disparities, and achieving access to preventive services for all. The keynote address to the 1990 meeting of the US Public Health Association (McBeath 1991) explores the origins of this work and its connections to HFA. By 1993, twenty-nine states had already set state health objectives for the year 2000, based on the national document, and a set of model standards detailing possible community implementation strategies for local action has also been produced (US Department of Health and Human Services 1994). The Minister of Health in the province of Ontario initiated a goal-setting process in 1985; this resulted in a preliminary paper published in 1987 (Spasoff *et al.* 1987). Australia published an HFA strategy in 1988 (HFA Committee 1988), and New Zealand in 1989 (Ministry of Health 1989).

Some countries have placed particular emphasis on parts of the strategy. In the European region, for example, Greece, Portugal, Spain, Morocco, and Turkey have put such emphasis on developing comprehensive PHC services (WHO/EURO 1986). Outside Europe, reports from Burkina Faso, China, Egypt, Indonesia, Malaysia, Mozambique, Nigeria, Papua New Guinea, Sri Lanka, and Thailand compiled at the end of the 1990s illustrate a wide range of different approaches to and successes in implementing PHC (Tarimo and Creese 1990). This emphasizes the importance of adapting primary care approaches to suit individual country circumstances, and also illustrates the wide range of difficulties still to be overcome. In many low-income countries, such difficulties are closely linked to the burdens imposed by interest repayments on external loans, by deteriorating terms of trade, and by economic adjustment policies, set within the context of deadlock in the North–South dialogue.

In some countries the ideas of HFA have been taken up at a more local level; for example, within the UK a number of regions and districts have adapted targets to the local level (Jacobson 1990; Rathwell 1992), and local HFA initiatives are also reported from Denmark, France, Spain, and Switzerland (WHO/EURO 1989a). The Healthy Cities and healthy communities work, discussed in Chapter 7, is yet another example of the ideas of HFA being taken up locally.

Farrant (1991) argues that the tensions between individually orientated biomedical approaches and more socially orientated perspectives that run throughout the HFA strategy, together with insufficient attention to the systematic analysis of power relations within and between communities and the difficulties this poses for achieving community participation, make it possible for the HFA targets to be interpreted extremely selectively to support a conventional medical model. Indeed, the UK government's policy paper *Health of the Nation* (Secretary of State for Health 1991) was also based on a target approach, but can be criticized for adopting exactly such a selective approach, and in particular for failing to address the health impact of government policies, the need to address social inequalities, and the importance of community participation (Radical Statistics Health Group 1991). Its policy stance is heavily based on individualism, and is dominated by a negative libertarian approach (Mills and Saward 1993) in which, although the need for some government action is acknowledged, the emphasis is placed on informed individuals taking responsibility for themselves. A study into responses to Health for All in Britain (Rathwell 1992) found exactly such selective use of targets at local level in many of those health authorities or health boards who had taken up Health for All.

In analysing developments related to HFA, one of the most important issues is distinguishing between rhetoric and reality. While many documents pay lip-service to important HFA principles such as intersectoral collaboration, consumer participation, PHC, and health promotion, it is still too early to establish whether the changes and mechanisms introduced will actually contribute in the desired direction, or produce exactly the opposite effect. A case in point is the most recent reorganizations of the health service within the UK, implemented on 1 April 1991, which were discussed in Chapter 5. Prior to implementation there was already considerable debate about the likely effects of the reforms and whether these would contribute to the stated aims; this debate still continues, with directly conflicting interpretations being common. The illumination of this debate and a clearer separation of rhetoric from reality will remain a key task for health service researchers for some years to come.

Outside Europe, while Bryant (1980) provides an optimistic view of the impact of HFA through a series of examples, a counterweight to this is provided by Mburu (1980), who in his critique of Bryant's paper seriously questions whether most low-income countries are prepared or ready to implement the ideas of HFA. An earlier study by Mburu (1979) of health policy and health services delivery in Kenya identified considerable differences between the policy rhetoric and the actual health programmes implemented. A study by Morley *et al.* (1983), carried out at a fairly early stage in the life of the HFA policy, examined evidence from a wide variety of different projects and programmes throughout a number of different low-income countries. They posed the question: 'is "Health for All" just an idle slogan, the latest fad in international development jargon?' (p. 319). They note a gap in many places between the rhetoric of high-level political commitment to HFA and the lack of action to tackle the necessary non-health factors, including political instability, the vested interests of the medical profession and drug-manufacturers (as well as other industrial interests), inequitable social structures, and uneven economic development, and conclude 'there are simply no purely technical solutions to community health problems. The political process intrudes everywhere – whether at national or community level' (p. 325). More positively however, the case studies they examine also illustrate how, in a democratic system with strong political commitments to equitable socio-economic development, for example the Indian state of Kerala, high levels of health can be achieved even on extremely modest levels of income.

The latest evaluation of the implementation of the global strategy for HFA (WHO 1993b), which covers the years 1985–90, provides a similarly mixed picture, as the following extract from the chapter assessing overall achievement illustrates (p. 139):

> Overall there has been strong political commitment to achieving health-for-all goals and most countries have endorsed at the highest level the necessary policies and strategies. . . . However in implementing the strategy the fundamental health-for-all policies and the principles applicable to health systems based on primary health care have not always been appropriately put into practice. . . . The factors that slowed progress in implementing the strategy were: (i) the slow pace at which existing disease-control programmes have been reoriented towards people's needs; (ii) problems of collaboration on a continuing basis through general health infrastructure; (iii) difficulties in involving all those concerned (individuals, communities and local non-governmental organizations as well as health personnel) in health care delivery; and (iv) weak management of health care delivery, especially at the operational level.

The report also notes resistance to HFA policies (see also Green 1991). In some countries this comes from within the health sector, mostly from the clinically-oriented medical professions; in others, non-health sectors are particularly resistant. Although some (e.g. Pannenborg 1991) argue that such factors represent major strategic weaknesses in the HFA paradigm, policy development work in a number of different areas, discussed in the next section, means that it is premature to conclude (as Pannenborg does) that PHC and HFA can no longer serve as guiding principles for international health. It does seem, however, that some elements of the HFA strategy may be in the process of being de-prioritized; this is illustrated slightly later in this chapter when we look at the 1993 World Bank report *Investing in health*, which was produced with close involvement from WHO, and consider some of the policy recommendations it puts forward. Before doing that, however, we shall review some important areas of policy development in support of HFA.

Policy development work in support of HFA: health promotion, Healthy Cities, and healthy public policy

Another, newer trend in the work of WHO has been policy development specifically aimed at supporting the implementation of HFA through achieving a better understanding of appropriate policy mechanisms. Three particularly important programmes of work have been those on health promotion, Healthy Cities, and healthy public policy. Aspects of the local implementation of these last two policy initiatives have already been discussed in detail in Chapter 7; in this chapter we concentrate on the policy origins.

A programme in health promotion was established in the WHO regional office for Europe in 1984, taking as its starting-point the need to clarify some of the concepts and principles involved in the promotion of health. The definition of health promotion adopted is:

> the process of enabling people to increase control over, and to improve, their health. This perspective is derived from a conception of 'health' as the extent to which an individual or group is able, on the one hand, to realise aspirations and satisfy needs; and, on the other hand, to change or cope with the environment. Health is, therefore, seen as a resource for everyday life, not the objective of living; it is a positive concept emphasising social and personal resources, as well as physical capacities. (WHO/EURO 1984: 653–4)

Along with this comes a recognition that health promotion requires action not only within the health services but within other sectors as well: 'health promotion . . . encompasses actions to protect or enhance health, including legal, fiscal, educational and social measures' (Whitehead 1989: 7). Elsewhere, this type of model of health promotion has been referred to as the 'empowerment model' (Wallerstein and Bernstein 1988; Whitehead 1989), with the former linking it to Freire's pedagogy. The key features of the empowerment model are that it is positive, dynamic, enabling, and participative. It aims to perform a delicate balancing act between recognizing the constraints on healthy choices faced by people due to the environments (in the widest sense) in which they live and work, and strengthening people's potential for taking action to improve their health (without slipping into victim-blaming). Roles for health promotion programmes and initiatives therefore involve a wide spectrum of activities including advocacy, mediation, enabling, etc.

This model of health promotion is the one underlying the European HFA targets (WHO/EURO 1985a, 1993a), and has also been taken up elsewhere in both high- and low-income countries (WHO Working Group on Health Promotion in Developing Countries 1991). Although WHO has provided much useful clarification in this area (WHO/EURO 1984; Ottawa Charter for Health Promotion 1986), there are still struggles over this model of health promotion (see e.g. Naidoo 1986). Much of current health promotion practice, however, is dominated by a different model which might be labelled the 'education model'. Here the role of health promotion is limited to education in the narrowest sense: providing information. The model assumes that changes in attitudes and practice naturally follow after the provision of information, a somewhat simplistic view of behaviour change which is not well supported by the research evidence.

Following on from work developing basic principles and concepts of health promotion, further work was initiated to explore mechanisms by which these might be put into practice. The first of these was a project aimed at exploring the concept of the 'healthy city' (Duhl 1986). This commenced in the European region of WHO in 1985, involving a small network of European cities (Tsouros 1990). By 1993 the project had grown rapidly, yielding eighteen national networks and hundreds of towns and cities actively involved in Europe, North America, and, increasingly, low-income countries (Hancock 1993); it had also become known as the Healthy Cities movement, or Healthy Communities in some parts of the world. Besides the European Region, WHO has supported Healthy Cities activities in the African region, the eastern Mediterranean region, and the western Pacific region (WHO 1994). The work has drawn explicitly on experience outside the health sector: from community development workers as well as that gained in other broad social movements with origins in the community/voluntary sector, such as feminist organizations and groups, black and minority ethnic groups and organizations, civil rights groups, green organizations and groups. A valuable emphasis of much of the successful Healthy Cities work has been in stressing the importance of the local context for the selection of priorities for action and of the importance of achieving widespread involvement in, and commitment to, local action. This has particularly involved work around community participation and intersectoral collaboration, two of the themes of HFA. The early work on Healthy Cities has since led on to the development of other approaches based in particular settings, such as health promotion in schools, workplaces, and hospitals (WHO/EURO 1993a).

Work in the third area to be discussed in this section, that on 'healthy public policy', built on the earlier experience and responded to the danger of reducing health promotion to a variety of victim-blaming strategies by emphasizing the importance of policy in all sectors in creating social, physical, and economic environments where 'healthy choices become the easy choices':

Healthy public policy is the policy challenge set by a new vision of public health. It refers to policy decisions in any sector or level of government that are characterised by an explicit concern for health

and an accountability for health impact. It is expressed through horizontal strategies such as intersectoral cooperation and public participation. (Adelaide Conference on Healthy Public Policy 1988)

The notion of healthy public policy is argued to provide a foundation for
promoting physical and social environments that support the adoption of
healthy patterns of living (WHO/EURO 1993a). Its aim is to ensure equitable access to the prerequisites for health, whether in the form of consumer
goods, supportive living environments, or services that contribute to healthy
living. It seeks to stimulate action and the development of specific mechanisms so that decision-makers at all levels and in all sectors are aware of the
consequences for health of their decisions, and are willing to accept their
share of responsibility for health in their communities.

Common elements in the process by which work in each of these three
areas (health promotion, Healthy Cities, healthy public policy) has been carried out include the development and clarification of key concepts and the
use of demonstration projects exploring mechanisms for implementation.
This type of policy work involved something of a new development for
WHO, in working in partnership with different groups at the local level,
rather than supplying expertise to the local level. Ensuring that components
like participation are not rendered as merely rhetoric has necessitated frequent intervention by activists, for example at national and international
Healthy Cities and health promotion conferences to ensure that issues pertinent to women, minority ethnic communities, and the disabled are kept on
the agenda. This type of work can be regarded as highly successful in clari-
fying important policy concepts and in stimulating a number of successful
demonstration projects and local initiatives, some of which have been discussed in Chapter 7. Policy development work carried out in all these three
areas was reflected in the 1991 revisions to the European HFA targets
(WHO/EURO 1993a), allowing a more sophisticated analysis to be offered
in these areas and much more specific strategies and mechanisms to be discussed.

Health and development: the World Bank and 'Investing in Health'

World Development Report 1993, the sixteenth in this annual series,
examines the interplay between human health, health policy, and economic development. The three most recent reports – on the environment, on development strategies, and on poverty – have furnished an
overview of the goals and means of development. This year's report on
health, like next year's on infrastructure, examines in depth a single

sector in which the impact of public finance and public policy is of particular importance. (World Bank 1993: iii)

The above is the opening paragraph in the foreword to the World Bank's 1993 *Development Report*. Subtitled *Investing in Health*, the report is notable for a number of reasons. It was produced in close collaboration with WHO, and offers a detailed analysis of health systems and their problems, including not only an analysis of the global burden of disease (discussed in Chapter 4) but also detailed analysis of the different policy options for reform in 'developing countries' and an assessment of their likely effects. It proposes an extensive agenda for action, not only covering international assistance for health but also discussing health policy reform requirements within what it refers to as 'developing countries'. In discussing 'developing countries', the report distinguishes three different groups: 'low-income countries', defined as those with 1991 per capita GNP not more than $635; 'middle-income countries', defined as countries with per capita GNP in 1991 of more than $635, but less than $7,911; and finally, the 'formerly socialist economies'. The report does not address policy recommendations to other, high-income countries, and as we shall see below, it adopts a somewhat uncritical attitude to aspects of health policy in such countries. In this section the policy components of the report are briefly discussed, particularly in relation to HFA, and linked to earlier discussions on HFA policy in countries.

The report proposes a three-pronged approach to government policies for improving health (these are summarized in Table 9.6). In discussing the three different groups of countries, further distinctions are made in terms of policy suggestions (see Table 9.7).

The second and third elements in the approach proposed (Table 9.6) are particularly interesting for their extremely mixed stance towards the value and role of markets in health policy. The second element, for example, starts with the statement: 'The challenge for most governments is to concentrate resources on compensating for market failures and efficiently financing services that will particularly benefit the poor' (World Bank 1993: 6) – a statement that seems to accept as inevitable, and also desirable, the existence of the market in health services. There is also extremely selective use of experience from high-income countries to support policy suggestions; for example, in discussions of cost containment, it is argued that 'other promising approaches are to allow government hospitals to compete with each other as semi-autonomous enterprises, as in the United Kingdom in recent years' (p. 161). The 'promise' of this approach in the UK is not well supported by research evidence from any evaluation of the effect of the reforms in the UK so far (see e.g. LeGrand and Bartlett 1993; Mullen 1995). This is perhaps especially curious, since the report also recognizes (e.g. pp. 5, 171) that there are considerable problems with market operations, identifying the need for government action to promote access to services by all and to

Table 9.6 Investing in health: World Bank suggestions for health policy reform in 'developing countries'

Fostering an environment that enables households to improve health	Improving government spending on health	Promoting diversity and competition
Promote macroeconomic policies that reduce poverty	Reduce government spending on tertiary facilities, specialist training, and interventions providing low health gain for money spent	Encourage social or private insurance
Expand investment in schooling, especially for girls	Finance and implement public health interventions to deal with substantial externalities surrounding infectious disease control, prevention of AIDS, environmental pollution	Encourage supplier competition in delivery of clinical services and provision of inputs such as drugs
Promote the rights and status of women through political and economic empowerment and legal protection against abuse	Finance and ensure delivery of a package of essential clinical services	Generate and disseminate information on provider performance, cost effectiveness and accreditation
	Improve management of government health services through measures such as decentralization and contracting out	

Source: World Bank (1993: 6, box 1).

reduce undesirable aspects of unregulated market operation (such as reductions in incentives to avoid risk, and escalating costs). This recognition that there are imperfections in the operation of markets in relation to health services does, however, represent a marked shift from the stance of earlier World Bank reports. Comparison of the policy suggestions offered to the three different country groups (Table 9.7) seems to show an increased emphasis on introduction of aspects of market operation as one moves from low-income to middle-income to formerly socialist countries.

The third element, promoting diversity and competition (Table 9.6), contains some particularly unsubstantiated appeals to the value of competition in driving down costs and improving quality, yet also argues that:

strong government regulation is crucial, including regulation of privately delivered health services to ensure safety and quality and of private insurance to encourage universal access to coverage and to discourage practices – such as fee-for-service payment to providers

Table 9.7 Investing in health: World Bank suggestions for key elements of health policy reform in different types of 'developing countries'

Low-income countries	Middle-income countries	Formerly socialist countries
Provide solid primary schooling for all children, especially girls	Phase out public subsidies to better-off groups	Improve efficiency of government health facilities and services, partly by reducing the size of the public system
Invest more resources in highly cost-effective public health activities that can substantially improve the health of the poor	Extend insurance cover more widely	Find new ways to finance health care
Shift health spending for clinical services from tertiary care facilities to district health infrastructure capable of delivering essential clinical care	Give consumers a choice of insurer	Encourage private supply of health services while strengthening public regulatory capacity
Reduce waste and efficiency in government health programmes	Encourage payment methods that control costs	
Encourage increased community control and financing of essential health care		

Source: World Bank (1993: 156–65).

reimbursed by a 'third-party' insurer – that leads to overuse of services and escalation of costs. (World Bank 1993: 7)

It remains unclear how this balance is to be struck. It is also suggested that quality and efficiency of government-provided services can be improved through a combination of decentralization, performance-based incentives, and training and development. This seems to ignore the considerable difficulties of ensuring equity under such circumstances, as illustrated by the discussion of the effects of decentralization in Chapter 8.

The report is certainly based implicitly, at least in part, on WHO's Health for All strategy, although this does not receive detailed discussion. Close links to the HFA strategy are found in the recognition of the importance of the effects of income and education on health, and of the need for reorientation of health services away from the tertiary sector. As emphasized earlier, socio-political change is also required; however, on this aspect the report is less explicit. Although the report recognizes the existence of an array of interest groups that stand to lose from change (cf. pp. 15, 165, 170–1), it is much less explicit than, for example, the European HFA strategy or the Alma-Ata declaration in exploring the issues involved. This lack of detail, together with the emphasis (discussed above) on market mechanisms, lays the report open to criticism on the grounds that this limited approach will not be effective, as it does not confront the various interests (class, cultural, capitalist, and industrial) bound up with the reproduction of mass poverty in low-income countries. In this respect, the report is very similar to the policy of previous World Bank development reports; see for example Burkett's (1991) analysis of the 1990 *World development report*, which concludes that World Bank policy effectively subordinates the needs of the poor, and ensures instead the maintenance of the basic structures underpinning uneven development.

There are other differences between the World Bank report and HFA that also seem significant. The World Bank report focuses on the provision of essential clinical services alongside a strengthening of public health interventions. There is no discussion of provision of comprehensive PHC services in the sense envisaged in the Alma-Ata report (and summarized in Table 9.1). This contraction to a selected range of services seems reminiscent of the concept of 'selective PHC' fostered by donor agencies and planners, which, as Rifkin and Walt (1986) have demonstrated, represents a fundamental departure from the key principles of PHC. It replaces a multisectoral approach, directed at the root causes of inequalities in health, by a strategy of tackling selected problems chosen according to cost-effectiveness criteria, and interprets community participation narrowly in terms of compliance with medically led advice. The danger, therefore, is that the selective nature of the measures called for here will produce similar problems. A further concern can be raised by the report's acceptance of the value of private provision in health services (see e.g. the final column of Table 9.7), provided only that it is subject to appropriate regulation.

Positively, on the other hand, the report can also be seen as an attempt to remove some of the obstacles to implementation of HFA identified in the discussions earlier, through its careful reframing of the arguments for HFA in the economistic language of 'investing in health'. The report attempts to express the debate about health expenditure in terms reminiscent of discussions of capital investment, etc., and thus attempts to move away from the view of the health sector as a drain on resources that could otherwise be spent elsewhere. The health sector is recognized as a productive sector of the economy, citing gains in worker productivity, improved utilization of natural resources, benefits to future generations through improved education, and reduced costs of medical care as key components in the argument for why health matters. This is in interesting contrast to the language of WHO in explaining why health matters, which grounds its arguments in the language of rights, for example, from the statements of WHO's Director-

General, Dr Hiroshi Nakajima, to the Executive Board and the World Health Assembly in 1993:

> Health is a fundamental human right. We must uphold this right. We must do so with peaceful but powerful weapons – care, compassion, mutual respect and education. Health is inseparable from individual rights and freedoms, and from the right to development. (Nakajima 1993: 7)

> Health is inseparable from individual liberties or from the right to development. Human beings and 'human security' must be the centre-piece of sustainable development. by promoting respect for ethical principles and for the human rights of individuals and communities alike – core values for the dedicated health profession worldwide – I intend with a new partnership to lead the World Health Organisation to the fulfilment of its constitutional mandate and moral mission: the building of universal peace through health for all. (p. 14)

The discussion in these quotations is clearly phrased in terms of the 'right to health'. Others have suggested that this formulation is unhelpful, and that it makes more sense to talk in terms of right of access to what Foucault terms the 'means of health' (1983: 170), which he defines as everything that society has at its disposal to remedy or alleviate ill health, and to which we might add, to protect against ill health and promote good health.

The World Bank report is also interesting in its suggestions that levels of aid should be explicitly linked to countries' willingness to undertake major changes in health policy (cf. pp. 16, 167–8). The World Bank itself has considerably increased its support for the health sector, as a share of new World Bank lending: projects for health, population, and nutrition grew from less than 1 per cent in 1987 to nearly 7 per cent in 1991 (World Bank 1993: 169). In its analysis of trends, it demonstrates that increasingly, World Bank lending for health is being focused on broad policy reforms in the health sector.

In conclusion: the role of international organizations in general and WHO in particular

The criticisms of perpetuation of colonial attitudes to low-income countries through the operation of international organizations no longer seem quite so justified, especially given the major shifts in WHO's stance in the 1970s and 1980s, particularly in relation to non-allopathic versus allopathic medicine, and in relation to the radical health reforms sought as a part of the HFA strategy and related policy initiatives discussed in this chapter.

However, there are still many critics of the role of WHO, including Navarro (1984), who argues for the importance of WHO supplying concrete support to liberation movements. Pragmatically speaking, this would be extremely difficult to achieve while retaining consensus of support, and is unlikely to be implemented, since international organizations remain clearly aware of the dangers of losing the support of powerful countries, as has already happened in the case of UNESCO, from which countries like the USA and the UK have withdrawn funding, not only in reaction against corruption/misuse of funds but also against political and ideological positions different from their own. Some of the issues surrounding these possibilities are discussed further by Taylor (1992) and Walt (1994). With specific reference to WHO, a further constraint on action is the argument about issues being 'outside the health sector', and thus outside WHO's remit. Central to this debate is the whole question of the most effective way to improve health – and indeed, whether improvements in health of certain sectors of the population and removal of inequalities are possible without fundamental social change. It is here that Navarro and other marxist critics argue that it is impossible for WHO to be non-political and, further, that its existing stance is not neutral, but in fact adopts the political perspective of the dominant capitalist classes. Their analysis of the potential of WHO is thus very pessimistic. Senior WHO staff members, however, argue that WHO still has a role as an advocate for reduction in inequities and inequalities in health both between and within countries, linking this specifically to the necessity for wider social change (see e.g. Kickbusch 1993).

On the other hand, many would agree that WHO still maintains its traditional, valuable role in information/technology exchange. In terms of health policy development, it has an extremely limited directive role, but (as this chapter and related discussions in Chapter 7 have demonstrated) it can be influential by persuasion and example. In recent decades we have seen effects in terms of: more widespread acceptance of need for reorientation of health services; a questioning of assumptions about the effectiveness of (allopathic) medical care; moves towards the integration of non-allopathic practitioners. In terms of WHO's role in detailed policy development around issues relating to equity and social justice, views differ on the overall effectiveness of programmes of work in areas such as HFA, health promotion, Healthy Cities/Healthy Communities, and healthy public policy. While some specific studies have shown success in achieving desired effects, longer-term evaluation of these initiatives will be required before the sceptics can be convinced.

One area of particular interest for the future will be how a number of different tensions that are currently apparent get played out. The first of these is the tension between global and regional policy and activities, manifest in several ways. First there is increasing doubt about the support for continued use of PHC and HFA policy mechanisms; see for example the discussion of the World Bank report earlier and its ambiguous relationship to HFA, as well as reports in the development press (*New Internationalist,* Dec. 1994), the public health literature (Pannenborg 1991), and series of articles based on a wide range of interviews published in the *British Medical Journal* (Godlee 1994a–d, 1995a–c). While some, like Pannenborg (1991), question the effectiveness of these mechanisms (although there is little in the way of supporting evidence cited in his report), others (e.g. *New Internationalist,* Dec. 1994; Godlee 1994a, 1994b) link the change in stance to the change in Director-General which took place in 1988. This is also reflected in the recent criticisms expressed in a *Lancet* editorial (1995), by Smith (1995), and by Godlee (1995c) about the capability of the current leadership at WHO.

A related issue is whether the organization will change its manner of working in the future. There is certainly scope for this; for example, Taylor (1992) argues that WHO has paid insufficient attention to the role that legislation can play in the Health for All strategy, specifically through the promotion of the 'right to health'. Her analysis attributes this to the internal dynamics and politics of the organization, which she finds dominated by an organizational culture established by the 'conservative medical profession', a conclusion also reached by Godlee (1994a). Some organizational shifts have already occurred recently, with the institution in 1989 by the Geneva headquarters of WHO of a series of intensified co-operation programmes (WHO 1994) with a number of selected countries which effectively bypass the regional office structure, raising questions about the relationship between the central and regional offices of WHO. A report by DANIDA (1992) into the effectiveness of WHO at country level in four countries (Kenya, Nepal, Sudan, and Thailand) found little strategy behind spending at country level. Godlee (1994c) argues that the regional offices give only the illusion of decentralization, with the regional directors having too much power. A study by the UN Joint Inspection Unit (Daes and Daoudy 1994) concluded that there are many problems with WHO's decentralized structure, and that it is not functioning as efficiently as it used to; they recommend a strategy of strengthening the system of country offices, with resources and responsibilities devolved from the regional offices.

There is also a continued tension between clinical medicine and public health, between the biomedical model and the socio-ecological model of health displayed in the work of WHO. An excellent example is provided by Stephenson and Wagner (1993), who analyse two WHO reports on *in vitro* fertilization (one from headquarters and one from the European region). The policy recommendations from these differed widely: while the European report is consistent with the public health principles underlying Health for All, the other report is more narrowly focused on technical issues of laboratory and clinical practice (representing almost a step back in time). Godlee (1995a) presents another example in her conclusion that 'eradicationitis' is still prevalent in WHO, and that a shift back to vertical programmes is undermining the effectiveness of more integrated efforts. In its critique of the current operation of WHO, a recent *Lancet* editorial (1995) diagnoses domination by the biomedical model as one of the reasons for what it sees as WHO's lack of effective leadership in health.

Another tension, within Europe, is between the role and influence of WHO and the European Union (EU). With the growth in membership of the EU, and its growth in concern with health matters through the adoption of the Maastricht treaty which contains a specific health component, an increased influence of the EU on health policy and health systems can be envisaged (Svensson and Stephenson 1992). Kokkonen and Kekomäki (1993) also argue that there is considerable potential for legal and economic measures affecting health. Others conclude, however, that an expansion of the EU's role in health-related policies is likely to proceed only erratically, arguing that the EU policy formulation process is dominated by incrementalism, bargaining, and compromise (Bomberg and Peterson 1993). It remains to be seen how the EU will link with the activities of the European regional office of WHO. The re-election in September 1994 of Dr Jo Asvall as regional director in the WHO European regional office offers at least some hope of continuity in the role adopted by the regional office. In a speech in Kiel in November 1992, Asvall stated: 'the European regional organization of WHO believes its role is to stay different building on its strength and specificity. Its role is to be in the forefront of health development thinking, a catalyst in the search for better solutions and an advocate of more sensible policies and effective methods and approaches'. Such a role would obviously require close links with the EU. Some commentators, however (e.g. Pannenborg 1991; Godlee 1995b), question the relevance of maintaining a WHO regional office in Europe, given the increased health mandate of the EU.

A final question is how WHO will respond to the new challenges posed in a context of profound global social, economic, and environmental change. It has already undertaken a major re-examination of its mission and strategies, including the constitution of a working group of the executive board on the WHO response to global change. In order to strengthen WHO's policy-making capacity, and specifically to respond to the recommendations of this working group, the Director-General of WHO has recently established a global policy council (WHO 1994), whose members are the Director-General, the regional directors, the Assistant Directors-General (based in Geneva), and the Director of IARC (International Agency for Research on Cancer). A major research question for the future will be the exploration of the policy development consequent on such changes.

|10|

Towards 2000: issues for a research agenda

If this book has a single theme running throughout it, it is that of illustrating the changes in perspectives that have taken place in the study of health and societies. With that in mind, as we explained in the preface, we have concentrated most heavily on substantive material, covering aspects of theory only where we feel it is essential or helpful, and severely restricting our coverage of methods. In Chapter 1 we illustrated the theme of changing perspectives through the consideration of the development within a single discipline; the example we chose was that of medical geography, and we demonstrated how the perspectives adopted within medical geographical studies have changed over time. The remainder of the book took up the theme of changing perspectives by examining selected material in each of three different sub-domains of the study of health and societies: changing perspectives on health and ill health, covered in Chapters 2, 3, and 4; changing perspectives on health services, covered in Chapters 5, 6, and 7; and changing perspectives on health policy, covered in Chapters 8 and 9. As a consequence of choosing to talk about these three different sub-domains, rather than concentrating on only one or two of them, we recognize that we have necessarily provided an even more selective treatment in each of the three areas than otherwise would have been possible.

The discussion so far has used several examples of perspectives on health, health services, and health policy in order to demonstrate the importance of diversity and change. Diversity arises partly from the different points of view held by individuals and social groups, associated with differences in their experience and their theoretical and ideological positions. Also important are the influences of different contexts in time, place, and socio-economic structure. In the future, we can expect that further changes in perspectives will be needed to continue to do justice to the local, specific, contingent, and dynamic nature of knowledge, particularly in the social sphere. The examples considered illustrate the dynamic nature of health systems, and in this final chapter we point to some aspects of change which seem likely to be important for future developments. As we have seen in the case of the World Health Organization's strategy of 'Health for All by the year 2000', a certain significance is invested in aims and objectives targeted for the millennium. Although, of course, this date is quite unlikely to mark an objective watershed in the development of health and health services (and the WHO targets have already been revised with this reality in mind), there is a certain symbolic importance often attached to consideration of what may be the important issues at the beginning of the twenty-first century. In this chapter we have chosen for discussion some trends which seem likely to contribute to the shape of research on health and societies over the next five years or so, up to and beyond the year 2000. Clearly, a very wide research agenda is implied by the discussion in this book, so that this chapter is necessarily selective. We focus here on the following aspects which seem to be good illustrations of the continuing changes we expect to see and the emphasis in research which is likely to be evident in the short to medium term:

- pluralism in approaches to health and societies
- globalization and interdependence
- the structure of health service systems
- power and resistance: health and health services in societies

Pluralism in approaches to health and societies

One of the important trends contributing to the diversity of perspectives in health and health services systems has been a growing acknowledgement of pluralism in professional, academic, and popular approaches to health issues.

For example, we have identified throughout this book the changing balance between the perspectives offered by different health professionals including those in conventional biomedicine, in the socio-ecological paradigm of health, or in the various forms of complementary medicine. Although in many quarters biomedicine still maintains a dominant hold, this is continually contested and, increasingly, a perspective of professional pluralism is required. Growing professional pluralism will have implications for our understanding of health, and the nature and scope of health services is also likely to change because of the different strategies adopted towards health promotion, illness prevention, and the treatment of illness. Already, as discussed in Chapter 5, national health services are beginning to incorporate 'alternative' therapies, and there will be continuing debate over how to decide which approaches are effective and how access to them for the population may need to be regulated in terms of quality and quantity. There also will be limits to how far these different views and approaches can be comfortably integrated together, and health policy may in future need to resolve growing professional disputes over potentially conflicting practices, demands for resources, and authority in health service systems. The scope and nature of work within the separate professions is also beginning to change to allow more collaborative working, and to incorporate elements and perspectives from across the interprofessional divides (as shown by examples of intersectoral working in Chapters 7 and 9).

We have also highlighted the growing pluralism which is evident in academic approaches to health and societies. Not only is there a tendency towards more interdisciplinary work on health; within each discipline a broadening diversity of theoretical standpoints and methodological strategies is also evident. (This was demonstrated, for example, in the case of medical geography in Chapter 1.) The discussion in this text has drawn from examples of the work of researchers situated in sociology, anthropology, geography, economics, epidemiology, medicine, and other disciplines besides. There has been a tendency in the past for each discipline to approach health problems in isolation from the others, giving emphasis exclusively to its own theoretical and methodological frameworks (biomedicine, in particular, has historically been predisposed towards this strategy). However, new perspectives on health, health services, and health policy show a marked trend towards interdisciplinarity and cross-fertilization. There is no single body of theory which is adequate to represent or explain the complexity of factors relevant to health. A range of theories is now being applied to research on health and health services, from micro-level theories, useful to understand individual health beliefs and behaviour, to meta-theory, offering accounts of the role of health services and health policy in the broad political and economic structure of society.

Similarly, a wider range of methodologies are now becoming recognized in research on health and health care. The complementarity of 'intensive' and 'extensive' strategies is particularly evident in the examples considered in this book. Intensive research, typically focused on small, theoretically selected samples and employing qualitative techniques, provides rich accounts of particular parts of the larger picture. It is especially revealing of the way that a complex range of factors come together to influence the health of individuals. Intensive studies can also suggest hypotheses which might be postulated for general populations. Extensive studies, usually based on large, statistically representative samples and employing more systematized, and often quantifiable, methods of data collection, are sometimes the means by which the findings of intensive studies can be tested for their generalizability. Extensive research is often more descriptive than explanatory, but sometimes the patterns discovered can later be investigated in more depth in intensive research. Thus there should be a symbiotic relationship between intensive and extensive methodology. In future, it may become more common for researchers to be comfortable working with both sorts of approach together, and certainly there is a tendency for researchers to collaborate to combine both strategies in joint projects.

Such changes in the perspectives used by the research community to investigate health progressively feed into health service practice and health policy. In Chapter 6, for example, we noted the use of a wider range of methods, producing qualitative as well as quantitative information, to evaluate the outcomes of health services, and Chapters 7 and 9 showed how theories of the socio-economic factors relating to health have informed the 'new' public health movement and international policy relating to health.

Globalization and interdependence

Several of the perspectives developed in this book demonstrate that the factors which influence health within particular countries, even at the very local level, are significantly influenced by processes operating at the world scale. One obvious case is the importance of environmental factors and the effects of environmental change and pollution, which are no respecters of international frontiers. The transmission of infectious diseases is also facilitated in part by the increasing volume of international mobility and communication. Furthermore, migration processes have resulted in growing ethnic and cultural diversity within many countries, with important implications for health service needs and health policy.

Another example of global processes affecting health and health services is the increasing international organization of the biomedical industries supplying pharmaceuticals and medical equipment, the commercial medical insurance companies, and other industries with considerable effects on health, such as tobacco and alcohol. These developments are linked to the globalization of economic systems, associated with the organization of capital at the world scale and facilitated by the emergence of multinational companies and international economic unions of nations (described in detail by e.g. Knox and Agnew 1994). Global economic systems bring with them economic and social interdependency of countries. High-income countries are typically much more powerful in the world economic system than low-income countries and, as we have shown in Chapter 4, the high-income countries command greater resources for health services and have better average population health status. However, no country is unaffected by global trends or able completely to control their impact.

Continuing change in the world economy places pressures on countries to make structural adjustments to their political economy, and health policy evolves in the context of such change. Thus, for example, we discussed in Chapter 5 how high-income countries are attempting to contain the costs of medical care in the context of slowing economic growth. In Eastern Europe, the former socialist countries struggle to adapt their health and welfare systems to accommodate more capitalist models of organization in their societies. Finally, in low- and middle-income countries, the battle to provide essential health services for a growing population in an increasingly urban and industrial setting is hampered by their burden of international debt. The ways in which such global pressures are felt will obviously be different in each country, and mediated by the local circumstances. Our discussion in Chapter 4 of the diversity of experience of the epidemiological transition, for example, shows that it would be wrong to assume that because some of the processes are global there is a single, international approach to health policy which will be relevant in all cases. However, the fact that many of the underlying processes affecting health are internationally linked means that it is essential for countries to work together in some areas of health policy. Our discussion in Chapter 9 of the role of the World Health Organization and the World Bank underlines the importance of such collaboration in the future. New groupings of countries, such as the expanded European Union, will also have their roles to play on the international health scene, and represent important areas for future research in terms of their impact on health policy, health services, and health. There will be a growing need in future to work out more coherent and co-ordinated international strategies to promote health gain and to resolve some of the current inconsistencies of international policies. This will require, among other things, a good deal of work on methods of research and information systems to allow us to make more meaningful international comparisons of health and health services. A continuing major challenge will be to ensure that sufficient priority is given to population health, alongside other international objectives, and, in particular, that steps are taken towards reducing global inequalities in health.

The structure of health service systems

Among the significant dimensions of change which we have discussed in this book, reorganization of health service systems stands out as particularly important, having considerable potential implications for health in many countries of the world. We noted in Chapter 5 a shift towards managed markets for health services in many parts of the world. This trend has raised new questions over whether responsibility for health services rests with the public or the private sector. Many of the recent reforms have been introduced wholesale into countries, with little attempt to pilot them in the local context. In some cases the rationale for introduction of market competition and market-style management techniques seems to have been based as much on ideological criteria as on sound evidence that they will produce the promised benefits of more efficient and effective health services. Saltman

(1994: 289) comments that 'the present period of health care reform can be characterized as one of dynamic learning'. He suggests that work to evaluate the outcomes of introduction of market-style models in health services has really only just begun, and that there is an enormous task ahead to establish which of the changes have been successful from the point of view of health policy objectives and which may require further adjustment. We noted in Chapter 6 that, while various market-style methods have been adopted to try to monitor and enhance the efficiency and effectiveness of health services, for the moment these are too simplistic to be very useful. A much better understanding of how managers can assess the health outcomes of health services and health policy is clearly necessary, and is becoming a major research area.

Saltman also points out that health service reforms have been more difficult and more costly to implement than expected in many countries. He underlines how experience of reform has demonstrated the importance of national regulatory structures for the successful introduction of managed markets. For some governments (e.g. in the UK and in Russia) this may seem perverse, since their reforms were partly intended to reduce the degree of involvement and control of government. Nevertheless, it seems likely that as market-style reforms to health service systems begin to mature there will be increasing emphasis on how best to carry out these regulatory functions.

Throughout the book we have highlighted gaps between policy rhetoric and the reality of policy implementation in health service reforms. This has illustrated the inevitably problematic and contested nature of the rhetoric underlying different policy themes and the contested nature of policy implementation. This links back to discussion in Chapter 3 about contested views of health, as we can see those refracted and reflected in many of the policy disputes; the examples discussed in Chapter 3 can each be reread in policy process terms. They can also be read in terms of Fraser's categorization of different types of struggle over needs (Fraser 1989b: 173). This can be seen in the issue of central versus local control and responsibility that permeates the debates over health service organization. As Chapter 8 showed, recent changes in Britain, for example, reflect two apparently conflicting tendencies, with, on one hand, a growth in central control, constraint, and monitoring of the resources for health services and, on the other hand, a shifting of responsibility for the practical implementation of reform to the local level, and a growing acceptance of variation in the mode of health service delivery in response to local needs. The same conflicting tendencies were also noted in the US system in Chapter 8. At the local level this has generated tensions which are difficult to resolve, and which require new approaches to health service planning. Local health planners need better information on how conditions in their area combine to produce patterns of need for health services specific to their populations. There also needs to be greater understanding at central level that it may be inappropriate or unrealistic to apply universal national standards or targets locally without adjustment for local conditions. There is a large potential here for further research on the effects of the local context on health status, on health services, and on their interaction.

Finally, health policy at the international and the national scales places growing emphasis on the development of primary health care, health promotion, and the provision of secondary care in community settings. Many of these changes in health service systems imply that greater responsibility for health care will be taken by individuals or by the informal carers who look after them. However, we need much clearer information on the current role of this sector, the burden of care it carries, and whether there is really scope to increase this burden in the ways that policy implies. In future, it seems likely that health policy will also need to be more responsive to the felt and expressed needs of informal carers and of individual patients themselves, and this will require improved methods for making these needs more apparent and more influential in the health policy process. Chapter 8 discussed the sorts of strategies for consultation and participation which hopefully will be much more prominent in health service systems of the future.

Power and resistance: health and health services in societies

The importance of the discussions in Chapters 2 and 3 of views about 'health', 'illness', and 'disease' is not to try to find the most satisfactory definition of a set of terms but to illustrate the multiple and complex ways in which these terms are used discursively. They are carried into, and shape, every interaction that takes place within health service settings, whether it be in the name of treatment, cure, care, prevention, or even health promotion. The recognition of this has important implications for considering the relationships between the parties involved in the delivery of health services to an individual, and we have not had sufficient space to explore this area in detail. Fox (1993) analyses how the organization of health services and the caring relationship as presently constituted frequently operate to re-inscribe existing relations of power between the health professional and individual in support of the status quo. However, these power relationships are also characterized by resistance, and there is scope to recast these relationships to support the empowerment of the individuals involved. This concern with the philosophy and ethics of professional practice represents an important area for future research.

What counts as knowledge about health and ill health is continually reshaped in the course of health service delivery as well as in the health policy arena. In addition to the studies discussed earlier in the book (in Chapters 3 and 8 particularly), there are examples such as that of Arney and Neill (1982), who examine how Western obstetrics transformed its view of its practice in the post-World War II period, partly in response to the challenge of the natural childbirth movement, and Nettleton (1988), who examines the constitution of knowledge in Western dentistry at the turn of the century. There are also wider interactions that need to be considered, between health, health services, and wider aspects of the social context. In Chapter 8 we explored some Foucauldian studies in the context of health policy development and implementation. In examining this area, we have emphasized how Western medicine, and the health services structured around its use, have operated in the past to serve a variety of social control functions (see discussions in Chapters 1, 3, and 9 in particular).

A further example is provided by the work of Illich (1976). He is one of the most extreme critics of the role of doctors, arguing that institutionalized medicine has become a threat to health as a consequence of industrialization, bureaucratization, and professionalization of medicine. Illich distinguishes three different types of 'iatrogenesis' or damage induced by the provider. The first of these, clinical iatrogenesis, refers to the pain, dysfunction, disability, and anguish resulting from technical medical intervention through ineffective treatment, side-effects, and medical mistakes or malpractice; some of the issues concerned here were covered in Chapter 6. Secondly there is social iatrogenesis, the process through which medical practice sponsors sickness by reinforcing a society that 'preserves defectives' and 'breeds demand for the patient role', encourages people to consume curative, preventive, industrial, and environmental medicine, and fosters dependence. Finally there is cultural iatrogenesis, through which the 'so-called health professions' destroy the potential of people to deal with their human weakness, vulnerability, and uniqueness in a personal and autonomous way, leading to a reduction of autonomy and independence, and a paralysis of healthy responses to suffering, sickness, and death. Illich sees the basic struggle as one in which individuals must regain their self-reliance and liberty from the oppressive nature of industrialization (one aspect of which is medical practice). His views have been widely criticized by marxists as reliant on a lifestyle theory of health (with a danger of 'blaming the victim'), and as fundamentally flawed in failing to recognize the key role of the capitalist economy; see for example Navarro's (1975) critique of Illich. We might note also the contrast between Illich's views and that implicit in Hart's inverse care law (discussed in Chapter 6), which assumes that more doctors are unequivocally good for you. Furthermore, we might also want to question some of the assumptions that go into the labelling of people as 'defectives'. Nevertheless, Illich's polemic is stimulating to those who seek to develop discourses which challenge the dominance of biomedicine.

There are also a body of studies based on an explicitly Foucauldian position, which argues for a reconceptualization of the notion of the state and for a view of 'power', not as something exercised from a central location by domination, but conceived of as circulating and saturating the social field, acting through multiple and dispersed locations, and whose constitution is intimately related to that of knowledge. Some of these studies were discussed in Chapter 8. Other relevant examples include Prior (1987) on the mortuary as a socio-medical institution and the discourse of pathology, which identifies a fusion of the interests of medicine and the wider politico-technological system within which it is embedded; while Nettleton (1991) examines the discourse and practices that surround dentistry to illustrate how these serve as a means by which political rule is exercised. Deleuze and Guattari (1984) offer a thorough critique of the mainstay of much Western psychoanalysis – the Oedipus complex – illustrating amply its culture-bound nature and its links to the maintenance of oppressive and repressive social relations within Western capitalism, and to the maintenance of fascism, at the level of society and of the individual. Armstrong (1983) includes an examination of the discipline of social medicine which emerged in the twentieth century as the medical gaze began to move beyond the microscopic details of the individual body to observe the undifferentiated spaces between bodies, and thus forge a new political anatomy based especially on the development of the population survey as a tool. Thus, while our discussion of the new public health movement in Chapter 7 showed an emancipation from a strongly biomedical professional view of health which had developed in high-income countries during the twentieth century, we can also see that, inasmuch as it is informed by social medicine, the movement has potential to introduce new disciplinary mechanisms in the field of health.

Arguments, such as those by Illich and Armstrong, about the social control and disciplinary functions of medicine are not new, but they may be applied in new ways in the future. We have emphasized in this book that more recent accounts of health and society move beyond a simple understanding of the functioning of medicine as acting to preserve the social relations between dominant and subordinated social groups. Such accounts emphasize how the disciplinary practices within health service systems are continually subject to contestation and dynamic redefinition. Examples were included in Chapters 3, 8, and 9, and we might also mention Bloor and McIntosh's (1990) study of different techniques of 'client resistance' in health visiting and therapeutic communities, including collective ideological dissent, individual ideological dissent, non-co-operation, and scope for avoidance and concealment. This area presents important challenges for multidisciplinary and interdisciplinary research in the future, particularly to help in the task of shaping health policy and health services so that they respond to the diversity in the groups and communities they are supposed to serve. In approaching the study of health in societies in the future, what we argue for is a critical approach, based on an appreciation of the value of health services in relieving suffering and preventing ill health, whilst at the same time recognizing that such services, and the professionals that provide them, also have other, potentially less desirable, impacts upon societies.

References

Aakster, C. W. (1986) Concepts in alternative medicine. *Social Science and Medicine* 22, 265–74.

Aaron, H. (1994) Economic analysis of the Clinton health care reform proposals. *Health Affairs*, 13(1), 57–68.

Aase, A. (1989) Regionalizing mortality data: ischaemic heart disease in Norway. *Social Science and Medicine* 29, 907–11.

Abbott, A. (1992) Europe tightens rules that govern homeopathic products. *Nature* 359, 469.

Adelaide Conference on Healthy Public Policy (1988) *Report on Second International Conference on Health Promotion, April 5–9 1988*. Adelaide, South Australia.

Aggleton, P. and Chambers, H. (1986) *Nursing models and the nursing process*. Basingstoke: Macmillan.

Ahmad, W. I. U. (ed.) (1992) *The politics of 'race' and health*. Bradford: Race Relations Unit, University of Bradford, and Bradford and Ilkley Community College.

——(ed.) (1993a) *'Race' and health in contemporary Britain*. Milton Keynes: Open University.

——(1993b) Making black people sick: 'race', ideology and health research. In Ahmad (1993a: 11–33).

——Kernohan, E. and Baker, M. (1989) Health of British Asians: a literature review. *Community Medicine* 11, 49–56.

Aiach, P. and Curtis, S. (1990) Social inequalities in self-reported morbidity: interpretation and comparison of data from Britain and France. *Social Science and Medicine* 31(3), 267–74.

Akhtar, R. (ed.) (1987) *Health and disease in tropical Africa: geographical and medical viewpoints*. Chur, Switzerland: Harwood.

——and Izhar, N. (1994) Spatial inequalities and historical evolution in health provision: Indian and Zambian examples. In Phillips and Verhasselt (1994: 216–33).

Alaszewski, A., Tether, P. and McDonnell, H. (1981) Another dose of managerialism? *Social Science and Medicine* 15A, 3–15.

Alcohol, Drug Abuse and Mental Health Administration Reorganisation Act (1992) Public Law 102–321, 106 Stat. 383.

Aldridge, D. (1989) Europe looks at complementary medicine. *British Medical Journal* 229, 1121–2.

Alford, R. (1975) *Health care politics*. Chicago: University of Chicago Press.

Alladin, W. J. (1992) Clinical psychology provision: models, policies and prospects. In Ahmad (1992: 117–41).

Allen, E. J. (1995) Child health care delivery in the inner city: perceived roles and responsibilities. Ph.D. thesis, Queen Mary and Westfield College, University of London.

Allen, I. (ed). (1988) *Hearing the voice of the consumer*. London: Policy Studies Institute.

Allsop, J. and May, A. (1993) Between the devil and the deep blue sea: managing the NHS in the wake of the 1990 Act. *Critical Social Policy* 38, 5–22.

Alston, P. (1979) *Making and breaking human rights: the UN's specialised agencies and implementation of the International Covenant on Economic, Social and Cultural Rights*. London: Anti-Slavery Society and Committee for Indigenous Peoples.

Altman, D. (1986) *AIDS and the new puritanism*. London: Pluto Press. Also published (with the same pagination) as *AIDS in the mind of America*. New York: Anchor Press.

Altman, S. (1994) Health system reform: let's not miss our chance. *Health Affairs* 13(1), 69–80.

American Psychiatric Association (1952) *Diagnostic and statistical manual of mental disorders*. Washington, DC: American Psychiatric Association. 2nd edn 1968; 4th edn 1994.

Anderson, B. and Silver, B. (1989) The changing shape of Soviet mortality, 1958–1985: an evaluation of old and new evidence. *Population Studies* 43, 242–65.

Anderson, J. M., Blue, C. and Lau, A. (1991) Women's perspectives on chronic illness: ethnicity, ideology and restructuring of life. *Social Science and Medicine* 33, 101–13.

Andersson, N. and Marks, S. (1989) The state, class and the allocation of health resources in southern Africa. *Social Science and Medicine* 28, 515–30.

Anti-Drug Abuse Act (1988) Public Law 100–690, 102 Stat. 4202.

Antonucci, T. (1990) Personal characteristics, social support and social behaviour. In Binstock, R., and George, L. (eds), *Handbook of aging and the social sciences*, 3rd edn. San Diego, Calif.: Academic Press, 205–26.

Arber, S. and Ginn, J. (1993) Gender and inequalities in health in later life. *Social Science and Medicine* 36(1), 33–46.

Armstrong, D. (1983) *Political anatomy of the body: medical knowledge in Britain in the twentieth century.* Cambridge: Cambridge University Press.

Arney, W. R. and Neill, J. (1982) The location of pain in childbirth: natural childbirth and the transformation of obstetrics. *Sociology of Health and Illness* 4(1), 1–24.

Arnold, D. (ed.) (1988) *Imperial medicine and indigenous societies.* Manchester: Manchester University Press.

Arnstein, S. R. (1969) A ladder of citizen participation. *Journal of the American Institute of Planners* 35, 216–24.

Aronson, J. (1993) Giving consumers a say in policy development: influencing policy or just being heard. *Canadian Public Policy* 19(4), 367–78.

Ashton, J. (1991) The Healthy Cities Project: a challenge for health education. *Health Education Quarterly* 18(1), 39–48.

——(1992) The origins of healthy cities. In Ashton, J. (ed.), *Healthy Cities.* Milton Keynes: Open University, 1–14 .

——Gray, P. and Barnard, K. (1986) Healthy cities: WHO's new public health initiative. *Health Promotion* 1, 319–24.

Asthana, S. (1994) Primary health care and selective PHC: community participation in health and development. In Phillips and Verhasselt (1994: 182–98).

Audit Commission for Local Auhorities in England and Wales (1986) *Making a reality of community care.* London: HMSO.

Ayoade, A., Harrison, I., Warren, D. and Ademuwagun, Z. A. (eds) (1978) *African therapeutic systems.* Waltham, Mass.: Crossroads Press.

Baggott, R. (1986) Alcohol, politics and social policy. *Journal of Social Policy* 15(4), 467–88.

——(1990) *Alcohol, politics and social policy.* Aldershot: Avebury.

——(1994) *Health and health care in Britain.* New York: St Martin's Press.

Baghust A. (1994) In line for a new look at queuing. *Health Service Journal,* 30 June, 17.

Bain, J. (1994) Fundholding: a two-tier system? *British Medical Journal* 309, 396–9.

Balarajan, R., Bulusu, L., Adelstein, A. and Shukla, V. (1984) Patterns of mortality among immigrants to England and Wales from the Indian subcontinent. *British Medical Journal* 289, 1185–87.

——Yuen, P. and Machin D. (1992) Deprivation and general practitioner workload. *British Medical Journal* 304, 529–34.

Bancroft, J. (1988) Editorial: Homosexuality—compatible with full health. *British Medical Journal* 297, 308–9.

Banerji, D. (1984) The political economy of western medicine in third world countries. In McKinlay, J. B. (ed.), *Issues in the political economy of health.* New York: Tavistock, 257–82.

Bannerman, R. H., Burton, J. and Chen, W. C. (1983) *Traditional medicine and health care coverage.* Geneva: WHO.

Banzhaf, M., Chris, C., Christensen, K., Danzig, A., Denenberg, R., Leonard,

Z., Lurie, R., Pearl, M., Saalfield, C., Thistlethwaite, P., Walker, J. and Weil, B. (The ACT UP/New York Women and AIDS Book Group) (1992) *Women, AIDS and activism*, 3rd edn. Boston: South End Press.

Bardsley, M., Venables, C., Watson, J., Goodfellow, J. and Wright, P. (1992) Evidence for validity of a health status measure in assessing short term outcomes of cholecystectomy. *Quality in Health Care* 1, 10–14.

Baron, M. (1993) Genetic linkage and male homosexual orientation: reasons to be cautious. *British Medical Journal* 307, 337–8.

Barrera, M. (1986) Distinctions between social support concepts, measures and models. *American Journal of Community Psychology* 14(4), 413–45.

Barrett, F. A. (1991) 'Scurvy': Lind's medical geography. *Social Science and Medicine* 33(4), 347–53.

——(1993) A medical geographical anniversary. *Social Science and Medicine* 37(6), 701–10.

——(forthcoming) The origins and development of medical geography and geographical medicine: from the mid-5th century BC up to the mid-20th century AD.

Bartlett, W. and Harrison, L. (1993) Quasi-markets and the National Health Service reforms. In Le Grand and Bartlett (1993: 68–92).

Baruch, E. H. and Serrano, L. J. (1988) *Women analyze women—in France, England and the United States*. New York: Harvester Wheatsheaf.

Bass, A. (1994) Being there. *Health Service Journal* 104(5417), 25.

Baum, F. (1993) Noarlunga Healthy Cities pilot project: the contribution of research and evaluation. In Davies, J. K. and Kelly, M. P. (eds), *Healthy Cities: research and practice*. London: Routledge, 90–111.

Bayer, R. (1981) *Homosexuality and American psychiatry: the politics of diagnosis*. New York: Basic Books.

Bell, M. (1993) 'The pestilence that walketh in darkness': imperial health, gender and images of South Africa c. 1880–1910. *Transactions of the Institute of British Geographers* N.S. 18, 327–41.

Bennegadi, R. and Bourdillon, F. (1990) La santé des travailleurs migrants en France: aspects médico-sociaux et anthropologiques. *Revue Européenne des Migrations Internationales* 6(3), 129–42.

Bennett, A. L. (1991) *International organizations: principles and issues*. London: Prentice-Hall.

Bentham, G. (1994) Global environmental change and health. In Phillips. and Verhasselt (1994: 33–49).

Bergler, E. (1951) *Neurotic, counterfeit sex*. New York: Grune & Stratton.

Berk, M., Cunningham, P. and Beauregard, K. (1991) The health care of poor persons living in wealthy areas. *Social Science and Medicine* 32(10), 1097–1103.

Berman, P., Kendall, C. and Bhattacharyya, K. (1994) The household production of health: integrating social science perspectives on micro-level health determinants. *Social Science and Medicine* 38(2), 205–15.

Bhardwaj, S. (1980) Medical pluralism and homeopathy: a geographic perspective. *Social Science and Medicine* 14B, 209–16.

Bhat, R. (1993) The private health care sector in India. In Berman, P. and Khan, M. (eds), *Paying for India's health care*. London: Sage, 161–96.

Bie Nio Ong and Humphris, G. (1994) Prioritizing needs with communities: rapid appraisal methodologies in health. In Popay, J. and Williams, G. (eds), *Researching the people's health*. London: Routledge, 58–82.

Biehal, N. (1993) Changing practice: participation, rights and community care. *British Journal of Social Work* 23, 443–58.

Black Health Workers and Patients Group (1983) Psychiatry and the corporate state. *Race and Class* 25(2), 49–64.

Black, N. (1992) The relationship between evaluative research and audit. *Journal of Public Health Medicine* 14(4), 361–6.

——and Thompson, E. (1993) Obstacles to medical audit: British doctors speak. *Social Science and Medicine* 36(7), 849–56.

Blane, D., Davey G. and Bartley, M. (1993) Social selection: what does it contribute to social class differences in health? *Sociology of Health and Medicine* 15(1), 1–15.

Blanpain, J., Delesie, L. and Nys, H. (1978) *National health insurance and health resources: the European experience*. Cambridge, Mass.: Harvard University Press.

Blaxter, M. (1990) *Health and lifestyles*. London: Routledge.

——and Patterson, E. (1982) *Mothers and daughters: a three generational study of health attitudes and behaviour*. London: Heinemann.

Bloch, S. and Reddaway, P. (1984) *The shadows over world psychiatry*. London: Gollancz.

Bloor, M. and McIntosh, J. (1990) Surveillance and concealment: a comparison of techniques of client resistance in therapeutic communities and health visiting. In: Cunningham-Burley, S. and McKeganey, N. P. (eds), *Readings in medical sociology*. London: Tavistock/Routledge, 159–81.

Blum, A. and Monnier, A. (1989) Recent mortality trends in the USSR: new evidence. *Population Studies* 43, 211–41.

Blum, H. (1974) *Planning for health*. New York: Human Sciences Press.

Bobidilla, J. and Possas, C. (1991) *Epidemiological transition and health policies in Latin America: an analysis of three middle income countries*. Workshop organized by the Committee on Population of the National Academy of Science. Washington DC: National Academy of Science.

——Sepulveda, J. and Cervantes, M. (1989) Health transition in middle-income countries: new challenges for heath care. *Health Policy and Planning* 4(1), 29–39.

Boffin, T. and Gupta, S. (eds) (1990) *Ecstatic antibodies: resisting the AIDS mythology*. London: Rivers Oram Press.

Bognar, B. J. (1994) Cooperation between academia and consumers to optimize research in policy and program development. In Rey, J. C. and

Tilquin, C. (eds), *SYSTED 94: Proceedings of the Fifth International Conference on Systems Sciences in Health and Social Services for the Elderly and the Disabled*, Geneva, 2–6 May 1994. Aarau: Institut Suisse de la Santé Publique, 257–64.

Bomberg, E. and Peterson, J. (1993) Prevention from above? the role of the European Community. In Mills (1993: 140–60).

Booth, C. (1892) *The life and labour of the people of London, ii.* London: Macmillan.

Boudreau, F. (1991) Partnership as a new strategy in mental health policy: the case of Québec. *Journal of Health Politics, Policy and Law* 16(2), 307–29.

Boufford, J. (1993) US and UK health care reforms: reflections on quality. *Quality in Health Care* 2, 249–52.

Bowling, A. (1993) *Measuring health: a review of quality of life measurement scales.* Milton Keynes: Open University Press.

Boys, R., Foster, D. and Jozan, P. (1991) Mortality from causes amenable and non-amenable to medical care: the experience of eastern Europe. *British Medical Journal* 303, 879–83.

Bradshaw, J. (1972) The concept of need. *New Society*, 30 Mar. 1972, 640–3.

Brandon, R., Podhorzer, M. and Pollk, T. (1991) Premiums without benefits: waste and inefficiency in the commercial health insurance industry. *International Journal of Health Services* 21(2), 265–83.

Brannstrom, I., Persson, L. and Wall, S. (1994) Towards a framework for outcome assessment of heath intervention: conceptual and methodological considerations. *European Journal of Public Health* 4, 125–30.

Brearley, S. (1990) *Patient participation: the literature.* RCN research series. London: Harrow Soutain Press.

Breeze, E., Maidment, A., Bennett, N., Hatley J. and Carey, S. (1994) *Health Survey for England 1992.* London: OPCS, HMSO.

Breheny, R. (1991) Cholera: forget vaccines this time around. *British Medical Journal*, 302, 1033–34.

Briggs, A. (1961) *Social thought and social action: a study of the work of Seebohm Rowntree, 1871–1954.* London: Longman.

British Medical Association (BMA) (1993) *Complementary medicine: new approaches to good practice.* Oxford: Oxford University Press.

Britton, M. (1990) Geographic variation in mortality since 1920 for selected causes. In Britton, M. (ed.), *Mortality and geography: a review in the mid-1980s, England and Wales.* London: HMSO.

Brown, L. D. (1992) Political evolution of federal health care regulation. *Health Affairs – Millwood* 11(4), 17–37

Bryant, J. H. (1980) WHO's programme of health for all by the year 2000: a macrosystem for health policy making—a challenge to social science research. *Social Science and Medicine* 14A, 381–6.

Bryce, C., Curtis, S. and Mohan, J. (1994) Coronary heart disease: trends in spatial inequalities and implications for health care planning in England. *Social Science and Medicine* 38(5), 677–90.

Brzezinski, Z. J. (1985) *Mortality in the European region.* Copenhagen: WHO Regional Office for Europe.

Bulmer, M. (1987) *The social basis of community care.* London: Allen & Unwin.

Burkett, P. (1991) Poverty crisis in the third world: the contradictions of World Bank policy. *International Journal of Health Services* 21(3), 471–9.

Burton, I. (1963) The quantitative revolution and theoretical geography. *Canadian Geographer* 7, 151–62.

Butler, K. and Forrest, M. (1990) Citizen advocacy for people with disabilities. In Winn, L. (ed.), *Power to the people: the key to responsive services in health and social care.* London: King's Fund Centre, 59–67.

Buxton, M. (1994) Achievements of audit in the NHS. *Quality in Health Care* 3, Supplement, S31–4.

Calnan, M. (1984) The politics of health: the case of smoking control. *Journal of Social Policy* 13(3), 279–96.

——(1987) *Health and illness: the lay perspective.* London: Tavistock.

Campling, E., Devlin, H. and Lunn J. (1990) *The report of the national confidential enquiry into perioperative deaths (NCEPOD).* London: NCEPOD.

Cantor, D. (1993) Cortisone and the politics of empire: imperialism and British medicine, 1918-1955. *Bulletin of the History of Medicine* 67(3), 463–93.

Carpenter, E. (1980) Children's health care and the changing role of women. *Medical Care,* 38(12), 1208–17.

Carr-Hill, R. (1994) Efficiency and equity implications of the health care reforms. *Social Science and Medicine* 39(9), 1189–1201.

——and Sheldon, T. (1991) Designing a deprivation payment for general practitioners: the UPA(8) wonderland. *British Medical Journal* 302, 393–6.

——Smith, P., Martin, S., Peacock, S., and Hardman, G. (1994) Allocating resources to health authorities: development of method for small area analysis of use of inpatient services. *British Medical Journal* 309, 1046–49.

Carter, E. and Watney, S. (eds) (1989) *Taking liberties: AIDS and cultural politics.* London: Serpent's Tail.

Cartwright, S. A. (1851) Report of the diseases and physical peculiarities of the negro race. *New Orleans Medical and Surgical Journal,* May, 691–715.

Caselli, G. and Egidi, V. (1988) Les variation géographiques de la mortalité. In Vallin, J. and Meslé, F. (eds), Les causes de décès en France de 1925 à 1978. Cahier no. 115. Paris: Institut National d'Études Démographiques.

CDC (Centers for Disease Control) (1983) *Morbidity and mortality weekly Report, Sept. 9.* Atlanta: CDC.

CETHV (Council for the Education and Training of Health Visitors) (1977) *An investigation into the principles of health visiting*. London: CETHV.

Chadwick, E. (1842) *Report from the Poor Law commissioners on an inquiry on the sanitary conditions of the labouring population of Great Britain*. London: HMSO (repr. 1965).

Chamberlain, J. (1988) *On our own: patient controlled alternatives to the mental health system*. London: MIND.

Chambers, M. and Clarke, A. (1990) Measuring admission rates. *British Medical Journal* 301, 1134–36.

Charles, C. and Demaio, S. (1993) Lay participation in health care decision making: a conceptual framework. *Journal of Health Politics, Policy and Law* 18(4), 881–904.

Charlton, J., Hartley, R., Silver, R. and Holland, W. (1983) Geographical variation in mortality from conditions amenable to medical intervention in England and Wales. *Lancet* 26(3), 691–6.

——Wallace, M. and White, I. (1994) Long-term illness: results from the 1991 census. *Population Trends*, 75, 19–25.

Chave, S. (1987) *Recalling the medical officer of health*. London: King Edward's Hospital Fund for London.

Checkoway, B. (ed.) (1981) *Citizens and health care: participation and planning for social change*. New York: Pergamon.

Chen, P. (1981) Traditional and modern medicine in Malaysia. *Social Science and Medicine* 15A, 127–36.

Chirimuuta, R. and Chirimuuta, R. (1989) *AIDS, Africa and racism*. London: Free Association Books.

Chow, E. (1984) Traditional Chinese medicine: a holistic system. In Salmon, W. (ed.) *Alternative medicines: popular and policy perspectives*. London: Tavistock, Ch. 4, 114–37.

Chruscz, D., Pamuk, E. and Lentzner, H. (1991) Life expectancies in eastern central Europe: components of change in six countries in the 1980s. *Population Network Newsletter* 20, 1–4.

Cixous, H. and Clément, C. (1975) *The newly born woman*, trans. B Wing. Manchester: Manchester University Press.

Clare, A. (1980) *Psychiatry in dissent: controversial issues in theory and practice*, 2nd edn. London: Tavistock.

Clarke, A. (1990) Are readmissions avoidable? *British Medical Journal* 301, 1136–38.

Clarke, M. and Wilson, A. (1986) Developments in planning models for health care policy analysis. In Pacione (1986: 248–83).

Cliff, A. D. and Haggett, P. (1988) *Atlas of disease distributions: analytic approaches to epidemiological data*. Oxford: Blackwell. (Paperback 1992.)

——Ord, J. and Versey, G. (1981) *Spatial diffusion: an historical geography of epidemics in an island community*. Cambridge: Cambridge University Press.

Clifford, T. (1984) *Tibetan Buddhist medicine and psychiatry*. Main: Samuel Weiser.

COMARE (Committee on Medical Aspects of Radiation in the Environment), first report (1986), second report (1988), third report (1989). London: HMSO.

Comas-Diaz, L. and Greene, B. (eds) (1994) *Women of color: integrating ethnic and gender identities in psychotherapy*. New York: Guilford Press.

Coombes, Y. J. (1993) A geography of the new public health. Ph.D. thesis, Queen Mary and Westfield College, University of London.

Cooter, R. (ed.) (1988) *Studies in the history of alternative medicine*. Basingstoke: Macmillan.

Cornwell, J. (1984) *Hard earned lives: accounts of health and illness from East London*. London: Tavistock.

Cotgrove, A., Bell, G. and Katona, C. (1991) Psychiatric admissions and social deprivation: is the Jarman underprivileged area score relevant? *Journal of Epidemiology and Community Health* 46, 245–7.

Cox, B., Blaxter, M., Bucle, A., Fenner, N., Golding, J., Gore, M., Huppert, F., Nickson, J., Roth, Sir M., Stark, J., Wadsworth, M. and Whichelow, M. (1987) *Health and lifestyle survey*. Cambridge: Health Promotion Trust.

——Huppert, F. and Whichelow, M. (eds) (1993) *The health and lifestyle survey: seven years on*. Aldershot: Dartmouth.

Crail, M. (1994) Blinking indicators. *Health Service Journal,* 4 Aug., 10–11.

Crawford, R. (1984) A cultural account of 'health': control, release and the social body. In McKinlay, J. B. (ed.), *Issues in the political economy of health care*. New York: Tavistock, 60–103.

Crimp, D. (1987) AIDS: Cultural analysis/cultural activism. Published in Crimp (1988: 3–16).

——(ed.) (1988) AIDS: *Cultural analysis/cultural activism*. Cambridge, Mass.: MIT Press.

Croft, S. and Beresford, P. (1990) *From paternalism to participation: involving people in social services*. London: Open Services Project/Joseph Rowntree Foundation.

——(1992) The politics of participation. *Critical Social Policy,* autumn (35), 20–44.

Cromley, E. (1993) Geographical aspects of rural hospital conversion in the United States: an analysis of the EACH program. Paper presented at the Institute of British Geographers Annual Conference, 5–8 Jan. 1993. Royal Holloway College, University of London.

——and Craumer, P. (1990) Physician supply in the Soviet Union, 1940–1985. *Geographical Review*, 80, 132–40

Cruikshank, J. and Beevers, D. (1989) *Ethnic factors in health and disease*. London: Wright.

Currer, C. and Stacey, M. (eds) (1986) *Concepts of health, disease and illness*. Leamington Spa: Berg.

Curtice, L. (1993) Strategies and values: research and the WHO Healthy Cities project in Europe. In Davies, J. K. and Kelly, M. P. (eds), *Healthy Cities: research and practice*. London: Routledge, 34–54.

Curtis, S., Bucquet, D. and Colvez, A. (1992) Sources of instrumental support for dependent elderly people in three parts of France. *Ageing and Society* 12, 329–54.

Curtis, S. and Ogden, P. (1986) Bangladeshis in London; a challenge to welfare. *Revue Europeenne des Migrations Internationales* 2(3), 135–49.

Curtis, S., Petukhova, N. and Taket, A. (1995) Health care reforms in Russia: the example of St Petersburg. *Social Science and Medicine* 40(6), 755–65.

Curtis, S. E. (1982) Spatial analysis of surgery locations in general practice. *Social Science and Medicine* 16, 303–13.

——(1987a) The patient's view of general practice in an urban area. *Family Practice* 4(3), 200–6.

——(1987b) Self reported morbidity in London and Manchester: inter-urban and intra-urban variations. *Social Indicators Research* 19, 255–72.

——(1990) Use of survey data and small area statistics to assess the link between individual morbidity and neighbourhood deprivation. *Journal of Epidemiology and Community Health* 44, 62–8.

Curtis, S. E., Taket, A. R., Prokhorskas, R., Shabanah, M. A. and Thuriaux, M. C. (1989) Vers la santé pour tous dans la région européenne de L'OMS: surveillance des progrès accomplis. 2: Conditions préalables: bilan démographique et sanitaire. *Revue d'Épidémiologie et Santé Publique* 37, 295–317.

Curto De Casas, S. (1993) Geographical inequalities in mortality in Latin America. *Social Science and Medicine* 36(10), 1349–55.

Dada, M. (1990) Race and the AIDS agenda. In Boffin and Gupta (1990: 85–95).

Daes, E. I. A. and Daoudy, A. (1994) *Decentralization of organizations within the UN system, iii: the World Health Organization*. Report of the Joint Inspection Unit. General Assembly Official Records, 48th session. New York: UN.

Daker-White, G. (1995) Perceptions of illicit drug use and the accessibility of treatment services in an inner-London borough. Ph.D. thesis, Queen Mary and Westfield College, University of London.

Dalal, F. (1988) The racism of Jung. *Race and Class* 29(3), 1–22.

Danciger, E. (1987) *The emergence of homoeopathy*. London: Century.

DANIDA (Danish International Development Agency) (1991) *Effectiveness of multilateral agencies at country level: WHO in Kenya, Nepal, Sudan and Thailand*. Copenhagen: DANIDA, Ministry of Foreign Affairs.

Darton, R. (1984) *Trends 1970–1981*. In Laming, H.(ed.), *Residential care for the elderly: present problems and future Issues*. Discussion paper no. 8. London: Policy Studies Institute.

Dauncey, K., Giggs, J., Baker, K. and Harrison, G. (1993) Schizophrenia in Nottingham: lifelong residential mobility of a cohort. *British Journal of Psychiatry* 163, 613–19.

Davies, A. (1994) Patient defined outcomes. *Quality in Health Care* 3, Supplement, S6–S9.

Davies, B. (1968) *Social needs and resources in local services*. London: Michael Joseph.

Dear, M. J. and Taylor, S. M. (1982) *Not on our street: community attitudes to mental health care*. London: Pion.

Deleuze, G. and Guattari, F. (1984) *Anti-Oedipus: capitalism and schizophrenia*. London: Athlone.

——(1988) *A thousand plateaus: capitalism and schizophrenia*, trans. B Massumi. London: Athlone.

Department of Health (1990a) *General practice in the National Health Service: the 1990 contract. The government's programme for changes to GPs' terms of service and remuneration system*. London: Health Departments of Great Britain.

——(1990b) *Confidential enquiry into stillbirths and deaths in infancy: report of a working group*. London: Department of Health.

——(1991) *The patient's charter*. London: HMSO.

Department of Health and Social Security (DHSS) (1972) *Management arrangements for the reorganised National Health Service*. London: HMSO.

——(1976) *Sharing resources for health in England: report of the Resource Allocation Working Party*. London: HMSO.

——(1986a) *Neighbourhood nursing: a focus for care*. London: HMSO.

——(1986b) *Primary health care: an agenda for discussion*. London: HMSO.

D'Houtaud, A. and Field, M. (1984) The image of health: variations in perception by social class in a French population. *Sociology of Health and Illness* 6(1), 30–60.

Dixon, J., Dinwoodie, M., Hodson, D., Dodd, S., Poltorak, T., Garrett, C., Rice, P., Doncaster, I., Williams, M. (1994) Distribution of NHS funds between fundholding and non-fundholding practices. *British Medical Journal* 309, 30–4.

Donabedian, A. (1980) *The definition of quality and approaches to its assessment: explorations in quality assessment and monitoring, i*. Ann Arbor, Mich.: Health Administration Press.

Donden Y. (1986) *Health through balance: an introduction to Tibetan medicine*. Ithaca, NY: Snow Lion.

Donnison, J. (1977) *Midwives and medical men: a history of interprofessional rivalries and women's rights*. London: Heinemann.

Donovan, J. (1984) Ethnicity and health: a research review. *Social Science and Medicine* 19, 663–70.

——(1986) *We don't buy sickness, it just comes*. Aldershot: Gower.

——and Coast, J. (1994) Public preferences in priority setting: unresolved issues In Malek, M. (ed.), *Setting priorities in health care*. Chichester: Wiley, 31–45.

Donzelot, J. (1979) *The policing of families: welfare versus the state*, trans. R. Hurley. London: Hutchinson.

Dowson, S. (1990) *Keeping it safe: self advocacy by people with learning difficulties and the professional response*. London: Values into Action.

Doyal, L. (1979) *The political economy of health*. London: Pluto.

——and Gough, I. (1991) *A theory of human need*. London: Macmillan.

Drummond, M. (1994) Evaluation of health technology: economic issues for heath policy and policy issues for economic appraisal. *Social Science and Medicine* 38(12), 1593–1600.

Dubois, R., Rogers, W., Mosely J., Draper, D. and Brook, R. (1987) Hospital inpatient mortality: is it a predictor of quality? *New England Journal of Medicine* 317, 1674–80.

Dubos, R. (1960) *The mirage of health*. London: Allen & Unwin.

——1965: *Man adapting*. New Haven, Conn.: Yale University Press.

Duhl, L. J. (1976) The process of recreation: the health of the 'I' and the 'Us'. *Ethics in Science and Medicine* 3, 33–63.

——(1986) The healthy city: its function and its future. *Health Promotion* 1(1), 55–60.

Duncan, C., Jones, K. and Moon, G. (1993) Do places matter? A multi-level analysis of regional variations in health-related behaviour in Britain. *Social Science and Medicine* 37, 725–33.

Dunn, J. L. Taylor, M., Elliott, S. and Walter, D. (1994) Psychosocial effects of PCB contamination and remediation: the case of Smithville, Ontario. *Social Science and Medicine* 39(8), 1093–104.

Eames, M., Ben-Shlomo, Y. and Marmot, M. (1993) Social deprivation and premature mortality: regional comparison across England. *British Medical Journal* 307, 1097–102.

Eckstein, H. (1960) *Pressure group politics: the case of the British Medical Association*. London: Allen & Unwin.

Ehrenreich, B. and English, D. (1973) *Complaints and disorders: the sexual politics of sickness*. London: Writers and Readers Publishing Cooperative.

——(1974) *Witches, midwifery and nurses: a history of women healers*. London: Compendium.

Eisenberg, D. M., Kessler, R. C., Foster, C., Norlock, F. E., Calkins, D. R. and Delbanco, T. L. (1993) Unconventional medicine in the United States. *New England Journal of Medicine* 328(4), 246–52.

Ellencweig, A. (1992) *Analysing health systems: a modular approach*. Oxford: Oxford University Press.

Elling, R. H. (1981) Political economy, cultural hegemony and mixes of traditional and modern medicine. *Social Science and Medicine* 15A, 89-99.

——(1982) Industrialization and occupational health in underdeveloped countries. In Navarro (1982: 207–33).

Elliot, S. (1995) Psychosocial stress, women and heart health: a critical review. *Social Science and Medicine* 40(1), 105–15.

Elstad, J. I. (1990) Health services and decentralized government: the case of primary health services in Norway. *International Journal of Health Services* 20(4), 545–59.

Engel, G. L. (1977) The need for a new medical model: a challenge for biomedicine. *Science* 196, 129–36.

Enthoven, A. (1985) *Reflections on the management of the National Health Service, Occasional Paper 5.* London: Nuffield Provincial Hospitals Trust.

Erben, R., Franzkowiak, P. and Wenzel, E. (1992) Assessment of the outcomes of health intervention. *Social Science and Medicine* 35(4), 359–65.

Ernst, S. and Goodison, L. (1981) *In our own hands: a book of self-help therapy.* London: The Women's Press.

Euroqol Group (1990) EuroQOL: a new facility for the measurement of health-related quality of life. *Health Policy* 16, 199–208.

Evers, A., Farrant, W. and Trojan, A. (1990) *Local healthy public policy.* Boulder, Colo.: Westview Press.

Exworthy, M. (1994) The contest for control in community health services: general managers and professionals dispute decentralisation. *Policy and Politics* 22, 17–29.

Eyles, J. (1987) *The geography of the national health.* London: Croom Helm.

——and Donovan, J. (1986) Making sense of sickness and care. *Trans. Institute of British Geographers* 11, 415–27.

——and Woods, K. (1983) Societal constraints and systems of health care provision. In *The social geography of medicine and health.* London: Croom Helm, 183–231.

Fabrega, H. (1989) Cultural relativism and psychiatric illness. *Journal of Nervous and Mental Disease* 177, 415–25.

Faderman, L. (1980) *Surpassing the love of men.* London: Junction.

Falola, T. and Ityavyar, D. (eds) (1992) *The political economy of health in Africa.* Athens, Oh.: Ohio University Center for International Studies.

Fang Ru-Kang (1993) The geographical inequalities of mortality in China. *Social Science and Medicine* 36(10), 1319–24.

Faris, R. E. L. and Dunham, H. W. (1939) *Mental disorders in urban areas – an ecological study of schizophrenia and other psychoses.* Chicago: University of Chicago Press.

Farmer, P. (1992) *AIDS and accusation: Haiti and the geography of blame.* Berkeley, Calif.: University of California Press.

Farrant, W. (1991) Addressing the contradictions: health promotion and community health action in the United Kingdom. *International Journal of Health Services* 21(3), 423–39.

Fauveau, V., Goenig, M. and Wojtyniak, B. (1991) Excess female deaths among rural Bangladeshi children: an examination of cause specific mortality and morbidity. *International Journal of Epidemiology* 20(3), 729–35.

Feacham, R. (1986) Preventing diarrhoea: what are the policy options? *Health Policy and Planning* 1, 109–11.

Feingold, E. (1977) Citizen participation: a review of the issues. In Rosen, H. M., Metsch, J. M. and Levey, S. (eds), *The consumer and the health care system: social and managerial perspectives*. New York: Spectrum, 142–66.

Fenner, F., Henderson, D. A., Arita, I., Jezek, Z. and Ladnyi, I. D. (1988) *Smallpox and its eradication*. Geneva: WHO.

Fernando, S. (1988) *Race and culture in psychiatry*. London: Croom Helm.
——(1991) *Mental health, race and culture*. London: Macmillan.

Field, D. (1976) The social definition of illness. In Tuckett, D. (ed.), *An introduction to medical sociology*. London: Tavistock, 334–66.

Field, M. (ed.) (1989) *Success and crisis in national health systems: a comparative approach*. London: Routledge.

Fisher, D. (1978) Rockefeller philanthropy and the British empire: the creation of the London School of Hygiene and Tropical Medicine. *History of Education* 7(2), 129–43.

Fisher, P. and Ward, A. (1994) Complementary medicine in Europe. *British Medical Journal* 309, 107–11.

Fishman, R. (1982) *Urban utopias in the 20th century*. Cambridge, Mass.: MIT Press.

Fitzpatrick, J. and Whall, A. (1983) *Conceptual models of nursing*. Bowie, Md.: Brady.

Fitzpatrick, R. (1984) Lay concepts of illness. In Fitzpatrick, R., Hinton, J., Newman, S., Scambler, G. and Thompson, J. (eds), *The experience of illness*. London: Tavistock, 11–31.
——(1987) Political science and health policy. In Scambler, G. (ed.), *Sociological theory and medical sociology*. London: Tavistock, 221–45.
——and Boulton, M. (1994) Qualitative methods for assessing health care. *Quality in Health Care* 3, 1107–13.

Flynn, B. C. (1993) Healthy Cities within the American context. In Davies, J. K. and Kelly, M. P. (eds), *Healthy Cities: research and practice*. London, Routledge, 112–26.

Ford, C. and Beach, F. (1951) *Patterns of sexual behaviour*. New York: Harper.

Foster, A., Ratchford, D. Taylor, D. (1994) Auditing for patients. *Quality in Health Care* 3, Supplement, S16–S19.

Foucault, M. (1967) *Madness and civilization: a history of insanity in the age of reason*, trans. R. Howard. London: Tavistock.
——(1973) *The birth of the clinic: an archaeology of medical perception*, trans. A. M. Sheridan. London: Routledge.
——(1975) Prison talk. Repr. in Foucault, M. (1980) *Power/knowledge*, ed. C. Gordon, trans. C. Gordon *et al*. Brighton: Harvester, 37–54.

——(1976a) *The history of sexuality i: An introduction.* trans. R. Hurley. London: Penguin.

——(1976b) Two lectures. Repr. in Foucault, M. (1980) *Power/knowledge,* ed. C. Gordon, trans. C. Gordon *et al.* Brighton: Harvester, 78–108.

——(1976c) The politics of health in the eighteenth century. Repr. in Foucault, M. (1980) *Power/knowledge,* ed. C. Gordon, trans. C. Gordon *et al.* Brighton: Harvester, 166–82.

——(1978) The dangerous individual. Address to the Law and Psychiatry Symposium, York University, Toronto. Repr. in Kritzman, L. D. (ed.) (1988) *Politics, philosophy, culture,* trans. A. Sheridan *et al.* New York: Routledge, 125–51.

——(1983) Social security. Repr. in Kritzman, L. D. (ed.) (1988) *Politics, philosophy, culture,* trans. A. Sheridan *et al.* New York: Routledge, 159–77.

Fox, A., Goldblatt, P. and Jones, D. (1985) Social class mortality differentials: artefact or life circumstances. *Journal of Epidemiology and Community Health* 38(1), 1–8.

——Jones, D. and Goldblatt, P. (1984) Approaches to studying the effect of socio-economic circumstances on geographic differences in mortality in England and Wales. *British Medical Bulletin* 40(4), 309–14.

Fox, D. M. and Leichter, H. M. (1991) Rationing care in Oregon: the new accountability. *Health Affairs,* summer, 7–27.

Fox, J. (ed.) (1989) *Health inequalities in European countries.* Aldershot: Gower.

Fox, N. J. (1993) *Postmodernism, sociology and health.* Buckingham: Open University Press.

Frankel, S. and West, R. (1993) *Rationing and rationality in the National Health Service: the persistence of waiting lists.* Basingstoke: Macmillan.

Fraser, D. (1973) *The evolution of the British welfare state.* London: Macmillan.

Fraser, N. (1989a) Women, welfare, and the politics of need interpretation. In Fraser, N. (ed.), *Unruly practices: power discourse and gender in contemporary social theory.* Cambridge: Polity Press, 144–60.

——(1989b) Struggle over needs: outline of a socialist-feminist critical theory of late capitalist political culture. In Fraser, N. (ed.), *Unruly practices: power discourse and gender in contemporary social theory.* Cambridge: Polity Press, 161–87.

Freeman, L. and Maine, D. (1993) Women's mortality: a legacy of neglect. In Koblinsky, M., Timyan, J. and Gay, J. (eds), *The health of women: a global perspective.* Boulder, Colo.: Westview Press, 147–70.

Freidson, E. (1970) *Professional dominance: the social structure of medical care.* New York: Aldine.

Freire, P. (1968) *Pedagogy of the oppressed.* New York: Seabury Press.

Freud, S. (1905) Fragment of analysis of a case of hysteria. Repr. in *The standard edition of the complete psychological works of Sigmund Freud,* vii. London: Hogarth Press, 1956, 15–122.

Friedmann, J. (1992) *Empowerment: the politics of alternative development.* Oxford: Blackwell.

Fryer, P. (1987) Oxford's aim of health for all. *Health Service Journal,* 5 Mar., 274–5.

Fuchs, V. (1994) The Clinton plan: a researcher examines reform. *Health Affairs* 13(1), 102–14.

Fujimura, J. H. and Chou, D. Y. (1994) Dissent in science: styles of scientific practice in the controversy over the cause of AIDS. *Social Science and Medicine* 38(8), 1017–36.

Fulder, S. and Monro, R. (1985) Complementary medicine in the United Kingdom: patients, practitioners and consultations. *The Lancet* September 1985, 542–45.

Gallop, J. (1982) *Feminism and psychoanalysis: the daughter's seduction.* Basingstoke: Macmillan.

Gardner, M. J., Winter, P. D. and Barker, D. J. P. (1983) *Atlas of cancer mortality in England and Wales, 1968–1978.* Chichester: Wiley.

Gay, E. and Kronenfeld, J. (1990) Regulation, retrenchment – the DRG experience: problems from changing DRG practice. *Social Science and Medicine* 31(10), 1103–18.

Gayle, J. A. (1987) AIDS education in Black America. *Health Education Journal* 46(2), 77–8.

George, J. B. (1990) *Nursing theories: the basis for professional nursing practice.* Norwalk: Appleton & Lange.

Gerhardt, U. (1989) *Ideas about illness: an intellectual and political history of medical sociology.* Basingstoke: Macmillan.

Gesler, W. M. (1984) *Health care in developing countries.* Washington DC: Association of American Geographers.

——(1991) *The cultural geography of health care.* Pittsburgh: University of Pittsburgh Press.

——(1992) Therapeutic landscapes, medical issues in the light of the new cultural geography. *Social Science and Medicine* 34(7), 735–46.

Gever, M. (1987) Pictures of sickness: Stuart Marshall's 'Bright Eyes'. Published in Crimp (1988: 109–26).

Giggs, J. (1973) The distribution of schizophrenics in Nottingham. *Trans. Institute of British Geographers* 59, 55–76.

Gilman, S. L. (1985) *Difference and pathology: stereotypes of sexuality, race and madness.* Ithaca, NY: Cornell University Press.

——(1987) AIDS and syphilis: the iconography of disease. Published in Crimp (1988: 87–107).

——(1988) *Disease and representation: images of illness from madness to AIDS.* Ithaca, NY: Cornell University Press.

Giminez, M. (1989) Latino/'Hispanic': who needs a name? The case against a standardized terminology. *International Journal of Health Services* 19(3), 557–71.

Gisbers Van Wijk, C., Kolk, A., Van Den Bosch, W. and Van Den Hoogh,

H. (1992) Male and female morbidity in general practice: the nature of sex differences. *Social Science and Medicine* 35(5), 665–78.

Gish, O. (1992) Malaria eradication and the selective approach to health care: some lessons from Ethiopia. *International Journal of Health Services* 22(1), 179–92.

Glaser, B. and Strauss, A. (1967) *The discovery of grounded theory*. Chicago: Aldine.

Gleave, R. (1993) Images of contracting for the mental health service. *European Journal of Public Health* 3, 92–6.

Glendenning, C. (1988) Dependency and interdependency: the incomes of informal carers and the impact of social security. *Journal of Social Policy* 19(4), 467–97.

Godlee, F. (1991) Cholera pandemic. *British Medical Journal* 302, 1039–40.

——(1994a) WHO in crisis. *British Medical Journal* 309, 1424–28.

——(1994b) WHO in retreat: is it losing its influence? *British Medical Journal* 309, 1491–95.

——(1994c) The regions: too much power, too little influence. *British Medical Journal* 309, 1566–70.

——(1994d) WHO at country level: a little impact, no strategy. *British Medical Journal* 309, 1636–39.

——(1995a) WHO's special programmes: undermining from above. *British Medical Journal* 310, 178–82.

——(1995b) WHO in Europe: does it have a role? *British Medical Journal* 310, 389–94.

——(1995c) The WHO: the director general's view. *British Medical Journal* 310, 583–7.

Goffman, E. (1963) *Stigma*. Harmondsworth: Penguin.

Gonda, M. A., Wong-Staal, F., Gallo, R. C., Clements, J. F., Narayan, O. and Gilden, R. V. (1985) Sequence homology and morphologic similarity of HTLV-III and Visna virus, a pathogenic lentivirus. *Science* 227, 173–7.

Good, C. M. (1991) Pioneer medical missions in colonial Africa. *Social Science and Medicine* 32(1), 1–10.

Gorter, A., Sandford, P., Davey Smith, G. and Pauw, J. (1991) Water supply, sanitation and diarrhoeal disease in Nicaragua: results from a case-control study. *International Journal of Epidemiology* 20(2), 527–33.

Goudie, A. (1993) *The human impact on the natural environment*, 4th edn. Oxford: Blackwell.

Gould, M. and Jones, K. (1994) Analysing perceived limiting long-term illness using UK Census microdata. Sixth International Medical Geography Symposium, Vancouver, British Colombia, Canada, 12–14 July 1994.

Gould, P. (1993) *The slow plague: a geography of the AIDS pandemic*. Oxford: Blackwell.

Gouvernement Du Quebec (Ministère de la Santé et des Services Sociaux)

(1992) *The policy on health and well being.* Quebec: Gouvernement du Québec.

Graham, H. (1994) Gender and class as dimensions of smoking behaviour in Britain: insights from a survey of mothers. *Social Science and Medicine* 38(5), 691–8.

Grant, W. (1989) *Pressure groups, politics and democracy in Britain.* New York: Philip Allan.

Great Britain, Committee on Homosexual Offences and Prostitution (1957) The Wolfenden Report. American edition 1963. New York: Stein & Day.

Great Britain, Parliament (1987) *Promoting better health.* Cm. 249. London: HMSO.

——(1989a) *Working for patients.* Cm. 555. London: HMSO.

——(1989b) *Caring for people* London: HMSO.

——(1992) *The Health of the Nation,* Cm. 1986. London: HMSO.

Green, G. (1992) Liverpool. In Ashton J. (ed.), *Healthy Cities.* Milton Keynes: Open University, 87–95.

Green, H. (1988) *Informal Carers: a study carried out on behalf of the Department of Health and Social Security as part of the 1985 General Household Survey.* London: HMSO.

Green, R. H. (1991) Politics, power and poverty: health for all in 2000 in the third world? *Social Science and Medicine* 32, 745–55.

Greene B. (1992) Black feminist psychotherapy. In Wright, E. (ed.), *Feminism and psychoanalysis.* Oxford: Blackwell, 34–5.

Greenfield, S. (1988) Flaws in mortality data: the hazards of ignoring comorbid disease. *Journal of the American Medical Association* 260, 2253–5.

Griffiths, R. (1983) *NHS management inquiry: report.* London: DHSS.

Griscombe, J. (1850) *The sanitary condition of the laboring population of New York.* New York: Harper.

Grol, R. (1994) Quality improvement by peer review in primary care: a practical guide. *Quality in Health Care* 3, 147–52.

Grover, Z. J. (1988) AIDS: Keywords. In Crimp (1988: 17–30).

Gudgin, G. (1978) Response to the distribution of schizophrenics in Nottingham. *Trans. Institute of British Geographers* 64, 148–9.

Guerra De Macedo, C. (1991) Health in the Americas 20th–21st centuries: perspectives in international health. *Asia-Pacific Journal of Public Health* 5(2), 163–9.

Habermas, J. (1987) *The theory of communicative action, ii: Lifeworld and system: a critique of functionalist reason,* trans. T. McCarthy. Cambridge: Polity Press.

Hadhorn, D. C. (1991) The role of public values in setting health care priorities. *Social Science and Medicine* 32(7), 773–81.

——(1992) The problem of discrimination in health care priority setting. *Journal of the American Medical Association* 268(11), 1454–9.

Haggett, P. (1994) Prediction and predictability in geographical systems. *Trans. Institute of British Geographers* 19(1), 6–20.

Hall, J. and Haas, M. (1992) The rationing of health-care: should Oregon be transported to Australia? *Australian Journal of Public Health* 16(4), 435–40.

Ham, C. (1992) Reforming the Swedish health services: the international context. *Health Policy* 21, 129–41.

Hambleton, R. (1988) Consumerism, decentralization and local democracy. *Public Administration* 66, 125–47.

Hamer, D. H., Hu, S., Magnuson, V. L., Hu, N. and Pattatucci, A. M. L. (1993) A linkage between DNA markers on the X chromosomes and male sexual orientation. *Science* 261, 321–7.

Hammarstrom, A. (1994) Health consequences of youth unemployment: review from a gender perspective. *Social Science and Medicine* 38(5), 699–709.

Hancock, T. (1986) Beyond Lalonde: looking back at 'A new perspective on the health of Canadians'. *Health Promotion* 1(1), 93–100.

——(1993) The evolution, impact and significance of the healthy cities/healthy communities movement. *Journal of Public Health Policy* 14(1), 5–18.

Hann, A. (1993) The decision to screen. In Mills (1993: 40–51).

Hansson, L. F., Norheim, O. F. and Ruyter, K. W. (1994) Equality, explicitness, severity, and rigidity: the Oregon plan evaluated from a Scandinavian perspective. *Journal of Medicine and Philosophy* 19(4), 343–66.

Haraway, D. J. (1991) The biopolitics of postmodern bodies: constitutions of self in immune systems discourse. In Haraway, D. J., *Simians, cyborgs and women: the reinvention of nature*. London: Free Association Books, 203–30.

Hardie, M. (1986) Community patchwork. *Social Services Insight* xxxx.

——and Morris, R. (1982) Patchwork in big cities. *Lancet* 1, 287–8.

——Harpham, T. (1994a) Cities and health in the third world. In Phillips and Verhasselt (1994: 111–21).

——(1994b) Urbanization and mental health in developing countries. *Social Science and Medicine* 39(2), 233–45.

——Lusty, T. and Vaughan, P. (eds) (1988) *In the shadow of the city: community health and the urban poor*. Oxford: Oxford University Press.

Harrison, S., Hunter, D. J., Marnoch, G. and Pollitt, C. (1992) *Just managing: power and culture in the National Health Service*. Basingstoke: Macmillan.

——and Pollitt, C. (1990) *The dynamics of British health policy*. London: Routledge.

Hart, G. (1993) Rural hospital in the United States: a critical review. Paper presented at Institute of British Geographers Annual Conference 5–8 Jan. 1993, Royal Holloway College, University of London.

Harvey, D. (1972) *Social justice and the city*. London: Edward Arnold.

——(1992) Social justice, postmodernism and the city. *International Journal of Urban and Regional Research* 16, 588–601.

Hatch, S. and Kickbusch, I. (1983) *Self help and health in Europe: new approaches in health care.* Copenhagen: WHO.

Hayes-Bautista, D. and Chapa, J. (1987) Latino terminology: a conceptual basis for a standardised terminology. (Different views.) *American Journal of Public Health* 77, 61–8.

Haynes, R. M. and Bentham, C. G. (1979) *Community hospitals and rural accessibility.* Farnborough: Saxon House.

Health Research Group (1982) *Contemporary perspectives on health and health care.* Occasional Paper no 20. London: Department of Geography, Queen Mary College.

Helander, E. (1993) *Prejudice and dignity: an introduction to community based rehabilitation.* New York: UN Development Programme.

Held, D. (1980) *Introduction to critical theory.* Berkeley, Calif.: University of California Press.

Helman, C. (1978) 'Feed a cold, starve a fever': folk models of infection in an English suburban community, and their relation to medical treatment. *Culture, Medicine and Psychiatry* 2, 107–37.

Henrard, J.-C. (1988) *Soins et aides aux personnes âgées: description, fonctionnement du système français.* Paris: Centre Technique National d'Études et de Recherches sur les Handicaps et les Inadaptations.

Herzlich, C. (1973) *Health and illness: a social psychological analysis.* trans. D. Graham. London: Academic Press.

——and Pierret, J. (1986) Illness: from causes to meaning. In Currer and Stacey (1986: 73–96).

——(1989) The construction of a social phenomenon: AIDS in the French press. *Social Science and Medicine* 29, 1235–42.

Hewitt, M. (1983) Bio-politics and social policy: Foucault's account of welfare. *Theory, Culture and Society* 2(1), 67–84.

Heyden, V. (1993) Never mind the quality. *Health Service Journal* 103(5352), 21.

HFA Committee (Health Targets and Implementation Committee) (1988) *Health for all Australians.* Canberra: Australian Government Publishing Service.

Hoggett, P. (1990) *Modernisation, political strategy and the welfare state: an organisational perspective.* Studies in Decentralization and Quasi-Markets no 2. Bristol: School for Advanced Urban Studies.

Holland, W. (ed). (1988) *European Community atlas of avoidable death.* Oxford: Oxford University Press.

Hooker, E. (1956) A preliminary analysis of group behavior of homosexuals. *Journal of Psychology* 42, 217–25.

——(1957) The adjustment of the male overt homosexual. *Journal of Projective Techniques* 21, 18–31.

——(1965) Male homosexuals and their 'worlds'. In Marmor, J. (ed.), *Sexual inversion.* New York: Basic Books, 83–107.

——(1968) Sexual behaviour: homosexuality. In *International*

Encyclopaedia of the Social Sciences, xiv. New York: Macmillan, 222–33.

Horner, A. and Taylor, A. (1979) Grasping the nettle: locational strategies for Irish hospitals. *Administration* 27, 348–70.

Howe, G. M. (1963) *A national atlas of disease mortality in Britain.* London: Nelson.

——(1972) *Man, environment and disease in Britain.* Newton Abbot: David & Charles.

Hoyes, L. and Means, R. (1993) Quasi-markets and the reform of community care. In Le Grand and Bartlett (1993: 93–124).

Hudson, R. (1989) Impact of the new Right. *Health Services Journal* 99, 1546–7.

Humblett, P., Lagasse, R., Moens, G., Vande Voorde, H., Wollast, E. (1986) *Atlas de la mortalité evitable en Belgique.* Leuven: École de Santé Publique; Brussels: Université Libre de Bruxelles and Scholl voor Maatschappelijke Gezondheidszorg, Katholieke Universiteit.

Hunt, S., McEwen, J. and McKenna, S. (1986) *Measuring health status.* London: Croom Helm.

Hunter, J. and Shannon, G. (1985) Jarvis revisited: distance decay in service areas of mid-19th century asylums. *Professional Geographer* 37(3), 296–302.

Hutchinson, A., Foy, C., Sandhu, B. (1989) Comparison of two scores for allocating resources to doctors in deprived areas. *British Medical Journal* 299, 1142–4.

ICD (1977) *International Classification of Diseases,* 9th edn.

Illich, I. (1976) *Medical nemesis: limits to medicine.* Harmondsworth: Penguin.

Illsley, R. (1986) Occupational class, selection and inequalities in health. *Quarterly Journal of Social Affairs* 2, 151–65.

——(1987) Occupational class, selection and inequalities in health: rejoinder to Wilkinson's reply. *Quarterly Journal of Social Affairs* 3, 213–23.

Ingleby, D. (1983) Mental health and social order. In Cohen, S. and Scull, A. (eds), *Social control and the state.* Oxford: Martin Robinson, 141–88.

Institute of Medicine (1988) *The future of public health.* Washington, DC: National Academy Press.

Irigaray, L. (1977) The poverty of psychoanalysis. Repr. in M. Whitford (ed.), *The Irigaray reader* . Oxford: Blackwell, 1991, 79–104.

Ityavyar, D. (1992) The colonial origins of health care services: the Nigerian example. In Falola, T. and Ityavyar, D. (eds), *The political economy of health in Africa.* Athens, Oh.: Ohio University Center for International Studies, 65–87.

Iyun, B. (1994) Health care in the third world: Africa. In Phillips and Verhasselt (1994: 249–58).

Jackson, P. (1989) *Maps of meaning: an introduction to cultural geography.* London: Unwin Hyman.

Jacobson, B. (1990) Planning an achievable strategy for 'health for all' in the inner city: the development of local targets. *Health Education Journal* 49(4), 171–5.

Janzen, J. M. (1978) The comparative study of medical systems as changing social systems. *Social Science and Medicine* 12B, 121–9.

——(1979) Ideologies and institutions in the precolonial history of equatorial African therapeutic systems. *Social Science and Medicine* 13B, 239–43.

——and Feierman, S. (1979) Introduction to special issue on medicine in Africa. *Social Science and Medicine* 13B, 239–43.

——and Prins, G. (eds) (1981) Causality and classification in African medicine and health. Special issue: *Social Science and Medicine* 15B(3), 169–429.

Jarman, B. (1981) *A survey of primary care in London*. Occasional Paper no. 16. London: Royal College of General Practitioners.

——(1983) Identification of underprivileged areas. *British Medical Journal* 286, 1705–9.

——(1984) Underprivileged areas: validation and distribution of scores. *British Medical Journal* 289, 1587–92.

Jencks, S. and Wilensky, G. (1992) The health care quality improvement initiative: a new approach to assurance in medicare. *Journal of American Medical Association* 268, 900–3.

Jenkins, C. (1991) Assessment of outcomes of health intervention. *Social Science and Medicine* 35(4), 367–75.

Jessop, E. (1992) Individual morbidity and neighbourhood deprivation in a non-metropolitan area. *Journal of Epidemiology and Community Health* 46, 543–6.

Jewell, J. (1983) Theoretical basis of Chinese traditional medicine. In Hillier, S. and Jewell, J. *Health care and traditional medicine in China 1800–1982*. London: Routledge & Kegan Paul, 221–41.

Johnson, E. (1991) Magic: my AIDS nightmare. *Sports Illustrated*, 18 Nov.

Johnston, R. J., Gregory, D. and Smith, D. M. (eds) (1994) *The dictionary of human geography*. Oxford: Blackwell.

Jones, H. (1994) *Health and Society in 20th Century Britain*. London: Longman.

Jones, I. R. (1995) Health care needs assessment: the case of renal services. Ph.D. thesis, Queen Mary and Westfield College, University of London.

Jones, K. and Moon, G. (1987) *Health, disease and society: an introduction to medical geography*. London: Routledge & Kegan Paul.

——(1992) Medical geography: global perspectives. *Progress in Human Geography* 16(4), 563–72.

——(1993) Medical geography, taking space seriously. *Progress in Human Geography* 17(4), 515–24

——and Clegg, A. (1991) Ecological and individual effects in childhood immunisation uptake: a multi-level approach. *Social Science and Medicine* 33(4), 501–8.

Jones, P. R. (1981) *Doctors and the BMA: a case study in collective action.* Farnborough: Gower.

Jones, R. (1994) Ready reckoners. *Health Service Journal* 10 Feb., 31.

Joseph, A. E. and Phillips, D. R. (1984) *Accessibility and utilization: geographical perspectives on health care delivery.* New York: Harper & Row.

Kakar, S. (1984) *Shamans, mystics and doctors: a psychological inquiry into India and its healing.* London: Unwin.

Kaminski, M., Bouvier-Colle, M.-H. and Blondel, B. (1986) *Mortalité des jeunes dans les pays de la communauté européenne (de la naissance à 24 ans).* Paris: Doin.

Kanji, N., Kanji, N. and Manji, F. (1991) From development to sustained crisis: structural adjustment, equity and health. *Social Science and Medicine* 33(9), 985–93.

Kaplan, R. M. (1994) Value judgement in the Oregon Medicaid experiment. *Medical Care* 32(10), 975–88.

Kaptchuk, T. J. (1983) *Chinese medicine: the web that has no weaver.* London: Rider.

Katayama, H., Anme, T. and Takayama, T. (1994) Interactive-media network system as a support technology for the elderly and the disabled. In Rey, J. C. and Tilquin, C. (eds), *SYSTED 94: Proceedings of the Fifth International Conference on Systems Sciences in Health and Social Services for the Elderly and the Disabled*, Geneva, 2–6 May 1994. Aarau: Institut Suisse de la Santé Publique, 568–71.

Katner, H. P. and Pankey, G. A. (1987) Evidence for a Euro-American origin of HIV. *Journal of the National Medical Association* 79, 1068-1072.

Katz, A. H. and Bender, E. I. (1976) *The strength in us: self-help groups in the modern world.* New York: New Viewpoints.

Kearns, R. A. (1991) The place of health in the health of the place: the case of the Hokianga special medical area. *Social Science and Medicine* 33(4), 519–30.

——and Joseph, A. E. (1993) Space in its place: developing the link in medical geography. *Social Science and Medicine* 37(6), 711–17.

Kelly, M., Davies, J. and Charlton, B. (1993) Healthy cities: a modern problem or a post modern solution. In Davies, J. and Kelly, M. (eds), *Healthy Cities: research and practice.* London: Routledge, 159–67.

Kickbusch, I. (1993) Health promotion and disease prevention. In Normand, C. E. M. and Vaughan, P. (eds), *Europe without frontiers: the implications for health.* London: Wiley, 47–54.

Kimberley, K. A. (1990) Regulating physician investment and referral behaviour in the competitive health care marketplace of the '90s: an argument for decentralization. *Washington Law Review* 65(3), 657–75.

Kindig, D. and Yan, G. (1993) Data Watch: Physician supply in rural areas with large minority populations. *Health Affairs*, summer, 177–84.

King, E. (1993) *Safety in numbers: safer sex and gay men.* London: Cassell.

King, G. (1994) Health care reform and the Medicare program. *Health Affairs* 13(4), 39–47.

King's Fund (1987) *Decentralising community health services.* London: King's Fund.

——(1988) *Can Cumberlege work in the inner cities?* London: King's Fund.

Kinsey, A., Pomeroy, W. B. and Martin, C. L. (1948) *Sexual behaviour in the human male.* Philadelphia: Saunders.

Kitzhaber, J. A. (1993) Prioritising health services in an era of limits: the Oregon experience. *British Medical Journal* 307, 373–7.

Kitzinger, C. (1987) *The social construction of lesbianism.* London: Sage.

Klein, R. (1974) Policy making in the National Health Service *Political Studies* 22(1), 1–14.

——(1984) Who makes the decisions in the NHS? *British Medical Journal* 288, 1706–8.

——(1992) Warning signals from Oregon: the different dimensions of rationing need untangling. *British Medical Journal* 304, 1457–8.

Kleinman, A. (1978) Concepts and model for the comparison of medical systems as cultural systems. *Social Science and Medicine* 12, 85–93.

——(1988) *The illness narratives.* New York: Basic Books.

——Kunstadter, P., Alexander, E. R. and Gale, J. L. (eds) (1975) *Medicine in Chinese cultures.* Washington: US Dept. of Health Education and Welfare, 75–653.

Kloos, H. and Zein, Z. A. (1993) *The ecology of health and disease in Ethiopia.* Boulder, Colo.: Westview.

Knox, P. (1978) The intra urban ecology of primary medical care: patterns of accessibility and their policy implications. *Environment and Planning A* 10, 415–35.

——and Agnew, P. (1994) *The geography of the world economy.* London: Edward Arnold.

——Bohland, J. and Shumsky, N. (1983) The urban transition and the evolution of the medical care delivery system in America. *Social Science and Medicine* 17, 37–43.

Kokkonen, P. T. and Kekomäki, M. (1993) Legal and economic issues in European public health. In Normand, C. E. M. and Vaughan, P. (eds), *Europe without frontiers: the implications for health.* London: Wiley, 35–43.

Kronenfeld, J. and Wasner, C. (1982) The use of unorthodox therapies and marginal practitioners. *Social Science and Medicine* 16, 1119–25.

Kuper, L. (1994) *Race, class and power: ideology and revolutional change in plural societies.* London: Duckworth.

Kurup, P. (1983) Ayurveda. In Bannerman, R., Burton, J. and Ch'en Wen-Chieh (eds), *Traditional medicine and health care coverage: a reader for health administrators and practitioners.* Geneva: WHO, 50–60.

Kurtz, Z. (1993) Better health for black and ethnic minority children and young people. In Hopkins, A. and Bahl, V. (eds), *Access to health care for people from black and ethnic minorities.* London: Royal College of Physicians.

Labonte, R. (1993) A holosphere of healthy and sustainable communities. *Australian Journal of Public Health* 17(1), 4–12.

Lado, C. (1992) Female labour participation in agricultural production and the implications for nutrition and health in rural Africa. *Social Science and Medicine* 34(7), 789–807.

Lagergren, M. (1987) Methodological considerations in long-range health planning: the Swedish experience. In Brouwer, J. J. and Schreuder, R. F. (eds), *Scenarios and other methods to support long term health planning.* Netherlands: Steering Committee on Future Health Scenarios, 77–87.

Lahelma, E. and Arber, S. (1994) Health inequalities among men and women in contrasting welfare states. *European Journal of Public Health* 4(3), 213–26.

Lalonde, M. (1974) *A new perspective on the health of Canadians.* Ottawa: Government of Canada.

Lambo, A. (1964) Patterns of psychiatric care in developing African countries. In Kiev, A. (ed.), *Magic, faith and healing, 4.* New York: Free Press of Glencoe, 443–53.

Lancet (1988) Editorial – Back to the Future: the reinvention of public health. *Lancet* 1 (23 Nov.), 157–9.

——(1995) Editorial – Fortress WHO: breaching the ramparts for health's sake. *Lancet* 345 (8944), 203–4.

Last, J. (1963) The iceberg: completing the clinical picture in general practice. *Lancet* 2, 28–31.

Le Grand, J. and Bartlett, W. (eds) (1993) *Quasi-markets and social policy.* Basingstoke: Macmillan.

Learmonth, A. (1988) *Disease ecology.* Oxford: Blackwell.

Leavitt, J. and Numbers, R. (eds) (1978) *Sickness and health in America: readings in the history of medicine and public health.* Madison: University of Wisconsin Press.

Lee, K. and Mills, A. (1982) *Policy-making and planning in the health sector.* London: Croom Helm.

Levitt, R. (1980) *The people's voice in the NHS.* London: King Edward's Hospital Fund for London.

Lewis R. (1952) *Edwin Chadwick and the public health movement 1832–1854.* London: Longmans Green & Co.

Li, J. Y. (1989) Cancer mapping as an epidemiologic research resource in China. *Recent Results in Cancer Research* 114, 115–36.

Lilienfeld, M. and Lilienfeld, D. (1980) *Foundations of epidemiology.* Oxford: Oxford University Press.

Limb, M. (1994) One for the record. *Health Service Journal* 104(5420), 12–13.

Lin Wei and Zhu Chow (1984) Guiding principles for health development and methods of health research in China. In Haifeng, C. and Chao, Z.

(eds) *Modern Chinese Medicine* Vol. 3. *Chinese Health Care*. Lancaster: MTP Press Ltd, Ch.1.

Lindblom, C. E. (1959) The science of muddling through. *Public Administration Review* 19(3), 79–88.

——(1979) Still muddling, not yet through. *Public Administration Review* 39(6), 517–26.

Lindgren, B. and Prütz, C. (1994) Care of the elderly in Sweden: is there a role for private producers and quasi-markets? In Rey, J. C. and Tilquin, C. (eds), *SYSTED 94: Proceedings of the Fifth International Conference on Systems Sciences in Health and Social Services for the Elderly and the Disabled*, Geneva, 2–6 May 1994. Aarau: Institut Suisse de la Santé Publique, 480–2.

Linn, M. W., Linn, B. S. and Stein, S. R. (1982) Beliefs about causes of cancer in cancer patients. *Social Science and Medicine* 16, 835–40.

Little, M. (1991) Imperialism, colonialism and the new science of nutrition: the Tanganyika experience 1925–1945. *Social Science and Medicine* 32(1), 11–14.

Littlewood, R. and Lipsedge, M. (1989) *Aliens and alienists: ethnic minorities and psychiatry*. London: Unwin Hyman.

Lohr, K. (ed.) (1990) *Medicare: a strategy for quality assurance*. Washington, DC: National Academic Press.

Loveman, G. and Sengenberger, W. (1991) The re-emergence of small-scale production: an international comparison. *Small Business Economics* 1, 1–38.

Lowry, S. (1988) Focus on performance indicators. *British Medical Journal* 296, 992–4.

——(1990) Health and homelessness. *British Medical Journal* 300, 32–4.

Loytonen, M. (1991) The spatial diffusion of human immunodeficiency virus type 1 in Finland, 1982-1997. *Annals of the Association of American Geographers* 81(1), 127–51.

Lundberg, O. (1991) Causal explanations for class inequality in health: an empirical analysis. *Social Science and Medicine* 32(4), 385–93.

Lyall, J. (1995) A world apart. *Health Service Journal* 5 Jan., 7–12.

Lyons, M. (1992) *The colonial disease: a social history of sleeping sickness in northern Zaire 1900–1940*. Cambridge: Cambridge University Press.

Mackenbach, J., Kunst, A. and Looman, C. (1991) Cultural and economic determinants of geographical mortality patterns in the Netherlands. *Journal of Epidemiology and Community Health* 45, 231–7.

——Looman, C. and Kunst, A. (1990) Geographic variation in the onset of decline of ischemic heart disease in the Netherlands: a comparison between men and women. Paper presented at the regional European meeting of the International Epidemiological Association, Granada, Spain, 14–16 Feb.

McBeath, W. H. (1991) Health for all: a public health vision. *American Journal of Public Health* 81(12), 1560–5.

McCracken, J., Pretty, J. and Conway, G. (1988) *An introduction to RRA for agricultural development*. London: International Institute of Environmental Development.

MacDonald, C. (1981) Political-economic structures–approaches to traditional and modern medical systems. *Social Science and Medicine* 15A, 101–8.

McFarlane, D. R. (1992) Restructuring federalism: the effects of decentralized federal policy on states' responsiveness to family planning needs. *Women and Health* 19(1), 43–63.

McGhee, S. and McEwan, J. (1993) Evaluating the Healthy Cities project in Drumchapel, Glasgow. In Davies, J. K. and Kelly, M. P. (eds), *Healthy Cities: research and practice*. London: Routledge, 148–58.

McGlashan, N. D. and Blunden, J. R. (eds) (1983) *Geographical aspects of health*. London: Academic Press.

McGrath, M. and Grant, G. (1992) Supporting 'needs-led' services: implications for planning and management systems. *Journal of Social Policy* 21(1), 71–97.

McIntosh, M. (1968) The homosexual role. *Social Problems* 16(2), 182–91.

McKenzie, K. and Crowcroft, N. (1994) Race, ethnicity, culture, and science. *British Medical Journal* 309, 286–7.

Mackenzie, W. J. M. (1979) *Power and responsibility in health care*. Oxford: Oxford University Press.

McKeown, K., Whitelaw, S., Hambleton, D. and Green, F. (1994) Setting priorities: science, art or politics? In Malek, M. (ed.), *Setting priorities in health care*. Chichester, Wiley, 19–29.

McLachan, G. and Maynard, A. (eds) (1982) The public/private mix for health: the relevance and effects of change. London: Nuffield Provincial Hospitals Trust

MacLeod, R. (1988) Introduction. In MacLeod, R. and Lewis, M. (eds), *Disease, medicine and empire*. London: Routledge, 1–18.

——and Lewis, M. (eds) (1988) *Disease, medicine and empire*. London: Routledge.

McLoone, P. and Boddy, F. (1994) Deprivation and mortality in Scotland, 1981 and 1991. *British Medical Journal* 309, 1465–70.

McMichael, A. (1993) Global environmental change and human population health: a conceptual and scientific challenge for epidemiology. *International Journal of Epidemiology* 22(1), 1–8.

Madhubuti, H. R. (1990) AIDS: the purposeful destruction of the black world. In Madhubuti, H. R., *Black men: obsolete, single, dangerous?* Chicago: Third World Press, 51–8.

Maff (Ministry of Agriculture, Fisheries and Food) (1987) *Survey of consumer attitudes to food additives*. London: HMSO.

Malmgren, R., Warlow, C., Bamford, J. and Sandercock, P. (1987) Geographical and secular trends in stroke incidence. *Lancet* (21 Nov.), 1196–200.

Mann, P. (1953) Octavia Hill: an appraisal. *Town Planning Review* 23(3), 223–37.

Marchant, C. (1992) *Radical ecology: the search for a liveable world.* London: Routledge.

Marks, S. and Andersson, N. (1988) Typhus and social control: South Africa, 1917–50. In Macleod, R. and Lewis, M. (eds), *Disease, medicine and empire.* London: Routledge, 257–83.

Marmor, J. (1975) Homosexuality and sexual orientation disturbances. In Freedman, A., Kaplan, H. and Sadock, B. (eds), *Comprehensive textbook of psychiatry.* Baltimore, Md: Williams & Wilkins, 1510–20.

Marmor, T. R. (1973) *The politics of Medicare.* Chicago: Aldine.

Marmot, M., Adelstein, A. and Bulusu, L. (1984) *Immigrant mortality in England and Wales 1970–1978.* Studies on Medical and Population Subjects no. 47. London: HMSO.

Martin, D. (1992) Postcodes and the 1991 Census of Population: issues, problems and prospects. *Trans. Institute of British Geographers* 17(3), 350–7.

Martin, E. (1987) *The woman in the body: a cultural analysis of reproduction.* Reprinted Milton Keynes: Open University Press, 1989.

Massam B. and Malczewski, J. (1991) The location of health centers in a rural region using a decision support system: a Zambian case study. *Geography Research Forum* 11, 1–24.

Masson, J. M. (1986) *Women, sexuality and psychiatry in the nineteenth century.* New York: Noonday.

Maxwell, R. (1984) Quality assessment in health. *British Medical Journal* 300, 919–22.

——(1993) Other cities, same problems. In Smith, J. (ed.), *London after Tomlinson: reorganizing big city medicine.* London: British Medical Journal Publishing Group, 101–7.

Mayer, J. (1992) Challenges to understanding spatial patterns of disease: philosophical alternatives to logical positivism. *Social Science and Medicine* 35(4), 579–87.

Mayhew, L. (1986) *Urban hospital location.* London: George Allen & Unwin.

Mays, N. (1987) Measuring morbidity for resource allocation. *British Medical Journal* 295, 764–7.

Mburu, F. M. (1979) Rhetoric-implementation gap in health policy and health services delivery for a rural population in a developing country. *Social Science and Medicine* 13A, 577–83.

——(1980) A critique of John H Bryant's paper. *Social Science and Medicine* 14A, 387–9.

Meade, M., Florin, J. and Gesler, W. (1988) *Medical geography.* New York: Guilford Press.

Mechali, D. (1990) Pathologie des étrangers ou des migrants: problèmes cliniques et thérapeutiques, enjeux de santé publique. *Revue Européenne des Migrations Internationales* 6(3), 99–127.

Mercer, K. (1986) Racism and transcultural psychiatry. In Miller, P. and Rose, N. (eds), *The power of psychiatry.* Cambridge: Polity Press, 112–42.

Meyer, C. (1994) Les critères et les procédures d'allocation géographique des ressources: bilan de quelques pratiques étrangères de référence. Paper presented to seminar 'Etudes des populations et épidémiologie en planification sanitaire', 12–14 Sept, 1994, Ecole de Santé Publique, Rennes.

Mezentseva, E. and Rimachevskaya, N. (1990) The Soviet country profile: health of the USSR population in the 70's and 80's – an approach to a comprehensive analysis. *Social Science and Medicine* 31(8), 867–77.

Miles, A. (1991) *Women, health and illness*. Milton Keynes: Open University Press.

Miller, N. and Rockwell, R. (eds) (1988) *AIDS in Africa: the social and policy impact*. Lewiston, NY: Edwin Mellen Press.

Mills, A., Vaughan, J. P., Smith, D. L. and Tabibzadeh, I. (1990) *Health system decentralization: concepts, issues and country experience*. Geneva: WHO.

Mills, M. (ed.) (1993) *Prevention, health and British politics*. Aldershot: Avebury.

——and Saward, M. (1993) Liberalism, democracy and prevention. In Mills (1993: 161–73).

Milio, N. (1981) *Promoting health through public policy*. Philadelphia: F. A. Davis.

——(1987) Making healthy public policy. *Health Promotion* 2(3), 263–74.

Ministry of Health (1989) *New Zealand health goals and targets*. New Zealand: Ministry of Health.

Minkler, M. and Cox, K. (1980) Creating critical consciousness in health: application of Freire's philosophy and methods to the health care setting. *International Journal of Health Services* 10(2), 311–22.

Mishler, E. G. (1981) Critical perspectives on the biomedical model. In Mishler, E. G., Amarsingham, L. R., Hauser, S. T., Liem, R., Osherson, S. D. and Waxler, N. E., *Social contexts of health, illness and patient care*. Cambridge: Cambridge University Press, 1–23.

Mitchell, J. (1974) *Psychoanalysis and feminism*. Harmondsworth: Penguin.

Mohan, J. (1988) Restructuring, privatization and the geography of health care provision in England 1983–1987. *Trans. Institute of British Geographers*, 13(4), 449–65

——(1989) Medical geography: competing diagnoses and prescriptions. *Antipode* 21(2), 166–77.

——(1990a) Health care policy and the state in 'austerity capitalism': Britain and the USA compared. In Simmie, J. and King, R. (eds) *The state in action: public policy and politics*. London: Pinter, 74–94.

——(1990b) Spatial implications of the National Health Service White Paper. *Regional Studies* 24(6), 553–9.

——(1995) *A National Health Service? Restructuring of health care in Britain since 1979*. London: Macmillan.

Mohan, J. F., Chambers, J., Johnson, K., Killoran, A. and McKenzie, J.

(1990a) *Mapping the epidemic: coronary heart disease in England.* London: Health Education Authority.

Mohan, J. M., Killoran, A., Johnson, K. and McKenzie, J. (1990b) Reducing coronary heart disease in England: targets and implications. *Health Education Journal* 49(4), 176–80.

Moon, G. (1990) Conceptions of space and community in British health policy. *Social Science and Medicine* 30(1), 165–71.

Mooney, G., Russell, E. and Weir, R. (1980) *Choices for health care.* Basingstoke: Macmillan.

Moore, T. V. (1945) The pathogenesis and treatment of homosexual disorders: a digest of some pertinent evidence. *Journal of Personality* 14 (Sept.), 47–83.

MORI (1994) *Evaluation of bi-lingual health care schemes in East London.* London: MORI.

Morin, S. F. (1977) Heterosexual bias in psychological research on lesbianism and male homsexuality. *American Psychologist* 32, 629–37.

Morley, D., Rohde, J. and Williams, G. (eds) (1983) *Practising health for all.* Oxford: Oxford Medical Publications.

Morone, J. and Marmor, T. R. (1981) Representing consumer interests: the case of American health planning. In Checkoway, B. (ed.), *Citizens and health care: participation and planning for social change.* New York: Pergamon, 25–48.

Mullen, P. M. (1995) The provision of specialist services under contracting. *Public Money and Management.*

Murrell, R. K. (1987) Telling it like it isn't: representations of science in 'Tomorrow's World'. *Theory, Culture and Society* 4, 89–106.

Naidoo, J. (1986) Limits to individualism. In Rodmell, S. and Watt, A. (eds), *The politics of health education.* London: Routledge, 17–37.

Nakajima, H. (1993) *Statements of Dr Hiroshi Nakajima, Director-General, to the Executive Board and the World Health Assembly.* Geneva: WHO.

National Health Planning and Resources Development Act (1974) Public Law 93–641.

Navarro, V. (1975) The industrialisation of fetishism or the fetishism of industrialisation: a critique of Ivan Illich. *International Journal of Health Services* 5, 351–71.

——(1978) *Class struggle, the state and medicine.* London: Martin Robertson.

——(ed.) (1982) *Imperialism, health and medicine.* London: Pluto Press.

——(1984) A critique of the ideological and political positions of the Brandt report and the Alma-Ata declaration. *Social Science and Medicine* 18, 467–74.

——(1986) Work, ideology and science: the case of medicine. In *Crisis, health and medicine: a social critique.* New York: Tavistock, 143–82.

——(1990) Race or class versus race and class: mortality differentials in the United States. *Lancet* 336, 1238–40.

——(1991) The relevance of the US experience to the reforms in the British National Health Service: the case of general practitioner fund holding. *International Journal of Health Services* 21(3), 381–7.

——(ed.) (1993) *Why the United States does not have a national health programme.* New York: Baywood.

Nelson, M. V., Bailie, G. R. and Areny, H. (1990) Pharmacists' perceptions of alternative health approaches: a comparison between US and British pharmacists. *Journal of Clinical Pharmacy and Therapeutics* 15, 141–6.

Nettleton, S. (1988) Protecting a valuable margin: towards an understanding of how the mouth came to be separated from the body. *Sociology of Health and Illness* 10(2), 156–69.

——(1991) Wisdom, diligence and teeth: discursive practices and the creation of mothers. *Sociology of Health and Illness* 13(1), 98–111.

New, S. J. and Senior, M. L. (1991) 'I don't believe in needles': qualitative aspects of a study into the uptake of infant immunisation in two English health authorities. *Social Science and Medicine* 33(4), 509–18.

Newson, L. (1993) Highland–lowland contrasts in the impact of old world diseases in early colonial Ecuador. *Social Science and Medicine* 36(9), 1187–95.

NHS Executive (1994) *Managing the new NHS: functions and responsibilities in the new NHS.* London: Department of Health.

NHS Management Executive (1991) *Hospital and Community Health Services: 1991–2 Cash Limits Exposition Booklet; Revenue and Capital.* London: Department of Health.

——(1992) *Local voices: the views of local people in purchasing for health.* London: Department of Health.

Nissel, M. and Bonnerjea, L. (1982) *Family care of the handicapped elderly: who pays?* London: Policy Studies Institute.

Noack, H. (1987) Concepts of health and health promotion. In Abelin, T. *et al.* (eds), *Measurement in health promotion and protection.* Copenhagen: WHO, 5–28.

Nocon, A. (1989) Forms of ignorance and their role in the joint planning process. *Social Policy and Administration* 23(1), 31–47.

Noyce, J., Smaith, A. and Trickey, A. (1974) Regional variations in the allocation of financial resources to the community health services. *Lancet* 1 (30 Mar.), 554–7.

Nugent, R. and Gramick, J. (1989) Homosexuality: Protestant, Catholic and Jewish issues – a fishbone tale. *Journal of Homosexuality* 18, 7–46.

Obeyesekere, G. (1977) The theory and practice of Ayurvedic medicine. *Culture, Medicine and Psychiatry* 1, 155–81.

Offe, C. (1984) *Contradictions of the welfare state.* London: Hutchinson.

Okafor, S. (1990) Distributive effects of location of government hospitals in Ibadan. *Area* 23(3), 128–35.

Okin, S. (1989) *Justice, gender and the family.* New York: Basic Books.

Oliver, M., Corney, R., Begum, N., Momen, A. & Kelly, M. P. (1994)

Final report on the evaluation of three projects in respect of primary care developments in Tower Hamlets. London: University of Greenwich.

Omran, A. (1971) The epidemiological transition: a theory of the epidemiology of population change. *Milbank Quarterly* 64, 355–91.

——(1983) The epidemiological transition theory: a preliminary update. *Journal of Tropical Pediatrics* 29, 305–16.

OPCS (1987) *General Household Survey, 1985.* London: HMSO.

——(1989) *General Household Survey, 1987.* London: HMSO.

——(1992) *General Household Survey, 1990.* London: HMSO.

——(1993) *1991 Census: limiting long term illness – Great Britain.* London: HMSO.

——(1994) *General Household Survey, 1992.* London: HMSO.

O'Shea E. and Blackwell, J. (1993) The relationship between the cost of community care and the dependency of old people. *Social Science and Medicine* 37(5), 583–90.

Ottawa Charter for Health Promotion (1986) *Ottawa Charter*, first International Conference on Health Promotion in Industrialized Countries, Ottawa, 17–21 Nov. Copenhagen: WHO Regional Office for Europe.

Ottowill, R. and Wall, A. (1990) *The growth and development of the community health services.* Sunderland: Business Education Publishers.

Ovretveit, J. (1993) Purchasing for health gain. *European Journal of Public Health* 3, 77–84.

Pacione, M. (ed.) (1986) *Medical geography, progress and prospect.* London: Croom Helm.

Packard, R. M. and Epstein, P. (1991) Epidemiologists, social scientists, and the structure of medical research on AIDS in Africa. *Social Science and Medicine* 33(7), 771–94.

PAHO (Pan-American Health Organization) (1991) Mortality due to intestinal infectious diseases in Latin America and the Caribbean, 1965–1990. *Epidemiologial Bulletin* 12(3), 1–6.

——(1993) *Implementation of the global strategy for health for all by the year 2000, second evaluation: eighth report on the world health situation, iii: Region of the Americas.* Washington, DC: Pan American Health Organization, Pan American Sanitary Bureau, Regional Office of the World Health Organization for the Americas.

Palmer, R., Donabedian, A. and Povar, G. (1991) *Striving for quality in health care: an inquiry into policy and practice.* Ann Arbor, Mich.: Health Administration Press.

Palumbo, D. J. and Calista, D. J. (eds) (1990) *Implementation and the policy process: opening up the black box.* New York: Greenwood.

Pannenborg, C. O. (1991) Shifting paradigms of international health. *Asia and Pacific Journal of Public Health* 5(2), 176–84.

Panos (1990) *The third epidemic: repercussions of the fear of AIDS.* London: Panos.

Paolisso, M. and Leslie, J. (1995) Meeting the changing health needs of women in developing countries. *Social Science and Medicine* 40(1), 55–65.

Parry, N. and Parry, J. (1976) *The rise of the medical profession: a study of collective social mobility*. London: Croom Helm.

Pater, J. E. (1981) *The making of the NHS*. London: King Edward's Hospital Fund for London.

Paton, C. (1992) *Competition and planning in the NHS: the danger of unplanned markets*. London: Chapman & Hall.

——(1993) Devolution and centralism in the National Health Service. *Social Policy and Administration* 27, 83–108.

Patricelli, R. (1994) Managed care industry perspectives: why do we need Health Alliances? *Health Affairs* 13(1), 239–42.

Patton, C. (1989) The AIDS industry: constructions of 'victims', 'volunteers' and 'experts'. In Carter, E. and Watney, S. (eds), *Taking liberties: AIDS and cultural politics*. London: Serpent's Tail, 112–25.

——(1990) *Inventing AIDS*. London: Routledge.

Payer, L. (1988) *Medicine and culture*. New York: Holt.

Peat, M. (1994) Community based rehabilitation: status and strategies for development. In Rey, J. C. and Tilquin, C. (eds), *SYSTED 94: Proceedings of the Fifth International Conference on Systems Sciences in Health and Social Services for the Elderly and the Disabled*, Geneva, 2–6 May 1994. Aarau: Institut Suisse de la Santé Publique, 669–74.

——and Boyce, W. (1993) Canadian community rehabilitation services: challenges for the future. *Canadian Journal of Rehabilitation* 6(4), 281–9.

Peffer, R. (1990) *Marxism, morality and social justice*. Princeton, NJ: Princeton University Press.

Pepper, D. (1993) *Eco-socialism: from deep ecology to social justice*. London: Routledge.

Pereira, J. (1990) The economics of inequality in health: a bibliography. *Social Science and Medicine* 31(3), 413–20.

Pettigrew, A., Ferlie, E. and McKee, L. (1992) *Shaping strategic change*. London: Sage.

Phillimore, P. and Morris, D. (1991) Discrepant legacies: premature mortality in two industrial towns. *Social Science and Medicine* 33(2), 139–52.

——and Reading, R. (1992) A rural disadvantage? Urban–rural health differences in Northern England. *Journal of Public Health Medicine* 14, 290–9.

Phillips, D. and Verhasselt, Y. (eds) (1994) *Health and development*. London: Routledge.

Phillips, D. L. (1967) Identification of mental illness: its consequences for rejection. *Community Mental Health Journal* 3, 262–6.

Phillips, D. R. (1981) *Contemporary issues in the geography of health care*. Norwich: Geo Books.

——(1990) *Health and health care in the third world.* Harlow: Longman.

Philo, C. (1986) 'The same and the other': on geographies, madness and outsiders. Occasional paper no 11. Loughborough: Loughborough University of Technology, Department of Geography.

——(1987) 'Fit localities for an asylum': the historical geography of the nineteenth-century 'mad-business' in England as viewed through the pages of the *Asylum Journal. Journal of Historical Geography* 13, 398–415.

——(1989) 'Enough to drive one mad': the organization of space in 19th century lunatic asylums. In Wolch, J. and Dear, M. (eds), *The power of geography.* Boston: Unwin Hyman, 258–90.

Picheral, H. (1989) Géographie de la transition epidémiologique. *Annales de Géographie* 546, 129–51.

Pile, S. and Rose, G. (1992) All or nothing? Politics and critique in the modernism–postmodernism debate. *Environment and Planning D: Society and Space* 10, 123–36.

Pilkington, E. (1994) Out-patients. *Search* 21 (winter), 28–32.

Pill, R. and Stott, N. C. H. (1982) Concepts of illness causation and responsibility: some preliminary data from a sample of working class mothers. *Social Science and Medicine* 16, 43–52.

Pineault, R., Lamarche, P. A., Champagne, F., Contandriopoulos, A. P. and Denis, J. L. (1993) The reform of the Quebec health care system: potential for innovation? *Journal of Public Health Policy* 14(2), 198–219.

Plaut, R. & Silvi, J. (1991) Communicable disease mortality: now you see it, now you don't. *Journal of Public Health Policy* (winter), 464–74.

Polednak, A. (1989) *Racial and ethnic differences in disease.* Oxford: Oxford University Press.

Polgar, S. (1968) Health. In Sills, D. L. (ed.), *International encyclopaedia of the social sciences,* vi. New York: Macmillan and Free Press, 330–6.

Pollitt, C. (1988) Bringing consumers into performance measurement: concepts, consequences and constraints. *Policy and Politics* 16(2), 77–87.

——(1993) The struggle for quality: the case of the National Health Service. *Policy and Politics* 21(3), 161–70.

Pollock, A. (1994) Debate on health service financing. *Health Services Journal* 104(5426), 12–13.

Popay, J., Bartley, M. and Owen, C. (1993) Gender inequalities in health: social position, affective disorders and minor physical morbidity. *Social Science and Medicine* 36(1), 21–32.

Ports, S. T. and Banzhaf, M. (1990) Many cultures, many approaches. Repr. in Banzhaf, M. *et al.* (the ACT UP/New York Women and AIDS Book Group), *Women, AIDS and activism,* 3rd edn. Boston: South End Press, 1992, 107–11.

Potter, J. (1988) Consumerism and the public sector: how well does the coat fit? *Public Administration* 66, 149–64.

Powell, M. (1990) Need and provision in the National Health Service: an inverse care law? *Policy and Politics* 18(1), 31–7.

Powles, J. (1992) Changes in disease patterns and related social trends. *Social Science and Medicine* 35(4), 377–87

Prior, L. (1987) Policing the dead: a sociology of the mortuary. *Sociology* 21(3), 355–76.

Puentes-Markides, C. (1992) Women and access to health care. *Social Science and Medicine* 35(4), 613–17.

Pyle, G. (1969) The diffusion of cholera in the United States in the nineteenth century. *Geographical Analysis* 1, 59–75.

——(1980) Geographical perspectives on influenza diffusion: the United States in the 1940s. In Meade, M. (ed.), *Conceptual and methodological issues in medical geography.* Studies in Geography no. 15. Chapel Hill, NC: University of North Carolina, Department of Geography.

——(1990) Regional inequalities in infant mortality within North Carolina, USA. *Espace, Populations, Sociétés* 3, 439–45.

Radical Statistics Health Group (1991) Missing: a strategy for health of the nation. *British Medical Journal* 303, 299–302.

Raeburn, J. and Rootman, I. (1989) Towards an expanded health field concept: conceptual and research issues in a new era of health promotion. *Health Promotion* 4(3), 386–7.

Raffel, M.W. and Raffel, N. K. (1992) Czechoslovakia's changing health care system. *Public Health Reports* 107(6), 636–43.

Rains, P. (1971) *Becoming an unwed mother: a sociological account.* Chicago: Aldine.

Raleigh, V. and Balarjan, R. (1994) Public health and the 1991 census: nonrandom underenumeration complicates interpretation. *British Medical Journal* 309, 287–8.

Ransome, O., Roode, H. and Reinach, S. (1986) Facilities for children in state and provincial hospitals. *South African Medical Journal* 69, 612–3.

Rathwell, T. (1992) Pursuing health for all in Britain: an assessment. *Social Science and Medicine* 34(2), 169–82.

——and Phillips, D. (eds) (1989) *Health, race and ethnicity.* London: Croom Helm.

Read, M. D. (1993) The failure to implement an anti-smoking policy. In Mills (1993: 85–104).

Rheinhardt, U. (1994) The Clinton plan: a salute to American pluralism. *Health Affairs* 13(1), 161–78.

Rhodes, P. (1985) *An outline history of medicine.* London: Butterworths.

Rhodes, R. A. W. (1987) Developing the public service orientation. *Local Government Studies,* May/June, 63–73.

Rice, M. and Rasmusson, E. (1992) Healthy cities in developing countries. In Ashton, J. (ed.) *Healthy Cities.* Milton Keynes: Open University Press, Ch. 9, 70–86.

Rider, M. and Flynn, B. (1992) Indiana. In Ashton, J. (ed.), *Healthy Cities*. Milton Keynes: Open University, 195–204.

Rifkin, S. B. and Walt, G. (1986) Why health improves: defining the issues concerning 'comprehensive primary health care' and 'selective primary health care'. *Social Science and Medicine* 23, 559–66.

Riley, J. and Alter, G. (1989) The epidemiological transition and morbidity. *Annales de Démographie Historique*, 199–213.

Riley, J. C. (1987) *The eighteenth-century campaign to avoid disease*. Basingstoke: Macmillan.

Rivett, G. C. (1986) *The development of the London hospital system 1823–1982*. London: King Edward's Hospital Fund for London.

Rodgers, J. (1994) Power to the people. *Health Service Journal* 104(5395), 28–9.

Rodwin, V and Sandier, S. (1993) Health care under French national health insurance. *Health Affairs,* fall, 111–31.

Roemer, M. (1976) *Health care systems in world perspective*. Ann Arbor, Mich.: Health Administration Press.

Root, A. (1995) Oxford blues. *Health Service Journal,* 19 Jan., 32–3.

Rose, J. (1986) *Sexuality in the field of vision*. London: Verso.

Rose, N. (1985) *The psychological complex: psychology, politics and society in England 1869 to 1939*. London: Routledge & Kegan Paul.

Rosen, G. (1979) The evolution of social medicine. In Freeman, H. E. *et al.* (eds), *Handbook of medical sociology*. Englewood Cliffs, NJ: Prentice Hall, 23–50.

Rosenberg, M. W. (1988) Linking the geographical, the medical and the political in analysing health care delivery systems. *Social Science and Medicine* 26(1), 179–86.

Rosenhan, D. L. (1973) On being sane in insane places. *Science* 179, 250–8.

Rosenthal, M. (1987) *Health care in the People's Republic of China: moving towards modernization*. Boulder, Colo.: Westview Press.

Rossi-Espagnet, A., Goldstein, G. and Tabibzadeh, I. (1991) Urbanizaton and health in developing countries: a challenge for health for all. *World Health Statistics Quarterly* 44(4), 187–224.

Rowland, A. and Cooper, P. (1983) *Environment and health*. London: Edward Arnold.

Royston, G., Hurst, J., Lister, E. and Stewart, P. (1992) Modelling the use of health services by populations of small areas to inform the allocation of central resources to larger regions. *Socio-Economic Planning Sciences* 26(3), 169–80.

Rumsey, S. (1990) AIDS issues for African-American and African-Caribbean women. In Banzhaf, M. *et al.* (the ACT UP/New York Women and AIDS Book Group), *Women, AIDS and activism*, 3rd edn. Boston: South End Press, 1992, 103–6.

Rutstein, D., Berenberg, W., Chalmers, T., Child, C., Fishman, A. and

Perrin, E. (1976) Measuring the quality of medical care. *New England Journal of Medicine* 11, 582–8.

——(1980) Measuring the quality of medical care: second revision of table of indexes. *New England Journal of Medicine* 302, 1146.

Ryan, M. (1978) *The organisation of Soviet medical care*. Oxford: Blackwell.

——(1993) Health care in Moscow. *British Medical Journal* 307, 782–4.

Rydell, C. (1994) Special report: an agenda for federalism from state leaders. *Health Affairs*, 13(5), 252–3.

Sabatier, R. (1988) *Blaming others: prejudice, race and worldwide AIDS*. London: Panos Institute.

Said, H. (1983) The Unani system of health and medicare. In Bannerman, R., Burton, J. and Ch'en Wen-Chieh (eds), *Traditional medicine and health care coverage: a reader for health administrators and practitioners*. Geneva: WHO, 61–7.

Salter, B. (1993) The politics of purchasing in the national health service. *Policy and Politics* 21(3), 171–84.

Saltman, R. (1994) A conceptual overview of recent health care reforms. *European Journal of Public Health* 4(4), 287–93.

——and von Otter, C. (1989) Public competition versus mixed markets: an analytic comparison. *Health Policy* 11, 43–55

——and——(1992) *Planned markets and public competition*. Milton Keynes: Open University Press.

Salvador, T. (1990) Intersectoral argumentation: the case of tobacco. In Taket (1990: 55–72).

Sanders, D. and Carver, R. (1985) *The struggle for health: medicine and the politics of underdevelopment*. Basingstoke: Macmillan.

Santé Québec (1987) *Et la santé, ça va? i: Rapport de l'enquête Santé Québec 1987*. Québec: Santé Québec.

——(1994) *Enquête sociale et de santé 1992–1993: faits saillants*. Québec: Santé Québec.

Sawicki, J. (1991) Disciplining mothers: feminism and the new reproductive technologies. In Sawicki, J., *Disciplining Foucault: feminism, power and the body*. New York: Routledge, 67–94.

Scarpaci, J. L. (1988a) Help-seeking behavior, use, and satisfaction among frequent primary care users in Santiago de Chile. *Journal of Health and Social Behaviour* 29, 199–213.

——(1988b) *Primary medical care in Chile: accessibility under military rule*. Pittsburgh, Penn.: University of Pittsburgh Press.

——(1990) *Health services privatisation in industrial societies*. New Brunswick, NJ: Rutgers University Press.

——(1991) Primary-care decentralization in the Southern Cone: shantytown health care as an urban social movement. *Annals of the Association of American Geographers* 81(1), 103–26.

——(1993) On the validity of language: speaking, knowing and understanding in medical geography. *Social Science and Medicine* 37(6), 719–24.

Scheper-Hughes, N. (1992) *Death without weeping: the violence of everyday life in Brazil.* Berkeley, Calif.: University of California Press.

Schieber, G., Pollier, J.-P. and Greenwald, M. (1994) Health system performance in OECD countries, 1980–1992. *Health Affairs*, 13(3), 100–12.

Schneider, D., Greenberg, M. R., Donaldson, M. H. and Choi, D. (1993) Cancer clusters: the importance of monitoring multiple geographic scales. *Social Science and Medicine* 37(6), 753–9.

Schoepf, B. G. (1991) Ethical, methodological and political issues of AIDS research in central Africa. *Social Science and Medicine* 33(7), 749–63.

Schofield, R., Reher, D. and Bideau, A. (eds) (1991) *The decline of mortality in Europe.* Oxford: Clarendon Press.

Schwartz, S. (1994) The fallacy of the ecological fallacy: the potential misuse of a concept and the consequences. *American Journal of Public Health* 84, 819–24.

Scott-Samuel, A. (1989) Building the new public health: a public health alliance and a new social epidemiology. In Maring, C. and McQueen, D. (eds) *Readings for a new public health*, Edinburgh: Edinburgh University Press.

Seale, J. (1988) Origins of the AIDS viruses HIV-1 and HIV-2: fact or fiction? *Journal of the Royal Society of Medicine* 81, 537–9.

Secretariat for Futures Studies (1984) *Time to care.* London: Pergamon.

Secretary of State for Health (1991) *The health of the nation: a consultative document for health in England.* London: HMSO, Cm. 1523.

Seedhouse, D. (1986) *Health: the foundations for achievement.* Chichester: Wiley.

Segal, J. and Segal, L. (1987) AIDS: Natur und Ursprung. In Kruse, K. (ed.), *AIDS: Erreger aus dem Genlabor.* Berlin: Simon & Leutner, 78–127.

Segal, S. P., Silvervan, C. and Temkin, T. (1993) Empowerment and self-help agency practice for people with mental disabilities. *Social Work* 38(6), 705–12.

Senior M. (1991) Deprivation payments to GPs: not what the doctor ordered. *Environment and Planning C* 19, 79–94.

——(1995) Updating or radically changing deprivation payments to GPs in England: experiments using 1991 census data. Paper presented to the Institute of British Geographers, Annual Conference, Newcastle upon Tyne, 3–6 Jan.

Senior, P. and Bhopal, R. (1994) Ethnicity as a variable in epidemiological research. *British Medical Journal* 309, 327–30.

Sermet, C. (1993) *Enquête sur la santé et les soins medicaux 1991–1992: méthodologie.* Paris: CREDES.

Shange, N. (1992) The love space demands: an interview. *Spare Rib* 238, 17–19.

Shannon, G. and Dever, G. (1974) *Health care delivery: spatial perspectives.* New York: McGraw Hill.

——Pyle, G. and Bashur, R. (1991) *The geography of AIDS: the origins and course of an epidemic.* New York: Guilford.

Shaper, A. (1984) Geographic variations in cardiovascular mortality in Great Britain. *British Medical Bulletin* 40, 366–73

Sharma, V. (1992) *Complementary medicine today.* London: Routledge.

Shattuck, L. (1850) *Report of the Sanitary Commission of Massachusetts.* Boston: Dulton & Wentworth.

Shaw, C. (1993) Quality assurance in the United Kingdom. *Quality Assurance in Health Care* 5(2), 107–18.

——and Brooks, T. (1991) Hospital accreditation in the United Kingdom. *Quality Assurance in Health Care* 3, 133.

Sheiman, I. (1994) Forming the system of health insurance in the Russian Federation. *Social Science and Medicine* 39(10), 1425–32.

Sheldon, T. (1994) Quality: link with effectiveness. *Quality in Health Care* 3, 41–5.

——and Parker, H. (1992) Race and ethnicity in health research. *Journal of Public Health Medicine* 14(2), 104–10.

——Smith, P., Borowitz, M., Martin, S. and Carr-Hill, R. (1994) Attempt at deriving a formula for setting general practitioner fundholding budgets, *British Medical Journal* 309, 1059–64.

Short, S. D. (1989) Community participation or community manipulation? a case study of the Illawarra Cancer Appeal-a-thon. *Community Health Studies* 13(1), 34–8.

Shouls, S., Congdon, P. and Curtis, S. (1994) Modelling inequality in reported long term illness: developing a strategy for combining individual and area characteristics. Paper presented at Multi-Level Modelling Workshop, University of Portsmouth, 23–4 Sept.

Showalter, E. (1991) *The female malady: women, madness and English culture 1830–1980.* London: Virago.

Skultans, V. (1979) *English madness: ideas on insanity 1580–1890.* London: Routledge & Kegan Paul.

Smallman-Raynor, M. and Cliff, A. (1990) Acquired Immune Deficiency Syndrome: literature, geographical origins and global patterns. *Progress in Human Geography* 14, 157–213.

——, ——and Haggett, P. (1992) London International Atlas of AIDS. Oxford: Oxford University Press.

Smi, V. (1993) *Environmental change and social flexibility.* London: Routledge.

Smith, D. M. (1977) *Human geography: a welfare approach.* London: Edward Arnold.

——(1994) *Geography and social justice.* Oxford: Blackwell.

Smith, L. K. (1991) Community participation in health: a case study of World Health Organization's Healthy Cities project in Barcelona and Sheffield. *Community Development Journal* 26(2), 112–17.

Smith, P. (1993) Outcome-related performance indicators and organizational control in the public sector. *British Journal of Management* 4, 135–51.

——Sheldon, T., Carr-Hill, R., Martin, S., Peacock, S. and Hardman, G. (1994) Allocating resources to health authorities: results and policy implications of small area analysis of inpatient services. *British Medical Journal* 309, 1050–4.

Smith, R. (1995) The WHO: change or die. *British Medical Journal*, 310, 543–4.

Smith, S. (1993) Tip of the iceberg. *Health Service Journal*, 25 Feb., 22–4.

Smith-Rosenberg, C. (1972) The hysterical woman: sex roles and role conflict in 19th century America. *Social Research* 39, 652–72.

Snow, J. (1855) *On the mode of communication of cholera*, 2nd edn. London: Churchill.

Social Science and Medicine (1990) Special issue: 'Health inequities in Europe'. *Social Science and Medicine* 31(3).

——(1993) Special issue: 'Geographical inequalities in Europe'. *Social Science and Medicine* 36(10).

Solórzano, A. (1992) Sowing the seeds of neo-imperialism: the Rockefeller Foundation's yellow fever campaign in Mexico. *International Journal of Health Services* 22(3), 529–54.

Sontag, S. (1979) *Illness as metaphor*. London: Allen Lane.

——(1989) *AIDS and its metaphors*. London: Allen Lane.

Spasoff, R. A., Cole, P. *et al.* (1987) *Health for all Ontario*. Report of the Panel on Health Goals for Ontario. Ontario: Ministry of Health.

St Leger, A., Schnieden, H. and Walsworth-Bell, J. (1992) *Evaluating health services effectiveness: a guide for health professionals, service managers and policy makers*. Milton Keynes: Open University Press.

Stacey, M. (1988) *The sociology of health and healing*. London: Routledge.

Stainton-Rogers, W. (1991) *Explaining health and illness: an exploration of diversity*. Hemel-Hempstead: Harvester Wheatsheaf.

Stephenson, P. A. and Wagner, M. G. (1993) WHO recommendations for IVF: do they fit with 'Health for All'. *Lancet* 341, 26 June, 1648–9.

Stimson, R. J. (1983) Research design and methodological problems in the geography of health. In McGlashan, N. D. and Blunden, J. R. (eds), *Geographical aspects of health*. London: Academic Press, 321–34.

Stone, G. P. and Faberman, H. A. (1981) *Social psychology through symbolic interaction*. Chichester, Wiley.

Strong, P. and Robinson, J. (1990) *The NHS under new management*. Milton Keynes: Open University Press.

Strong, P. M. (1986) A new-modelled medicine? Comments on the WHO's regional strategy for Europe. *Social Science and Medicine* 22, 193–9.

Strosberg, M. A., Wiener, J. M., Baker, R. and Fein, I. A. (eds) (1992) *Rationing America's medical care: the Oregon plan and beyond*. Washington, DC: Brookings Institute.

Stubbs, P. (1993) 'Ethnically sensitive' or 'anti-racist'? Models for health research and service delivery. In Ahmad, W. (ed.), *'Race' and health in contemporary Britain*. Milton Keynes: Open University Press, 34–50.

Summer, L. (1994) The escalating number of uninsured in the United States. *International Journal of Health Services* 24(3) 409–13.

Susser, M. (1994) The logic in ecological, i: The logic of analysis. *American Journal of Public Health* 84, 825–9.

Svensson, P.-G. and Stephenson, P. (1992) Health care consequences of the European Economic Community in 1993 and beyond. *Social Science and Medicine* 35(4), 525–9.

Sweeting, H. (1995) Reversals of fortune? Sex differences in health in childhood and adolescence. *Social Science and Medicine* 40(1), 77–90.

Sykes, W., Collins, M., Hunter, D. J., Poppay, J. and Williams, G. (1992a) *Listening to local voices: a guide to research methods, i: Summary of main issues.* Leeds: Nuffield Institute for Health Service Studies and the Public Health Research and Resource Centre, Salford.

——(1992b) *Listening to local voices: a guide to research methods, ii: An introduction to available research methods.* Leeds: Nuffield Institute for Health Service Studies and the Public Health Research and Resource Centre, Salford.

——(1992c) *Listening to local voices: a guide to research methods, iii: The research process.* Leeds: Nuffield Institute for Health Service Studies and the Public Health Research and Resource Centre, Salford.

Szasz, T. S. (1971) *The manufacture of madness.* London: Routledge & Kegan Paul.

——(1974) *The myth of mental illness.* New York: Harper & Row.

Taket, A. R. (1989) Equity and access: exploring the effects of hospital location on the population served. *Journal of the Operational Research Society* 40(11), 1001–9.

——(ed.) (1990) *Making partners: intersectoral action for health.* The Hague: WHO/Ministry for Welfare, Health and Cultural Affairs.

——(1993) Mixing and matching. *OR Insight* 6(4), 18–23.

——(1995) Methodological issues in the evaluation of health promotion and their relevance to the modelling of noncommunicable diseases. In Morgenstern W *et al.* (eds), *Modelling of noncommunicable diseases: methodological issues.* Berlin: Springer.

——and Curtis, S. E. (1989) Locality planning in health care: a case study in east London. *Area* 21(4), 357–64.

——, —— and Thuriaux, M. C. (1990) Vers la santé pour tous dans la région européenne de L'OMS: surveillance des progrès accomplis, iii: Modes de vie et milieu physique. *La Revue d'Épidémiologie et Santé Publique* 38, 3–18.

——Mayhew, L. D., Gibberd, R. W., Hall, N. M., Bevan, R. G., Waring, D., Bertuglia, C. S., Tadei, R. and Rising, E. J. (1986) RAMOS: a model of the spatial allocation of health care resources. In *Health projections in Europe: methods and applications.* Copenhagen: WHO Regional Office for Europe, 218–36.

——and Parkinson, J. (1990a) Helping hands. *Health Service Journal* 100, 1474–5.

——, ——Malcolm, G., Redmond, B., Crosti, L., Green, D. and Bass, A. (1990b) *Report on the Child Protection Advisers Project in Tower Hamlets Health Authority.* QMW Working Paper in Geography no. 5. London: Queen Mary and Westfield College, University of London.

Tarimo, E. and Creese, A. (eds) (1990) *Achieving health for all by the year 2000: midway reports of country experiences.* Geneva: WHO.

Taylor, A. L. (1992) Making the World Health Organization work: a legal framework for universal access to the conditions for health. *American Journal of Law and Medicine* 18(4), 301–46.

Taylor, R. (1984) Alternative medicine and the medical encounter in Britain and the United States. In Salmon, W. (ed.), *Alternative medicines: popular and policy perspectives.* London: Tavistock, 191–228.

Taylor, S., Elliott S., Frank, J., Haight, M., Steiner, D., Walter, S., White, N. and Willms, D. (1991) Psychosocial impacts in populations exposed to solid waste facilities. *Social Science and Medicine* 33(4), 441–7.

Taylor-Gooby, P. (1992) Who has the best tunes? Rights, needs and moral hazard. In Manning, N. and Page, R. (eds) *Social Policy Review 4.* London: Social Policy Association, 261–79.

Telyukov, A. (1991) A concept of health financing reform in the Soviet Union. *International Journal of Health Services* 21(3), 493–504.

Temerlin, M. K. (1968) Suggestion effects in psychiatric diagnosis. *Journal of Nervous and Mental Disease* 147, 349–53.

Terris, M. (1980) Three world systems of medical care: trends and prospects. *World Health Forum* 1, 78–86.

Tesh, S. N. (1988) *Hidden arguments: political ideology and disease prevention policy.* New Brunswick, NJ: Rutgers University Press.

Tomlinson, B. (1992) *Report of the inquiry into London's health services, medical education and research.* London: HMSO.

Tonnelier, F. (1990) Espace economique et acces aux soins: les zones d'étude pour l'emploi. *Cahier de sociologie et de démographie médicales* 30(2), 193–208.

Torkington, N. P. K. (1991) Consultative paper 6: mental health. In *Black health: a political issue.* Liverpool: Catholic Association for Racial Justice and Liverpool Institute for Higher Education, 112–57.

Townsend, P. and Davidson, N. (1982) *Inequalities in health: the Black report.* Harmondsworth: Penguin.

——, ,——and Whitehead, M. (1988) *Inequalities in health.* Harmondsworth: Pelican.

——Phillimore, P. and Beattie, A. (1988) *Health and deprivation: inequality and the north.* London: Croom Helm.

Treichler, P. A. (1988) AIDS, homophobia and biomedical discourse: an epidemic of signification. In Crimp (1988: 31–70).

——(1989) AIDS and HIV infection in the third world: a first world chronicle. In Kruger, B. and Mariani, P. (eds), *Remaking history*. Seattle, Wash.: Bay Press, 31–86.

Tsouros, A. D. (1990) *World Health Organization Healthy Cities Project: a project becomes a movement. Review of progress 1987–1990*. Milan: SOGESS.

Tudor-Hart, J. (1971) The inverse care law. *Lancet* 1 (27 Feb.), 405–12.

Turshen, M. (1984) *The political ecology of disease in Tanzania*. New Brunswick, NJ: Rutgers University Press.

UK Department of Health and Social Security (1975) *Sharing resources for health in England and Wales*. London: HMSO.

Umhau, T. H., Umhau, J. C. and Morgan, R. E. Jr (1991) National and international health agencies: profile of key players. *Infectious Disease Clinics of North America* 5(2), 197–220.

Unschuld, P. (1986) The conceptual determination (*uberformung*) of individual and collective experiences of illness. In Currer and Stacey (1986: 51–70).

US Department of Health and Human Services (1980) *Promoting health/preventing disease: Year 1990 objectives for the nation*. Washington, DC: US Department of Health and Human Services.

——(1989a) *Promoting health/preventing disease: Year 2000 objectives for the nation*. Washington, DC: US Department of Health and Human Services.

——(1989b) *National health interview survey, United States, 1989*. Vital and Health Statistics series 10, no. 176. Washington, DC: US Government Printing Office.

——(1990) *Healthy People 2000: national health promotion and disease prevention objectives*. Washington, DC: US Goverment Printing Office.

——(1991) *Health United States, 1990*. Hyattsville, Md.: US Department of Health and Human Services.

——(1994) *Healthy People 2000: fact sheet, March 1994*. Washington, DC: US Goverment Printing Office.

Ussher, J. (1991) *Women's madness: misogyny or mental illness?* New York: Harvester Wheatsheaf.

Van Dalen, H., Williams, A. and Gudex, C. (1994) Lay people's evaluations of health: are there variations between different subgroups? *Journal of Epidemiology and Community Health* 48, 248–53.

Van Os, J., Galdos, P., Lewis, G., Bourgeois, M. and Mann, A. (1993) Schizophrenia sans frontières: concepts of schizophrenia among French and British psychiatrists. *British Medical Journal* 307, 489–92

Van Oyen, H. J. (1990) *Health for all in Europe: an epidemiological review*. Copenhagen: WHO Regional Office for Europe.

Veith, I (1965) *Hysteria: the history of a disease*. Chicago: University of Chicago Press.

Virganskaya I. and Dimitriev V. (1992) Some problems of medicodemographic development in the former USSR. *World Health Statistics Quarterly* 45(1), 4–14.

Von Korff, M., Koepsell, T., Curry, S., Diehr, P. (1992) Multi-level analysis in epidemiological research on health behaviours and outcomes. *American Journal of Epidemiology* 135(10), 1077–82.

Von Thurn, D. R., Moore, J. C. and Marting, E. A. (1993) National health interview survey redesign: an anthropological investigation of mental health concepts. Paper presented at the Annual Conference of the American Association for Public Opinion Research, St. Charles, Ill., 20–3 May.

Vuori, H. (1982) The World Health Organization and traditional medicine. *Community Medicine* 4, 129–37.

Wagstaff, A., Doorslaer, E.V., Paci, P. (1989) Equity in the finance and delivery of health care; some tentative cross-country comparisons. *Oxford Review of Economic Policy* 5(1), 89.

Waite, G. (1988) The politics of disease: the AIDS virus and Africa. In Miller and Rockwell (1988: 145–64).

Walker, S. and Rosser, R. (1993) *Quality of life: key issues in the 1990s.* London: Kluwer Academic.

Wallace, R. (1990) Urban desertification, public health and public order: 'planned shrinkage', violent death, substance abuse and AIDS in the Bronx. *Social Science and Medicine* 31, 801–13.

——, R., Fullilove, M., Fullilove, R., Gould, P. and Wallace, D. (1994) Will AIDS be contained within US minority urban populations? *Social Science and Medicine* 39(8), 1051–62.

Wallerstein, N. (1993) Empowerment and health: the theory and practice of community change. *Community Development Journal* 28, 218–27.

——and Bernstein, E. (1988) Empowerment education: Freire's ideas adapted to health education. *Health Education Quarterly* 15(4), 379–94.

Walt, G. (1993) WHO under stress: implications for health policy. *Health Policy* 24, 125–44.

——(1994) *Health policy: an introduction to process and power.* London: Zed Press.

Walters, V. (1993) Stress, anxiety and depression: women's accounts of their health problems. *Social Science and Medicine* 36(4), 393–402.

Walzer, M. (1983) *Spheres of justice: a defense of pluralism and equality.* Oxford: Blackwell.

Wardwell, W. (1994) Alternative medicine in the United States. *Social Science and Medicine* 38(8), 1061–68.

Ware, J. and Sherbourne, C. (1992) The MOS 36-item short-form health survey (SF-36). Conceptual framework and item selection. *Medical Care* 30, 473–83.

Wareham, N. (1994a) External monitoring of quality of health care in the United States. *Quality in Health Care* 3, 97–101.

——(1994b) Changing systems of external monitoring of quality of health care in the United States. *Quality in Health Care* 3, 102–106.

Warner, M., Winkler, F. and Gritzner, C. (1991) Best for Britain? *Health Service Journal*, 22 Aug., 19.

Warren, K. S. (1988) The evolution of selective primary health care. *Social Science and Medicine* 26(9), 891–8.

Watney, S. (1989) Taking liberties. In Carter and Watney (1989: 11–57).

——(1990) Missionary positions: AIDS, Africa and race. In Ferguson, R. *et al.* (eds), *Out there: marginalization and contemporary cultures.* Cambridge, Mass.: MIT Press, 89–103.

Watson, G. and Williams, J. (1992) Feminist practice in therapy. In Ussher, J. M. and Nicholson, P. (eds), *Gender issues in clinical psychology.* London: Routledge, 212–36.

Wedderburn-Tate, C., Bruster, S., Broadley, K., Maxwell, E. and Stevens, L. (1995) What do patients really think? *Health Service Journal* 105(5435), 12 Jan., 18–20.

Weeks, J. (1989) *Sex, politics and society.* London: Longman.

Wei, L. and Chao, Z. (1984) Principles for health development and methods of health research in China. In Haifeng, C. and Chao, Z. (eds) *Modern Chinese medicine, iii: Chinese health care.* Lancaster: MTP Press.

West, P. (1991) Rethinking the health selection explanation for health inequalities. *Social Science and Medicine* 32(4), 373–84.

West, R. and Lowe, C. (1976) Regional variations in need for and provision and use of child health services in England and Wales. *British Medical Journal* 272, 843–6.

While, A. (ed.) (1989) *Health in the inner city.* London: Heinemann Medical.

White, L. A. and Taket, A. R. (1994) Using rapid appraisal for evaluating the voluntary sector's delivery of health care programmes. In Rey, J. C. and Tilquin, C. (eds), *SYSTED 94: Proceedings of the Fifth International Conference on Systems Sciences in Health and Social Services for the Elderly and the Disabled,* Geneva, 2–6 May 1994. Aarau: Institut Suisse de la Santé Publique, 812–17.

Whitehead, M. (1989) *Swimming upstream: trends and prospects for education in health.* London: King's Fund.

WHO (1957) *Measurement of levels of health: report of the WHO study group.* Technical Report Series no. 137. Geneva: WHO.

——(1977) *The selection of essential drugs: report of the WHO expert committee.* Technical Report series no. 615. Geneva: WHO.

——(1978) *The promotion and development of traditional medicine.* Technical Report Series no. 622. Geneva: WHO.

——(1980a) *Sixth report on the world health situation, 1973–1977.* Geneva: WHO.

——(1980b) *Guiding principles for the managerial process for national health development.* Geneva: WHO.

——(1981) *Global strategy for Health for All by the year 2000.* Geneva: WHO.

——(1983) *Primary health care: the Chinese experience.* Geneva: WHO.

——(1986) *Evaluation of the global strategy for health for all by the year 2000: seventh report on the world health situation.* (7 vols.) Geneva: WHO.

——(1988) *Basic documentation, 37th edn.* Geneva: WHO.

——(1993a) *Implementation of the global strategy for health for all by the year 2000, second evaluation: eighth report on the world health situation, i: Global review.* Geneva: WHO.

——(1993b) *A global strategy for malaria control.* Geneva: WHO.

——(1994) *The work of WHO 1992–1993: biennial report of the Director General.* Geneva: WHO.

WHO/EURO (1983) *Targets in support of the regional strategy for HFA2000.* Document EUR/RC33/9, produced for the 1983 European Regional Committee, agenda item 10. Copenhagen: WHO Regional Office for Europe.

——(1984) *Summary report of the working group on concepts and principles of health promotion.* Copenhagen: WHO Regional Office for Europe. Repr. in Abelin, T. *et al.* (eds) 1987: *Measurement in health promotion and protection.* Copenhagen: WHO, 653–8.

——(1985a) *Targets for health for all, 1985.* Copenhagen: WHO Regional Office for Europe.

——(1985b) *Progress towards HFA in the European Region.* Copenhagen: WHO Regional Office for Europe.

——(1986) *Evaluation of the strategy for health for all by the year 2000: seventh report on the world health situation. v: European region.* Copenhagen: WHO Regional Office for Europe.

——(1987) *Indicators for primary health care.* Copenhagen: WHO Regional Office for Europe.

——(1988) *Priority research for health for all.* Copenhagen: WHO Regional Office for Europe.

——(1989a) *Monitoring of the strategy for health for all by the year 2000, i: The situation in the European Region, 1987/1988.* Copenhagen: WHO Regional Office for Europe.

——(1989b) *Monitoring of the strategy for health for all by the year 2000, ii: Monitoring by country, 1988/1989.* Copenhagen: WHO Regional Office for Europe.

——(1993a) *Health for all targets: the health policy for Europe.* rev. edn. Sept. 1991. Copenhagen: WHO Regional Office for Europe.

——(1993b) *Implementation of the global strategy for health for all by the year 2000, second evaluation: eighth report on the world health situation, v: European Region.* Copenhagen: World Health Organization, WHO Regional Publications, European Series no 52.

WHO/SEARO (1993) *Implementation of the global strategy for health for all by the year 2000, second evaluation: eighth report on the world health situation, iv: South-East Asia region.* New Delhi: WHO Regional Office for South-East Asia.

WHO/UNICEF (1978) *Primary health care: report of the International Conference on Primary Health Care, Alma-Ata.* Geneva: WHO.

WHO Working Group on Health Promotion in Developing Countries

(1991) A call for action: promoting health in developing countries. *Health Education Quarterly* 18(1), 5–15.

WHO/WPRO (1993) *Implementation of the global strategy for health for all by the year 2000, second evaluation: eighth report on the world health situation, vii: Western Pacific region.* Manila: WHO Regional Office for the Western Pacific.

Wilensky, G. (1994) Health reform: what will it take to pass? *Health Affairs* 13(1), 179–91.

Wilkinson, R. (1986a) *Class and health.* London: Tavistock.

——(1986b) Occupational class, selection and inequalities in health: a reply to Raymond Illsley. *Quarterly Journal of Social Affairs* 2, 415–22.

——(1987) A rejoinder to Raymond Illsley. *Quarterly Journal of Social Affairs* 3, 225–8.

——(1992) Income distribution and life expectancy. *British Medical Journal* 304, 165–8.

Willcocks, A. (1967) *The creation of the National Health Service.* London: Routledge & Kegan Paul.

Williams, F. (1992) Somewhere over the rainbow: universality and diversity in social policy. In Manning, N. and Page, R. (eds), *Social Policy Review iv.* Canterbury: Social Policy Association, 200–19.

Williams, R. (1983) Concepts of health: an analysis of lay logic. *Sociology* 17, 185–205.

Williamson, J. (1989) Every virus tells a story: the meanings of HIV and AIDS. In Carter and Watney (1989: 69–80).

Wilson, A. G. (1974) *Urban and regional models in geography and planning.* London: Wiley.

Wimberley, T. (1980) *Toward national health insurance in the United States: an historical outline 1910–1979. Social Science and Medicine* 14C, 13–25.

Wing, S., Casper, M., Hayes, C., Dargent-Molina, P., Riggan, W. and Tyroler, H. (1987) Changing association between community occupational structure and ischaemic heart disease mortality in the United States. *Lancet* 7, 1067–70.

——, ——, Davis, W., Hayes, C., Riggan, W., Tyroler, H. (1990) Trends in the geographic inequality of cardiovascular disease mortality in the United States 1962–1982. *Social Science and Medicine* 30, 261–6.

Winkler, F. (1987) Consumerism in health care: beyond the supermarket model. *Policy and Politics* 15(1), 1–8.

Wlodarczyk, C. and Sabbat, J. (1993) Regional integration of health services in Poland: an ambitious pilot project. *Health Policy* 23, 229–45.

Wolinsky, F. (1980) *The sociology of health.* Boston, MA. Little, Brown & Co.

Woodhead, D. (1995) 'Surveillant gays': HIV, space and the constitution of identities. In Bell, D. and Valentine, G. (eds), *Landscapes of desire: geographies of sexualities.* London: Routledge.

Woolhandler, S., Himmelstein, D. and Young, Q. (1993) High noon for US health care reform. *International Journal of Health Services* 23(2), 193–211.

Working Group on Managed Competition (1994) Managed competition: an analysis of consumer concerns. *International Journal of Health Services* 24(1), 11–24.

World Bank (1993) *World development report 1993: investing in health.* Oxford: Oxford University Press.

——(1994) *Human Development Report.* Oxford: Oxford University Press.

Yang, P., Lin, V. and Lawson, J. (1991) Health reform in the People's Republic of China. *International Journal of Health Services* 21(3), 481–91.

Yankauer, A. (1987) Hispanic/Latino: what's in a name? *American Journal of Public Health* 77(1), 15–17.

Yanow, D. (1993) The communication of policy meanings: implementation as interpretation and text. *Policy Sciences* 26(1), 41–61.

Yates, J. and Davidge, M. (1984) Can you measure performance? *British Medical Journal* 288, 1935–36.

Zelman, W. (1994) The rationale behind the Clinton health care reform plan. *Health Affairs* 13(1), 9–29.

Author index

Subject index